NanoScience and Technology

NanoScience and Technology

Series Editors:

P. Avouris B. Bhushan D. Bimberg K. von Klitzing H. Sakaki R. Wiesendanger

The series NanoScience and Technology is focused on the fascinating nano-world, mesoscopic physics, analysis with atomic resolution, nano and quantum-effect devices, nanomechanics and atomic-scale processes. All the basic aspects and technologyoriented developments in this emerging discipline are covered by comprehensive and timely books. The series constitutes a survey of the relevant special topics, which are presented by leading experts in the field. These books will appeal to researchers, engineers, and advanced students.

Bharat Bhushan
Harald Fuchs

Applied Scanning Probe Methods XIII

Biomimetics and Industrial Applications

With 101 Figures and 10 Tables
Including 29 Color Figures

 Springer

Editors
Prof. Dr. Bharat Bhushan
Ohio State University
Nanoprobe Laboratory for Bio- & Nanotechnology
& Biomimetics (NLB^2)
201 W. 19th Ave
Columbus, Ohio 43210-1142
USA
bhushan.2@osu.edu

Prof. Dr. Harald Fuchs
Universität Münster
FB 16
Physikalisches Institut
Wilhelm-Klemm-Str. 10
48149 Münster
Germany
fuchsh@uni-muenster.de

Series Editors
Professor Dr. Phaedon Avouris
IBM Research Division
Nanometer Scale Science & Technology
Thomas J.Watson Research Center, P.O. Box 218
Yorktown Heights, NY 10598, USA

Professor Bharat Bhushan
Nanoprobe Laboratory for Bio- & Nanotechnology
and Biomimetics (NLB^2)201 W. 19th Avenue
The Ohio State University
Columbus, Ohio 43210-1142, USA

Professor Dr. Dieter Bimberg
TU Berlin, Fakutät Mathematik,
Naturwissenschaften,
Institut für Festkörperphysik
Hardenbergstr. 36, 10623 Berlin, Germany

Professor Dr., Dres. h. c. Klaus von Klitzing
Max-Planck-Institut für Festkörperforschung
Heisenbergstrasse 1, 70569 Stuttgart, Germany

Professor Hiroyuki Sakaki
University of Tokyo
Institute of Industrial Science,
4-6-1Komaba, Meguro-ku, Tokyo 153-8505
Japan

Professor Dr. RolandWiesendanger
Institut für Angewandte Physik
Universität Hamburg
Jungiusstrasse 11, 20355 Hamburg, Germany

ISBN: 978-3-540-85048-9

e-ISBN: 978-3-540-85049-6

NanoScience and Technology ISSN 1434-4904

Library of Congress Control Number: 2008934059

© Springer-Verlag Berlin Heidelberg 2009

Cover design: WMXDesign GmbH, Heidelberg

Printed on acid-free paper

9 8 7 6 5 4 3 2 1

springer.com

Preface for Applied Scanning Probe Methods Vol. XI–XIII

The extremely positive response by the advanced community to the Springer series on Applied Scanning Probe Methods I–X as well as intense engagement of the researchers working in the field of applied scanning probe techniques have led to three more volumes of this series. Following the previous concept, the chapters were focused on development of novel scanning probe microscopy techniques in Vol. XI, characterization, i.e. the application of scanning probes on various surfaces in Vol. XII, and the application of SPM probe to biomimetics and industrial applications in Vol. XIII. The three volumes will complement the previous volumes I–X, and this demonstrates the rapid development of the field since Vol. I was published in 2004. The purpose of the series is to provide scientific background to newcomers in the field as well as provide the expert in the field sound information about recent development on a worldwide basis.

Vol. XI contains contributions about recent developments in scanning probe microscopy techniques. The topics contain new concepts of high frequency dynamic SPM technique, the use of force microscope cantilever systems as sensors, ultrasonic force microscopy, nanomechanical and nanoindentation methods as well as dissipation effects in dynamic AFM, and mechanisms of atomic friction.

Vol. XII contains contributions of SPM applications on a variety of systems including biological systems for the measurement of receptor–ligand interaction, the imaging of chemical groups on living cells, and the imaging of chemical groups on live cells. These biological applications are complemented by nearfield optical microscopy in life science and adhesional friction measurements of polymers at the nanoscale using AFM. The probing of mechanical properties by indentation using AFM, as well as investigating the mechanical properties of nanocontacts, the measurement of viscous damping in confined liquids, and microtension tests using in situ AFM represent important contributions to the probing of mechanical properties of surfaces and materials. The atomic scale STM can be applied on heterogeneous semiconductor surfaces.

Vol. XIII, dealing with biomimetics and industrial applications, deals with a variety of unconventional applications such as the investigations of the epicuticular grease in potato beetle wings, mechanical properties of mollusc shells, electrooxidative lithography for bottom-up nanofabrication, and the characterization of mechanical properties of biotool materials. The application of nanomechanics as tools for the investigation of blood clotting disease, the study of piezo-electric polymers, quantitative surface characterization, nanotribological characterization of carbonaceous materials, and aging studies of lithium ion batteries are also presented in this volume.

We gratefully acknowledge the support of all authors representing leading scientists in academia and industry for the highly valuable contribution to Vols. XI–XIII. We also cordially thank the series editor Marion Hertel and her staff members Beate Siek and Joern Mohr from Springer for their continued support and the organizational work allowing us to get the contributions published in due time.

We sincerely hope that readers find these volumes to be scientifically stimulating and rewarding.

August 2008   Bharat Bhushan
                                                                    Harald Fuchs

Contents – Volume XIII

Contents – Volume XI

5 Contact Resonance Force Microscopy Techniques for Nanomechanical Measurements

Contents – Volume XII

14   Mechanical Properties of Metallic Nanocontacts
         G. Rubio-Bollinger, J.J. Riquelme, S.Vieira, N. Agraït 121

**15   Dynamic AFM in Liquids: Viscous Damping and Applications to the Study of Confined
         Liquids**
         *Abdelhamid Maali, Touria Cohen-Bouhacina, Cedric Hurth, Cédric
         Jai, R. Boisgard, Jean-Pierre Aimé* 149

Contents – Volume I

Contents – Volume II

Contents – Volume III

Contents – Volume IV

Contents – Volume V

Contents – Volume VI

Contents – Volume VII

Contents – Volume VIII

Contents – Volume IX

Contents – Volume X

List of Contributors – Volume XIII

Francois Barthelat

Department of Mechanical Engineering, McGill University, Macdonald Engineering Building, Rm 351, 817 Sherbrooke Street West, Montreal, Quebec H3A 2K6
e-mail: francois.barthelat@mcgill.ca

Bharat Bhushan

Nanotribology Laboratory for Information Storage and MEMS/NEMS (NLIM), Ohio State University, Columbus, OH 43210, USA

Sophie Bistac

Université de Haute-Alsace, 15, rue Jean Starcky, BP 2488, 68057, Mulhouse Cedex, France
e-mail: Sophie.Bistac-Brogly@uha.fr

Horacio D. Espinosa

Department of Mechanical Engineering, Northwestern University, 2145 Sheridan Rd., Evanston, IL 60208-3111, USA
e-mail: espinosa@northwestern.edu

Stanislav Gorb

Evolutionary Biomaterials Group, Max-Planck-Institut für Metallforschung, Heisenbergstrasse 3, 70569 Stuttgart, Germany
e-mail: s.gorb@mf.mpg.de

Stephanie Hoeppener

Laboratory of Macromolecular Chemistry and Nanoscience, Eindhoven University of Technology, P.O. Box 513, 5600 MB Eindhoven, The Netherlands Center for NanoScience, Lehrstuhl für Photonik und Optoelektronik Luduig - Maximilians - Universität München, Geschwister-Scholl Platz 1, 80333 München, Germany
e-mail: s.hoeppener@tue.nl

Ingomar L. Jäger

Department of Materials Science, University of Leoben, Jahnstrasse 12, 8700 Leoben, Austria
e-mail: ingomar@unileoben.ac.at

Taekwon Jee

Mechanical Engineering, Texas A&M University, College Station, TX 77843-3123
e-mail: taekwonjee@gmail.com

Hyungoo Lee

Department of Mechanical Engineering, Texas A&M University, College Station, TX 77843, USA
e-mail: thanku7@gmail.com

Hong Liang

Department of Mechanical Engineering, Texas A&M University, College Station, TX 77843-3123, USA
e-mail: hliang@tamu.edu

Helga C. Lichtenegger

Institute of Materials Science and Technology E308, Vienna University of Technology, Favoritenstrasse 9-11, 1040 Wien, Austria
e-mail: helga.lichtenegger@tuwien.ac.at

Shrikant C. Nagpure

Nanotribology Laboratory for Information Storage and MEMS/NEMS (NLIM), Ohio State University, Columbus, OH 43210, USA
e-mail: nagpure.1@osu.edu

H. Peisker

Evolutionary Biomaterial Group, Max-Plank-Institut für Metallforschung, Heisenbergstrasse 3, 70569 Stuttgart, Germany

Jee E. Rim

Mechanical Engineering, Northwestern University, 2145 Sheridal Road, Technological Institute B224, Evanton, IL 60208
e-mail: j-rim@northwestern.edu

Maria Cecília Salvadori

Institute of Physics, University of São Paulo, C.P. 66318, CEP 05315-970, São Paulo, SP, Brazil
e-mail: mcsalvadori@if.usp.br

Marjorie Schmitt
Université de Haute-Alsace, 15, rue Jean Starcky, BP 2488, 68057, Mulhouse
Cedex, France
e-mail: Marjorie.Schmitt@uha.fr

Matthias Schneider
University of Augsburg, Experimental Physics I, Universitätsstr. 1, 86159 Augsburg,
Germany
e-mail: matthias.schneider@physik.uni-augsburg.de

Thomas Schöberl
Erich Schmid Institute of Materials Science of the Austrian Academy of Sciences,
Jahnstrasse 12, 8700 Leoben, Austria
e-mail: schoeber@unileoben.ac.at

Ulrich S. Schubert
Friedrich-Schiller-Universität Jena
Institute für Organische Chemie and Makromolekulare
Chemie, Humbolattstr. 10, 07743 Jena, Germany

Daniel Steppich
University of Augsburg, Experimental Physics I, Universitätsstr.
1, 86159 Augsburg, Germany
e-mail: daniel.steppich@physik.uni-augsburg.de

Stefan Thalhammer
GSF-Institut für Strahlenschutz, Neuherberg, Germany
e-mail: stefan.thalhammer@gsf.de

D. Voigt
Evolutionary Biomaterials Group, Max-Planck-Institut für
Metallforschung, Heisenbergstrasse 3, 70569 Stuttgart,
Germany
e-mail: voigt@mf.mpg.de

Ke Wang
Department of Mechanical Engineering, Texas A&M University,
College Station, TX 77843, USA
e-mail: ke.phwk@gmail.com

Achim Wixforth
University of Augsburg, Experimental Physics I, Universitätsstr.
1, 86159 Augsburg, Germany
e-mail: achim.wixforth@physik.uni-augsburg.de

List of Contributors – Volume XI

Houssein Awada

Université Catholique de Louvain, Unité de chimie et de physique des hauts polymères (POLY), Croix du Sud 1 – 1348 Louvain-la-Neuve – Belgique (B)
e-mail: houssein.awada@uclouvain.be

Elmar Bonaccurso

Max-Planck-Institute for Polymer Research, Ackermannweg 10, D-55128 Mainz, Germany
e-mail: bonaccur@mpip-mainz.mpg.de

Paolo Bonanno

Department of Biophysical and Electronic Engineering, Unversity of Genova, Via all'Opera Pia 11a, I-16145 Genova, Italy
e-mail: paolo.bonanno@unige.it

Maurice Brogly

Université de Haute Alsace (UHA), Equipe Interfaces Sous Contraintes (ICSI - CNRS UPR 9069), 15 rue Jean Starcky – 68057 Mulhouse Cx – France (F)
e-mail: maurice.brogly@uha.fr

Hans-Jürgen Butt

Max-Planck-Institute for Polymer Research, Ackermannweg 10, D-55128 Mainz Germany
e-mail: butt@mpip-mainz.mpg.de

Lorenzo Calabri

CNR-INFM – National Research Center on nanoStructures and bioSystems at Surfaces (S3), Via Campi 213/a, 41100 Modena, Italy
e-mail: calabri.lorenzo@unimore.it.

M. Teresa Cuberes

Laboratorio de Nanotécnicas, UCLM, Plaza Manuel de Meca 1, 13400 Almadén, Spain
e-mail: teresa.cuberes@uclm.es

Dmytro S. Golovko
Max-Planck-Institute for Polymer Research, Ackermannweg 10, D-55128 Mainz,
Germany
e-mail: golovkod@mpip-mainz.mpg.de

Mykhaylo Evstigneev
Fakultät für Physik, Universität Bielefeld, Universitätsstr. 25, 33615 Bielefeld,
Germany
e-mail: Mykhaylo@Physik.Uni-Bielefeld.De

Harald Fuchs
Physikalisches Institut and Center for Nanotechnology (CeNTech), Universität
Münster, Wilhelm-Klemm-Str. 10, Münster D48149, Germany
e-mail: fuchsh@uni-muenster.de

Thomas Haschke
University of Siegen, Faculty 11, Department of Simulation, Am Eichenhang 50,
D-57076 Siegen, Germany
e-mail: haschke@simtec.mb.uni-siegen.de

Donna C. Hurley
National Institute of Standards & Technology, 325 Broadway, Boulder, Colorado
80305 USA
e-mail: hurley@boulder.nist.gov

Johann Jersch
Physikalisches Institut, Universität Münster, Wilhelm-Klemm-Str. 10, Münster
D48149, Germany
e-mail: jersch@uni-muenster.de

Olivier Noel
Université du Maine, Laboratoire de Phjysique de l'Etat Condensé (CNRS UMR
6087), Avenue Olivier Messiaen – 72085 Le Mans Cx 9 – France (F)
e-mail: olivier.noel@univ-lemans.fr

Stefano Piccarolo
Dipartimento di Ingegneria Chimica dei Processi e dei Materiali, Università di
Palermo, Viale delle Scienze, 90128 Palermo, Italy and INSTM Udr Palermo
e-mail: piccarolo@unipa.it

Nicola Pugno
Department of Structural Engineering, Politecnico di Torino, Corso Duca degli
Abruzzi 24, 10129 Torino, Italy, National Institute of Nuclear Physics, National
Laboratories of Frascati, Via E. Fermi 40, 00044, Frascati, Italy
e-mail: nicola.pugno@polito.it

Roberto Raiteri
Department of Biophysical and Electronic Engineering, Unversity of Genova, Via all'Opera Pia 11a I-16145 Genova, Italy
e-mail: rr@unige.it

Davide Tranchida
Dipartimento di Ingegneria Chimica dei Processi e dei Materiali, Università di Palermo, Viale delle Scienze, 90128 Palermo, Italy and INSTM Udr Palermo

Sergio Valeri
CNR-INFM – National Research Center on nanoStructures and bioSystems at Surfaces (S3), Via Campi 213/a, 41100 Modena, Italy. Department of Physics, University of Modena and Reggio Emilia, via Campi 213/a 41100 Modena, Italy
e-mail: sergio.valeri@unimo.it

Wolfgang Wiechert
University of Siegen, Faculty 11, Department of Simulation, Am Eichenhang 50, D-57076 Siegen, Germany
e-mail: wolfgang.wiechert@uni-siegen.de

List of Contributors – Volume XII

N. Agraït
Departamento de Física de la Materia Condensada C-III, Universidad Autonoma de Madrid, Madrid 28049, Spain
e-mail: nicolas.agrait@uam.es

Jean-Pierre Aimé
CPMOH, Université Bordeaux1, 351 Cours de la Libération, 33405 Talence Cedex, France
e-mail: jp.aime@cpmoh.u-bordeaux1.fr

David Alsteens
Unité de Chimie des Interfaces, Université Catholique de Louvain, Croix du Sud 2/18, B-1348 Louvain-la-Neuve, Belgium
e-mail: david.alsteens@uclouvain.be

Guillaume André
Unité de Chimie des Interfaces, Université Catholique de Louvain, Croix du Sud 2/18, B-1348 Louvain-la-Neuve, Belgium
e-mail: guillaume.andre@uclouvain.be

Sophie Bistac
Université de Haute-Alsace, CNRS, 15, rue Jean Starcky, BP 2488, 68057 Mulhouse Cedex, France
e-mail: sophie.bistac-brogly@uha.fr

Massimiliano Bocciarelli
Politecnico di Milano, Dipartimento di Ingegneria Strutturale, piazza Leonardo da Vinci 32, 20133 Milano, Italy
e-mail: bocciarelli@stru.polimi.it

Rodolphe Boisgard
CPMOH, Université Bordeaux1, 351 Cours de la Libération, 33405 Talence Cedex, France
e-mail: r.boisgard@cpmoh.u-bordeaux1.fr

Gabriella Bolzon
Politecnico di Milano, Dipartimento di Ingegneria Strutturale, piazza Leonardo da
Vinci 32, 20133 Milano, Italy
e-mail: gabriella.bolzon@polimi.it, bolzon@stru.polimi.it

Enzo J. Chiarullo
Politecnico di Milano, Dipartimento di Ingegneria Strutturale, piazza Leonardo da
Vinci 32, 20133 Milano Italy
e-mail: chiarullo@stru.polimi.it

Touria Cohen-Bouhacina
CPMOH, Université Bordeaux1, 351 Cours de la Libération, 33405 Talence Cedex,
France
e-mail: t.bouhacina@cpmoh.u-bordeaux1.fr

Etienne Dague
Unité de Chimie des Interfaces, Université Catholique de Louvain, Croix du Sud
2/18, B-1348 Louvain-la-Neuve, Belgium
e-mail: etienne.dague@laas.fr

Jurg Dual
ETH Zentrum, IMES – Institute of Mechanical Systems, CLA J23.2, Department of
Mechanical and Process Engineering, 8092 Zürich, Switzerland
e-mail: juerg.dual@imes.mavt.ethz.ch

Yves F. Dufrêne
Unité de Chimie des Interfaces, Université Catholique de Louvain, Croix du Sud
2/18, B-1348 Louvain-la-Neuve, Belgium
e-mail: dufrene@cifa.ucl.ac.be

Vincent Dupres
Unité de Chimie des Interfaces, Université Catholique de Louvain, Croix du Sud
2/18, B-1348 Louvain-la-Neuve, Belgium
e-mail: vincent.dupres@uclouvain.be

Robert H. Eibl
Plainburgstr. 8, 83457 Bayerisch Gmain, Germany
e-mail: robert_eibl@yahoo.com

Grégory Francius
Unité de Chimie des Interfaces, Université Catholique de Louvain, Croix du Sud
2/18, B-1348 Louvain-la-Neuve,
Belgium
e-mail: gregory.francius@uclouvain.be

Hongjun Gao
Nanoscale Physics & Devices Laboratory, Institute of Physics, Chinese Academy of Sciences, P. O. Box 603, Beijing 100080, China
e-mail: hjgao@aphy.iphy.ac.cn

Pietro Giuseppe Gucciardi
CNR-Istituto per i Processi Chimico-Fisici, Salita Sperone c.da Papardo, I-98158 Messina, Italy
e-mail: gucciardi@me.cnr.it

Haiming Guo
Nanoscale Physics & Devices Laboratory, Institute of Physics, Chinese Academy of Sciences, P. O. Box 603, Beijing 100080, China
e-mail: hmguo@aphy.iphy.ac.cn

Cedric Hurth
CPMOH, Université Bordeaux1, 351 Cours de la Libération, 33405 Talence Cedex, France
e-mail: cedric.hurth@asu.edu

Cédric Jai
CPMOH, Université Bordeaux1, 351 Cours de la Libération, 33405 Talence Cedex, France
e-mail: c.jai@free.fr

Udo Lang
ETH Zentrum, IMES – Institute of Mechanical Systems, Department of Mechanical and Process Engineering, 8092 Zürich, Switzerland
e-mail: udo.lang@imes.mavt.ethz.ch

Abdelhamid Maali
CPMOH, Université Bordeaux1, 351 Cours de la Libération, 33405 Talence Cedex, France
e-mail: a.maali@cpmoh.u-bordeaux1.fr

J.J. Riquelme
Departamento de Física de la Materia Condensada C-III, Universidad Autonoma de Madrid, Madrid 28049, Spain
e-mail: juanjo.riquelme@uam.es

Gabino Rubio-Bollinger
Departamento de Física de la Materia Condensada C-III, Universidad Autonoma de Madrid, Madrid 28049, Spain
e-mail: gabino.rubio@uam.es

Marjorie Schmitt
Université de Haute-Alsace, CNRS, 15, rue Jean Starcky, BP 2488, 68057 Mulhouse
Cedex, France
e-mail: Marjorie.Schmitt@uha.fr

Claire Verbelen
Unité de Chimie des Interfaces, Université Catholique de Louvain, Croix du Sud
2/18, B-1348 Louvain-la-Neuve, Belgium
e-mail: claire.verbelen@uclouvain.be

S. Vieira
Departamento de Física de la Materia Condensada C-III, Universidad Autonoma de
Madrid, Madrid 28049, Spain
e-mail: sebastian.vieira@uam.es

Yeliang Wang
Nanoscale Physics & Devices Laboratory, Institute of Physics, Chinese Academy of
Sciences, P. O. Box 603, Beijing 100080, China
e-mail: ylwang@aphy.iphy.ac.cn

18 Visualization of Epicuticular Grease on the Covering Wings in the Colorado Potato Beetle: A Scanning Probe Approach

D. Voigt · H. Peisker · S. Gorb

Abstract. Insects and spiders are supposed to release a greasy layer on their body surface, which may be involved in chemical and physical interactions between the organisms and their environment. In mating events, males frequently adhere to the female's dorsal body site by means of their feet, whereas grease should play an important role at the feet-attachment substrate interface. The properties and thickness of epicuticular grease have been diversely reported, but no definite visualizations and measurements have been previously carried out. Using the Colorado Potato beetle as a model species, we visualized the epicuticular grease on covering wings and characterized its adhesive properties. In this study, three different AFM modes (contact, tapping, and phase contrast) were applied. Obtained data were compared with the results of the Cryo-SEM. The grease layer thickness is about 8 nm on elevated sites of the epicuticle. A strong adhesion on the beetle epicuticle due to the presence of the grease layer was measured. The influence of a semi-fluid greasy layer on male adhesion to female's wings during copulation is discussed.

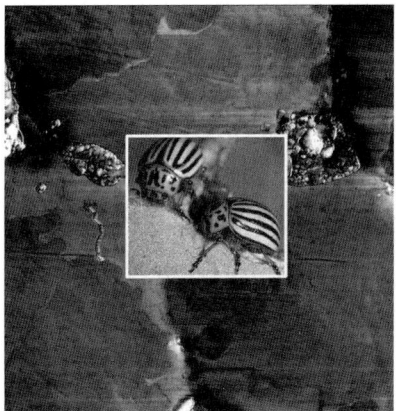

Key words: Adhesion, AFM, Chrysomelidae, Coleoptera, Covering wings, Cuticle, Elytra, Insect, *Leptinotarsa decemlineata*, SPM, Surface

Abbreviations

AFM   Atomic force microscopy
Cryo-SEM Cryo scanning electron microscopy
GR   Epicuticular grease

18.1
Introduction

18.1.1
Epicuticular Grease: Introductory Remarks

The covering wings (elytra) in the Colorado potato beetle *Leptinotarsa decemlineata* appear shiny, smooth, water-repellent and slightly slippery. These properties are due to the presence of epicuticle, the outermost layer of the insect integument, covered by a wax-like lipid surface layer called grease. Surface waxes have been previously reported in a variety of conditions, from liquid viscous coatings to crystalline structures in the form of plates, rods and filaments from many insects (adults and larvae) and arachnids [1]. Beament [2] and Wigglesworth [3] considered an outer thin layer of solid wax, whereas Lewis [4] rather assumed a lipophilous oil film on the epicuticle surface to be widespread throughout insects. According to Noble-Nesbitt [5], epicuticular lipids are found to be either waxes or mobile greases at ambient temperatures. Outermost greasy layers on the epicuticle have been reported in cockroaches [2, 6, 7], beetles [8], flies [9], true bugs [10], crickets [11], locusts [12], ticks [8, 13, 14], and spiders [15, 16].

The consistence of grease has been controversially discussed. Beament [2] extracted hard wax grease from the fresh cuticle of *Periplaneta* cockroaches, later described as mobile grease extracted from the same species [6]. In spite of observations that wax secretion films in cockroach grease remain stable [17], the cockroach's secretions spread over their entire surface by the pressure caused by this polar material and strongly reducing agent [13]. It may be a relatively fluid lipid layer hardening due to chemical reactions [7, 18].

The thickness of the superficial greasy wax layer covering the epicuticle is of molecular dimensions [17] varying from less than a few nm to several μm [11]. The bulk of lipids on an insect's surface form a layer of 0.1–1.0-μm thickness, probably in less well-oriented layers permeating the cement layer of the epicuticle [19]. But a precise definition of grease thickness has only been reported for the hunting spider *Cupiennius salei* Keys (Ctenidae), where the thickness of the surface viscous layer was measured within 20–40 nm [16].

The greasy material on the cuticle surface is considered to play a fundamental role in preventing water loss [e.g. 20, 21]. Furthermore, cuticular lipids may serve as chemical cues used by insects for signalling in olfactory communication [22].

Thus, previous research on grease mainly considered chemical and physicochemical approaches. Chloroform extractions and chemical analyses have indicated the presence of lipids in the epicuticular layers of several arthropod species [2, 7, 12, 15, 23]. In the bug *Rhodnius*, a viscous semi-liquid protein layer rich in polyphenols was found [17]. In *L. decemlineata*, a protective, yellow, oily fluid of strong odour and acrid taste, secreted from hypodermal glands located along the margins of the dark stripes on the elytron's surface, has been described [24].

Both fluid and solid coverage, located on the surface, may promote or prevent adhesive interactions at the interface between attachment devices of insects and the surface [25]. Males of the Colorado potato beetle *L. decemlineata* firmly adhere to the female surface during copulation using adhesive tarsal setae. Tarsal secretion has been previously supposed to be an adhesive agent in leaf beetles (Chrysomelidae) [26, 27] and for this reason one may conjecture that the grease of the female elytra may play an important role in male attachment on the female surface during copulation.

18.1.2
Covering Wings and Mating Behaviour

The female covering wings (elytra) of *L. decemlineata* are a common attachment substrate for male feet [24, 28–31]. The copulation process consists of several phases [28–32]. The male mounts the female laterally and backwards, at first wobbling on the convex female elytra (Fig. 18.1A, B). The male repeatedly touches the smooth surface of the elytra with shearing movements of his feet. After reaching an optimal position, the male may stay motionless for several hours, adhering to the female's elytra with tarsi of fore- and mid-legs. With the hind-legs, the male is mostly grasping terminal sclerites of the female's abdomen without touching the surface of the elytra. The male's feet, in contact with the female's surface, seem to be attached with the entire area of the attachment system, consisting of tenent setae located on the ventral surface of tarsomeres (Fig. 18.1D). Observation of the copulation posture lets us assume a role of the female elytra epicuticular grease on male adhesion (Fig. 18.1C).

Leptinotarsa decemlineata possess oval, triangular, convex elytra with a pale yellow to flavous coloured base [33]. Elytra give the insect a hemispherical appearance. Elytra are about 7.8 mm long and 6.8 mm wide [33]. The convex, anterior margin of the elytron tapers downward abruptly into the apophysis, whereas the two other sides of each elytron taper gradually into a point [34]. Each elytron has five black stripes (striae, vittae) extending from its base to its apex. The sutural margin of the elytron is black, the first vitta nearly joins it at the apex. Vittae 2 and 3 join at the apex. Punctation is course, patterned in the form of irregular double rows along the vittae. With two rows of punctuation between each two veins, and four in the subcostal space, they correspond exactly in position to the edges of the stripes [24]. In addition to these larger punctuations, smaller ones are also distributed without any discernible scheme of arrangement. Compound hypodermal glands are connected to the pores of the larger punctuation [24].

To understand mechanical interactions between male attachment devices and female elytra, detailed information on layer thickness, release, distribution, and physical properties of the grease are required. Previous morphological and ultrastructural studies using conventional electron microscopy methods (SEM, TEM) resulted in the removal of surface films, because samples were either dried and/or washed in organic solvents, such as ethanol, acetone, and propylene oxide, according to the conventional preparation procedures [15].

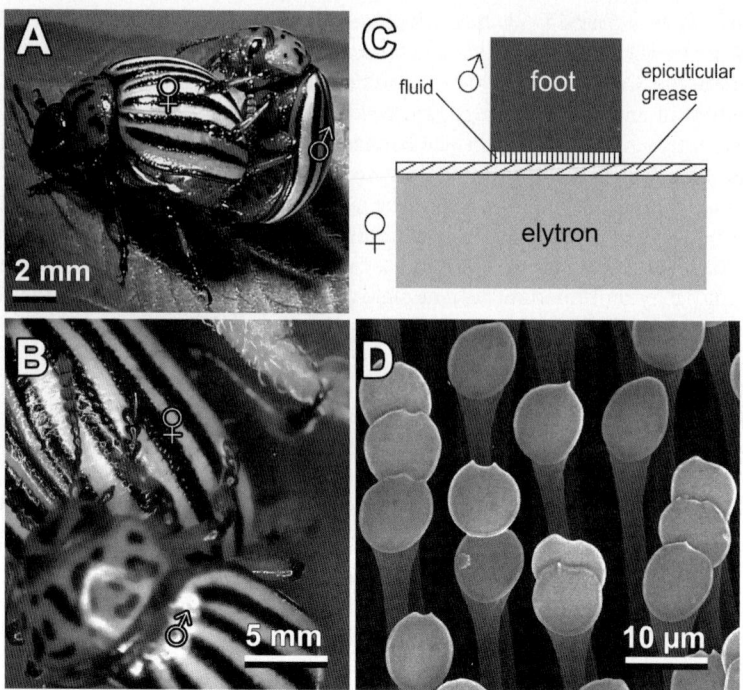

Fig. 18.1. Role of attachment in the mating process of the beetle *Leptinotarsa decemlineata*. **A, B** Typical mating posture demonstrating position of male attachment pads on the female surface. **C** Very simplified hypothetical diagram of the contact between the male tenent seta (foot) and female covering wing (elytron). Two fluid layers are presumably present between the cuticle of the seta and elytron. **D** Male tenent setae, Cryo-SEM image

The present study was undertaken to characterize the thickness of the epicuticular grease layer and its adhesive properties in beetle elytra. For this purpose, atomic force microscopy (AFM) was applied, which has been previously demonstrated to be an excellent approach for geometrical characterization of insect and spider cuticles and for estimation of their adhesive properties on a local scale [16, 35, 36]. We have compared AFM data of the present study with previous results obtained in cryo-SEM [30,31].

18.2
Methods

18.2.1
Insects and Sample Preparation

The beetles, *Leptinotarsa decemlineata* Say (Coleoptera, Chrysomelidae), were taken from a stock culture (25 °C, 60% RH, 16-h photoperiod) at Bayer Cropscience AG (Monheim, Germany).

Elytra were removed from anaesthetized alive females using micro-scissors and tweezers. For measurements, two types of samples were used: (1) fresh elytra without treatment and (2) fresh elytra washed in 50-ml chloroform shaking for 1 min in a vibrator, rinsed with aqua Millipore and dried using compressed air. Samples of the second type were treated with chloroform in order to remove grease and obtain a clean control surface without surface coverage. Four pieces ($5\,mm^2$) were cut out from the dorsal part of a single elytron, using a razor blade, and immediately mounted on a glass slide using double-sided adhesive tape. Prior to measurements, samples where kept in clean Petri dishes with a small piece of moist filter paper to prevent contamination and desiccation. In all, five female beetle individuals (ten elytra) were studied.

18.2.2
Scanning Probe Microscopy

We used a NanoWizard® atomic force microscope (JPK Instruments AG, Berlin, Germany) mounted on an inverted microscope Zeiss Axiovert 135 (Carl Zeiss MicroImaging GmbH, Göttingen, Germany). NanoWizard® image acquisition software 3.1.13 (JPK Instruments AG, Berlin, Germany) was used to obtain AFM images and NanoWizard® image processing software 3.1.6 was used to process images and calculate all necessary geometrical parameters of the surface. The most-common operating modes in the AFM are tapping mode and contact mode. The latter provides a better resolution, whereas, in the tapping mode, both the normal force and shear force applied to the sample are minimized. Visualization of the epicuticular grease in both modes provided us with complementary information. The force applied to the surface, was set to a minimum to ensure the integrity of the sample in both operating modes.

18.2.2.1
Tapping Mode

Scans were carried out in air at a scan rate of 1 Hz and a resolution of 512×512 pixels using standard, non-contact, high frequency cantilevers with reflex coating (NST-NCHF-R, Nascatec Technologies GmbH, Stuttgart, Germany). AFM images were taken from four pieces of the same elytron, two fresh untreated ones and two chloroform treated ones (see Sect. 18.2.1). Five scans ($20 \times 20\,\mu m$) per specimen were carried out. A total of 20 images per sample were taken.

18.2.2.2
Contact Mode

In contrast to the tapping mode, the contact mode always exerts a much larger lateral force, making it more difficult to properly visualize sensitive surfaces like insect epicuticular grease. However, load force must be sufficient to properly resolve the surface topography. The optimal balance between these two parameters was determined by scanning the same surface repeatedly while optimizing the setpoint, scan rate, and gains of the feedback loop.

Scans with a size of $20 \times 20\,\mu m$ were obtained at 1-Hz line rate and a resolution of 512×512 pixels using standard contact mode cantilevers (NST-CM, Nascatec Technologies GmbH, Stuttgart, Germany). A total of 20 images per sample were taken (four specimens used, five images per specimen).

18.2.2.3
Adhesion Measurements

The attraction and repulsion forces acting on the standard contact-mode cantilever tip (NST-CM, Nascatec Technologies GmbH, Stuttgart, Germany) were measured in pull-off experiments. The spring constant of each cantilever was determined according to the thermal noise technique [36, 37]. The idea of the technique is that while the tip is far from the sample, its motion is due only to thermal fluctuations, and measurements of this motion at frequencies near the resonant frequency of the spring allow estimating the spring constant [37].

Adhesive properties of fresh samples of beetle elytra (see Sect. 18.2.1) were compared with those of a silanized Si-wafer ($114°$ contact angle of aqua Millipore). The Si-wafer was used as a defined hydrophobic surface with the surface roughness comparable to that of single polygonal cells observed on the dorsal surface of the elytron (unpublished data). To exclude the influence of sites with stronger adhesion located in the vicinity of pores due to the accumulated secretion, force-distance curves were taken on the tops of pre-scanned, convex elytron cells. The pre-scanning procedure was carried out to identify the proper site at the top of the hexagonally shaped cell for taking force-distance curves.

In total, 127 force-distance curves were collected on both types of surfaces. The adhesion force was evaluated from the curves with NanoWizard® image processing software 3.1.6 (JPK Instruments AG, Berlin, Germany). The Mann–Whitney Rank Sum test was used to estimate statistic differences in the adhesion force acting on the cantilever tip in contact with the beetle elytra and with the silanized Si-wafer (SigmaStat 3.1.1® software, Systat Software, Inc., Richmond, California, USA).

18.2.3
Cryo-SEM

Cryo-SEM allows observation of fresh specimens and provides a unique possibility to prove our data obtained from AFM scans. Female elytra surfaces were analyzed using a cryo-SEM Hitachi S 4800 (Hitachi High-Technologies Corp., Tokyo, Japan) equipped with a Gatan ALTO 2500 cryo-preparation system (Gatan Inc., Abingdon, UK). Previously, this method has been successfully applied for visualization of tarsal adhesive fluid in flies [38]. Pieces of elytra from live, anaesthetized beetles were cut off with a fine knife, mounted on metal holders, frozen on a cryo-stage at $-140°C$, sputter-coated with gold-palladium (3 nm) in the preparation chamber, and examined in the SEM at $-120°C$ and with an accelerating voltage of 5 kV. Also, elytra treated with chloroform were examined for comparison.

18.3
Results

18.3.1
Elytra Topography and Grease Visualization

Cryo-SEM images of fresh, shock-frozen elytra demonstrated large pores (diameter $75.1 \pm 6.98 \,\mu m$, mean $\pm$ SD, n = 10) located at margins of dark stripes (Fig. 18.2). Additionally, there are some mid-sized pores unevenly distributed over the surface (diameter $29.5 \pm 3.32 \,\mu m$, mean $\pm$ SD, n = 10). The elytron surface consists of a hexagonal shaped, cell-like pattern. Smaller pores (diameter $3.9 \pm 0.02 \,\mu m$, mean $\pm$ SD, n = 10) are often situated at the corners of hexagons. Several tiny pores of 0.2 ± 0.02-μm diameter (mean $\pm$ SD, n = 10) were found on the top of cells. The pores observed are presumably responsible for delivering secretory substances to the cuticle surface. At higher magnification, an amorphous layer smeared over the surface can be detected (Fig. 18.2D). This layer often covers small pores and gaps between hexagons and thus makes the surface uniformly smooth.

The gathered AFM tapping-mode height images of fresh samples provided 3D information about hexagonal cells at the nanoscale (Fig. 18.3). However, no grease was visible.

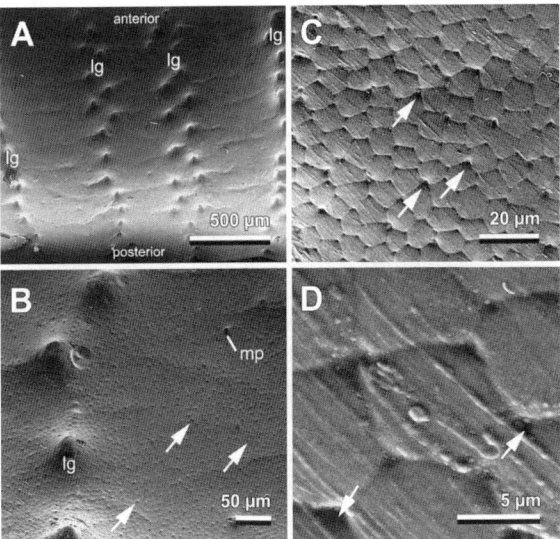

Fig. 18.2. Cryo SEM micrographs of the elytra surface in a fresh female *Leptinotarsa decem-lineata*. **A** Overview of the posterior elytra surface showing irregular double rows of pores corresponding to large compound hypodermal glands. **B** Detail of **A**, a row of large pores on the left side, hexagonal surface pattern, regularly dispersed small pores (*arrows*) and a single mid-sized pore is visible. **C** Detail of **B**, hexagonal-shaped cells and small pores (*arrows*). **D** Small pores (*arrows*) and single cells of the hexagonal surface pattern. Grease smeared over the surface is visible. lg, large glands; mp mid-sized pores

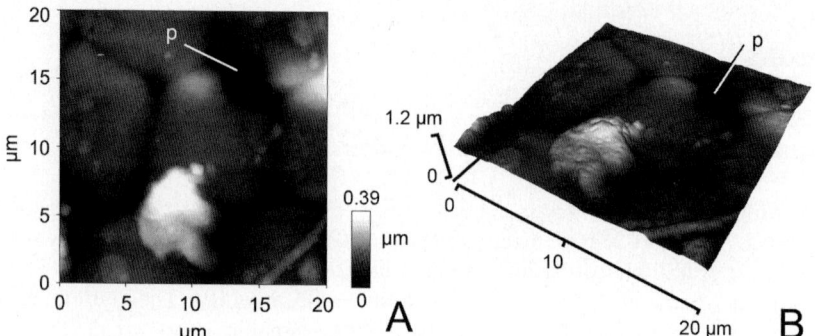

Fig. 18.3. AFM tapping-mode images of a fresh female elytron of *Leptinotarsa decemlineata*. **A** Height image. **B** 3D image. p, pores

On surfaces washed with chloroform, irregular runlet-like patches of presumably waxy residues were found in AFM tapping-mode images (Fig. 18.4). These residues demonstrate that chloroform can dissolve the grease, but after evaporation of the solvent, they form a specific pattern on the surface. The typical width of such runlets in a cross section was determined to be in the range of 250 nm, and their height about 12 nm.

In order to obtain a more detailed view of fresh surfaces, phase images were taken and compared to height images (Fig. 18.5). Height images alone did not allow a

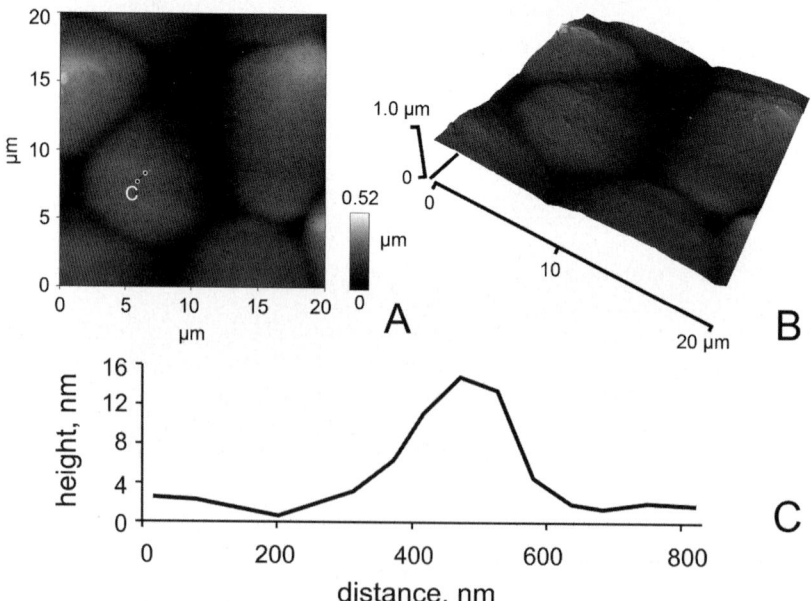

Fig. 18.4. AFM tapping-mode images of a chloroform-treated female elytron of *Leptinotarsa decemlineata* showing residues of grease. **A** Height image. **B** 3D image of the surface. **C**. Profile diagram of a representative cross section along the *line* indicated between two dots in **A**

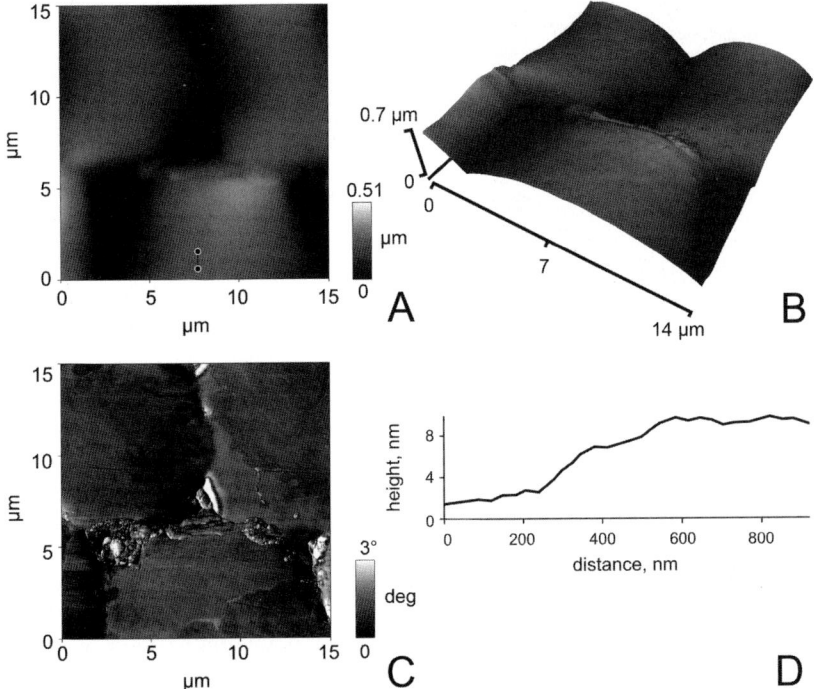

Fig. 18.5. AFM tapping-mode images of a fresh female elytron of *Leptinotarsa decemlineata*. **A** Height image. **B** Phase image of **A** indicating differences in surface properties/geometry by the use of phase-shift visualization. Grease distribution (*darker sites*) is clearly visible. **C** 3D phase image showing the grease boundary. **D** Profile diagram of the height image representing a cross section of the grease boundary along the *line* indicated between two dots in **A**

definition of the grease distribution over the surface (Fig. 18.5A), whereas phase images (Fig. 18.5B, C) showed distinct differences in material properties and/or geometry indicated by a phase shift. The grease often accumulated in the vicinity of small pores, within the pores themselves, or in the regions between cells. The grease layer was about 8 nm thick, shown by the profile crossing the boundary between a grease patch and the surface of the epicuticle (Fig. 18.5D).

Topographical images of the fresh elytra, obtained in the contact mode, showed a heavy, smeary, striped appearance especially in the proximity of cuticle pores (Fig. 18.6). The visualization of surface fine structure was disturbed in the vicinity of the smear. An attempt to decrease the force applied to the sample surface resulted in almost non-interpretable images of surface topography.

18.3.2
Adhesive Properties of the Elytra Surface

Having a thin layer of viscous grease on the surface of female elytra, one can expect an increase in adhesion (due to an increase of capillary forces), when male tarsal setae contact the surface. To prove this idea, we compared pull-off forces measured

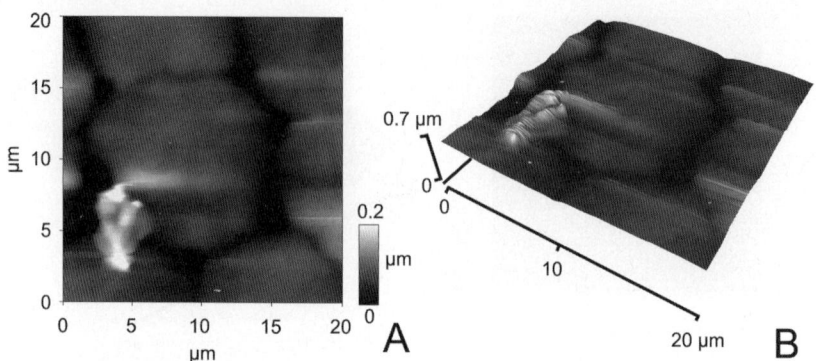

Fig. 18.6. AFM contact-mode scans of a fresh female elytron of *Leptinotarsa decemlineata*. **A** Height image. **B** 3D image. Images show smearing artefacts, presumably caused by pushing cuticle grease with the cantilever tip

on fresh female elytra with those obtained on the silanized Si-wafer. A relatively low adhesion force, ranging from 14 to 16 nN, was measured presumably due to the low capillary interactions on the hydrophobic surface (Fig. 18.7A). On fresh elytra, a considerably stronger pull-off force was detected (up to 28 nN). A relatively sharp contact breakage event was observed when retracting the cantilever tip from the elytra surface (Fig. 18.7B). The contact breakage was smoother in absolutely fresh elytra and became sharper when elytra began to dry.

18.4
Discussion

The present paper is a case study demonstrating the application of the AFM for visualization of the epicuticular grease in females of the Colorado potato beetle. The location and thickness of the grease was demonstrated, showing its non-uniform distribution over the elytra surface. The grease residues also remain on the top of the hexagonal cells in the form of patches, whereas the rest runs down the curvature and accumulates within the pores. Also, grease accumulation between the cells is clearly visible.

The data, obtained with the AFM, were partly proven with the use of Cryo-SEM, which supported the presence and geometry of the grease layer. These data, providing an important extension to the previous data of chemical analyses, were obtained for the first time with each of both microscopy methods.

The combined use of both AFM scanning modes (tapping and contact) shows the semi-fluid nature of the grease layer: the contact mode resulted in smeared images of the elytron surface, corresponding to fluid properties, whereas the tapping mode provided clear images of the surface, corresponding to solid properties.

Waxy residues found on the chloroform-treated samples indicate that lipid-like cuticle materials take a different shape after solvent evaporation, but are not completely removed from the surface. That is why chloroform treatment seems not to be

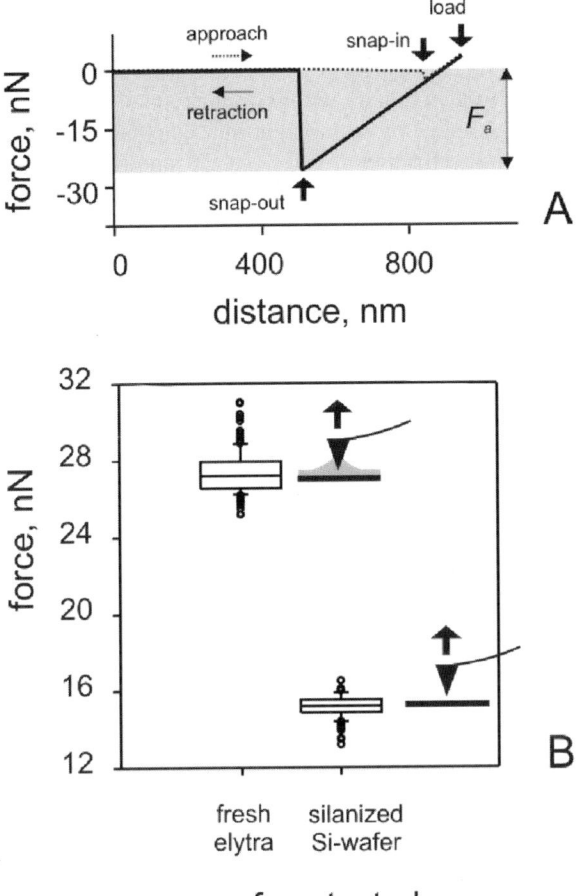

Fig. 18.7. Results of adhesion measurements. **A** Example of a force-distance curve obtained on a fresh female elytron of the beetle *Leptinotarsa decemlineata*. The *dotted line* represents data obtained while the cantilever approaches the surface and has to be read from left to right. The *black line* represents data acquired while the cantilever retracts from the sample showing adhesion of 27 nN. This line has to be read from right to left. The right side of the curve, where the force is positive, corresponds to the load applied to the sample (about 4 nN). The left part of the graph indicates the change in direction of the cantilever movement while not in contact. **B** Box-and-whiskers diagram of the adhesion force measured on fresh female elytra, and on a silanized Si-wafer. Statistically significant differences between the groups are revealed ($T = 24, 257$, $p \leq 0.001$, Mann–Whitney Rank Sum test, n = 127 per group)

a sufficient method to clean the elytra surface from cuticle fluids/secretions/grease. In addition, we cannot be sure that lipids situated in the depth of the cuticle are not dissolved and deposited on the surface after fluid evaporation. Thus, the method of chloroform treatment should also be critically evaluated in other kinds of experiments, for example in chemical analyses. The proper way of material collection for analysis of the cuticle grease is the preparation of surface prints on glass with further washing out of substances from the glass surface, as previously shown for tarsal adhesive secretion in the locust [12].

The phase image, used in this study for visualization of grease, has demonstrated very good results in combination with height images obtained in the tapping mode. Phase images aided in localization of places on the smooth epicuticle covered with the grease. In spite of uncertainty of the origin of such a phase contrast (caused either by (1) surface geometry, (2) differences in local material properties, or (3) both), it seems to be a powerful tool for detection and visualization of submicron-thick layers on the smooth surface of biological objects.

Hydrocarbons are a major component of the surface lipids in *L. decemlineata* [39]. There were quantitative compositional differences in epicuticular hydrocarbons found between different sexes: females contain a higher amount of branched methylalkanes [39,40]. It has been previously supposed that quantitative differences observed in the hydrocarbon profile of both sexes may be used in sex recognition [39]. Indeed, behaviour experiments on mating preferences have demonstrated that compounds of the female elytra stimulate mating responses in male, whereas those found on male elytra act as mating response inhibitors [41]. Taking into account the mating posture of both sexes, one can suggest a possible role of the female epicuticular grease in male attachment on the female surface.

Since roughness of the Si-wafer was comparable with the roughness of single hexagonal cells of elytra (Ra = 7.3 ± 0.77 nm, RMS = 9.1 ± 0.77 nm; mean $\pm$ SD, n = 15, scanning area 2×2 μm, unpublished data), detected differences in the pull-off force were not due to the surface topography. Also, epicuticle of the tanned cuticle is rather stiff [42] compared to the spring constant of the cantilever beam ($k = 0.0627$ N/m). That is why the differences in the material compliance between epicuticle and Si-wafer cannot explain differences in adhesion. We tend to explain differences in adhesive properties between these two substrates by the presence of the fluid layer on the beetle epicuticle. However, the influence of surface energy, which is presumably lower in silanized Si-wafer compared to clean epicuticle, cannot be completely excluded. The sharp contact breakage event, while retracting the cantilever tip from the elytra surface can be explained by highly viscous properties of the grease, especially in non-fresh specimens (unpublished data).

The sharp character of pull-off events that occurred when the cantilever tip was retracted from the sample surface was typical of a silicon substrate. Considering the assumed fluid-like properties of the cuticle grease, a less abrupt return of the cantilever to the zero deflection was expected (see our previous AFM data on the adhesive fluid of fly pulvilli [43–46]). The sharp character of the pull-off part of force-distance curves on the surface of the elytra may be explained by a very small radius of the cantilever tip (less than 10 nm).

As mentioned above, having a layer of greasy semi-fluid material on the elytra surface, one can expect its influence on male adhesion during copulation. The insect adhesive system relies on a certain amount of fluid in the contact area [47–50]. An increase in the fluid layer due to the presence of the grease presumably leads to an aquaplaning effect when, instead of an increase in friction forces, a lubrication (slipping) effect occurs. Video recordings of copulations of Colorado potato beetles [32] demonstrate the difficulties male beetles have in getting a firm grip on the female elytra surfaces. Male tarsi continuously slide along the female elytra. This means that, even having the proper foot position on the female elytra, male attachment systems fail on the female surface. It is known that females in many species of the beetle family Dytiscidae are dimorphic bearing smooth or structured elytra surfaces caused by sexual selection and sexual conflict. Female elytra and male attachment systems underlie continuous adaptations and counter-adaptations determined by the reproductive mechanisms and mate choice of females in a mix of cooperative and conflict interactions with males [51]. Whereas most males are adapted to the smooth surface of elytra, only the fittest ones succeed by proving their adhesion to rough surfaces [52]. Polyandry, mating and long mate guarding events have been suggested to be costs influencing female fitness [53, 54], also known in *L. decemlineata* [55]. They may drive the evolution of female adaptations to interfering with the male attachment system. Furthermore, females of *L. decemlineata* have been recorded to mate size-assortively [24] and could show clear preference for individual males [29]. All of these female adaptations may be related to the presence and properties of the grease. If female grease thickness changes during their life cycle, and if females are able to control the grease thickness, one may hypothesize that females may potentially use the grease thickness/properties for mating control.

18.5
Conclusions

No information about the distribution and thickness of cuticle grease on the cuticle surface of the Colorado potato beetle has been found in the literature. This study demonstrates three different AFM applications (contact mode, tapping mode, and phase contrast) not only for visualization of cuticle grease in living insect cuticle, but also for probing its properties. Cryo-SEM was applied as a control method to prove data obtained with AFM. The grease layer was about 8 nm thick on elevated sites of the epicuticle. Grease accumulation was observed between the hexagonally shaped cells of the cuticle and within the pores. We have demonstrated strong adhesion on the beetle epicuticle due to the presence of the grease layer. The possible influence of the presence of a layer of greasy semi-fluid material on male adhesion during copulation is discussed.

Acknowledgments. Peter Meisner, Bayer Cropscience AG, Research Insecticides/Insecticide Biology (Monheim, Germany) kindly provided the insects. Valuable discussions with Michael Varenberg (Technion, Haifa, Israel) on contact mechanics and tribology are greatly acknowledged. Cristian Löbbe (JPK, Berlin, Germany) helped with the data processing. Victoria Kastner (Tübingen, Germany) provided linguistic corrections on an early draft of the manuscript. This study was partly funded by the Federal Ministry of Education and Research, Germany (project InspiRat 01RI0633D).

References

1. Hadley NF (1981) Cuticular lipids of terrestrial plants and arthropods: a comparison of their structure, composition, and waterproofing function. Biol Rev 56:23–47
2. Beament JWL (1945) The cuticular lipoids of insects. J Exp Biol 21:115–131

3. Wigglesworth VB (1945) The insect cuticle. Biol Rev 23:408–451
4. Lewis CT (1962) Diffusion of oil films over insects. Nature 183:904
5. Noble-Nesbitt J (1992) Cuticular permeability and its control. In: Binnington K, Retnakaran A (eds) Physiology of the insect epidermis. CSIRO Australia, pp 252–283
6. Beament JWL (1958) The effect of temperature on the waterproofing mechanism of an insect. J Exp Biol 35:494–519
7. Gilby AR, Cox ME (1963) The cuticular lipids of the cockroach *Periplaneta americana* (L.). J Insect Physiol 9:671–681
8. Richards AG (1951) The integument of arthropods. The chemical components and their properties, the anatomy and development, and the permeability. University of Minnesota Press, Minneapolis
9. Wolfe LS (1954) Studies of the development of the imaginal cuticle of *Calliphora erythro- cephala*. J Cell Sci s3-95:67–78
10. Wigglesworth VB (1933) The physiology of the cuticle and of ecdysis in *Rhodnius prolixus* with special reference to the function of the oenocytes and of the dermal glands. Quart J Micr Sci 76:269
11. Hendricks GM, Hadley NF (1983) Structure of the cuticle of the common house cricket with reference to the location of lipids. Tissue Cell 15:761–779
12. Vötsch W, Nicholson G, Müller R, Stierhof Y-D, Gorb S, Schwarz U (2002) Chemical com- position of the attachment pad secretion of the locust *Locusta migratoria*. Insect Biochem Mol Biol 32:1605–1613
13. Lees AD, Beament JWL (1948) An egg-waxing organ in ticks. Quart J Micr Sci 89:291–332
14. Gilby AR (1957) Studies of cuticular lipids of Arthropods. III. The chemical composition of the wax from *Boophilus microplus*. Arch Biochem Biophys 67:320–324
15. Hadley NF (1981) Fine structure of the cuticle of the black widow spider with reference to surface lipids. Tissue Cell 13:805–817
16. McConney ME, Schaber CF, Julian MD, Barth FG, Tsukruk VV (2007) Viscoelastic nanoscale properties of cuticle contribute to the high-pass properties of spider vibration receptor (*Cupiennius salei* Keys). J R Soc Interface 4:1135–1143
17. Beament JWL (1955) Wax secretion in the cockroach. J Exp Biol 32:514–538
18. Gilby AR (1962) Absence of natural volatile solvents in cockroach grease. Nature 195:729
19. Locke M (1964) The structure and formation of the integument in insects. In: Rockstein M (ed) The physiology of Insecta. Academic Press, New York, pp 123–213
20. Ramsay JA (1935) The evaporation of water from the cockroach. J Exp Biol 12:373
21. Neville AC (1975) Biology of the arthropod cuticle. Springer, Berlin
22. Espelie KE, Bernays EA, Brown JJ (1991) Plant and insect cuticular lipids serve as behavioural cues for insects. Arch Insect Biochem Physiol 17:223–233
23. Lockey KH (1988) Lipids of the insect cuticle: origin, composition and function. Comp Biochem Physiol 89B:595–645
24. Tower WL (1906) An investigation of evolution in chrysomelid beetles of the genus *Lep- tinotarsa*. Carnegie Institution of Washington, Publication No. 48. Papers of the station of Experimental Evolution at Cold Spring Harbor, New York, No. 4, Press of Judd & Detweiler, Inc., Washington, DC
25. Gorb E (2007) Plant surfaces preventing insect adhesion. In: Brickwedde EF, Erb R, Lefèvre J, Schwake M (eds) Bionik und Nachhaltigkeit-Lernen von der Natur, 12. Internationale Sommerakademie St. Marienthal. Erich Schmidt Verlag GmbH & Co., Berlin, Initiativen zum Umweltschutz 68:103–110
26. Stork NE (1983) The adherence of beetle tarsal setae to glass. J Nat Hist 17:583–597
27. Eisner T, Aneshansley DJ (2000) Defence by foot adhesion in a beetle (*Hemisphaerota cyanea*). Proc Nat Acad Sci 97:6568–6573

28. Thibout E (1982) Le comportement sexuel du doryphore, *Leptinotarsa decemlineata* Say et son possible controle par l'homone juvenile et les corps allates. Behaviour 80:199–217

29. Szentesi Á (1985) Behavioral aspects of female guarding and inter-male conflict in the Colorado potato beetle. Proceedings of the Symposium on the Colorado potato beetle, XVIIth international congress of Entomology, Research Bull 704:127–137

30. Voigt D, Schuppert JM, Dattinger S, Gorb SN (2008) Sexualdimorphismus der Haftfähigkeit an rauen Oberflächen bei *Leptinotarsa decemlineata* Say (Coleoptera, Chrysomelidae). Mitt Dtsch Ges Allg Angew Ent 16:431–434

31. Voigt D, Schuppert JM, Dattinger S, Gorb SN (2008) Sexual dimorphism in the attachment ability of the Colorado potato beetle *Leptinotarsa decemlineata* (Coleoptera, Chrysomelidae) to rough substrates. J Insect Physiol 54:765–776

32. Wyss U (2005) Lebensweise und Entwicklung des Kartoffelkäfers *Leptinotarsa decemlineata*. Video documentation. Institut für Phythopatologie, Christian-Albrechts-Universität zu Kiel, Germany, www.entofilm.com

33. Jaques RL (1988) The potato beetles. The genus *Leptinotarsa* in North America (Coleoptera: Chrysomelidae). Flora & Fauna Handbook, no. 3, E. J. Brill, Leiden, New York, Kobenhavn, Köln

34. Rivnay E (1928) External morphology of the Colorado beetle (*Leptinotarsa decemlineata* Say). J New York Entomol Soc 26:25–145

35. Scherge M, Gorb SN (2001) Biological micro- and nanotribology. Springer, Berlin Heidelberg New York

36. Langer MG, Ruppersberg JP, Gorb S (2004) Adhesion forces measured at the level of a terminal plate of the fly's seta. Proc R Soc Lond B 271:2209–2215

37. Hutter JL, Bechhoefer J (1993) Calibration of atomic-force microscope tips. Rev Sci Instrum 64:1868–1873

38. Gorb SN (2006) Fly microdroplets viewed big: a Cryo-SEM approach. Microsc Today N9:38–39

39. Dubis E, Malínski E, Dubis A, Szafranek J, Nawrot J, Poplawski J, Wróbel JT (1987) Sex-dependent composition of cuticular hydrocarbons of the Colorado beetle, *Leptinotarsa decemlineata* Say. Comp Biochem Physiol A 87:839–843

40. Nelson DR, Adams TS, Fatland CL (2003) Hydrocarbons in the surface wax of eggs and adults of the Colorado potato beetle, *Leptinotarsa decemlineata*. Comp Biochem Physiol B 134:447–466

41. Mpho M, Seabrook WD (2003) Functions of antennae and palpi in the mating behaviour of the Colorado potato beetle *Leptinotarsa decemlineata* (Coleoptera: Chrysomelidae). Bull Entomol Res 93:91–95

42. Barbakadze N, Enders S, Gorb S, Arzt E (2006) Local mechanical properties of the head articulation cuticle in the beetle *Pachnoda marginata* (Coleoptera, Scarabaeidae). J Exp Biol 209:722–730

43. Stadler H, Mondon M, Then D, Jiao Y, Gorb SN, Ziegler C (2000) Scanning force microscopy measurements of the viscosity force of fly pad secretion. In: Scanning-probe microscopes and organic materials IX, Workshop, Hannover, 9.10.-11.11.2000, Abstract Booklet, edited by Kolb H-A, Enders O, Guckenberger R, Heckl WM, Hörber JKH, Rabe JP, and Ziegler C, Hannover.

44. Stadler H, Mondon M, Wallentin J, Jiao Y, Gorb S, Ziegler C. (2001) Viscosity force of the fly's pad secretion measured by atomic force microscopy. Technische Biologie und Bionik 5, ed. by Nachtigall W. Mainz: Akademie der Wissenschaften und der Literatur. Biona Report 15:340–344

45. Wallentin J, Mondon M, Stadler H, Gorb SN, Ziegler C (1999) Rasterkraftspektroskopische Untersuchung der Adhäsionseigenschaften von Fliegensekret. In: Deutsche Physikalische

Gesellschaft e.V., Frühjahrstagung des Arbeitskreises Festkörperphysik bei der DPG, Regensburg, März 27–31, 2000, Abstract Booklet, Regensburg, p 807

46. Wallentin J, Mondon M, Stadler H, Gorb SN, Ziegler C (1999) The secretes of fly secretes: adhesion properties probed by force-distance curves. In: Scanning-probe microscopes and organic materials VIII, Workshop, Basel, October 4–6, 1999, Abstract Booklet, ed. by Müller DJ, and Knapp HF, Basel, 1999, p 14

47. Gorb S (2001) Attachment devices of insect cuticle. Kluwer Academic Publishers, Dordrecht, p 305

48. Ishii S (1987) Adhesion of a leaf feeding ladybird *Epilachna vigintioctomaculata* (Coleoptera: Coccinellidae) on a vertically smooth surface. Appl Entomol Zool 22:222–228

49. Walker G, Yule AB, Ratcliffe J (1985) The adhesive organ of the blowfly, *Calliphora vomitoria*: a functional approach (Diptera: Calliphoridae). J Zool Lond A 205:297–307

50. Attygalle AB, Aneshansley DJ, Meinwald J, Eisner T (2000) Defence by foot adhesion in a chrysomelid beetle (*Hemisphaerota cyanea*): characterization of the adhesive oil. Zoology 103:1–6

51. Alcock, J (2006) Animal behavior. An evolutionary approach, 8th edn. Elsevier Spektrum Akademischer Verlag, p 577

52. Härdling R, Bergsten J (2006) Nonrandom mating preserves intrasexual polymorphism and stops population differentiation in sexual conflict. Am Nat 167:401–409

53. Arnquist G, Nilsson T (2000) The evolution of polyandry: multiple mating and female fitness in insects. Anim Behav 60:145–164.

54. Miller KB (2003) The phylogeny of diving beetles (Coleoptera: Dytiscidae) and the evolution of sexual conflict. Biol J Linn Soc 79:359–388

55. Orsetti DM, Rutowski RL (2003) No material benefits, and a fertilization cost, for multiple mating by female leaf beetles. Anim Behav 66:477–484

19 A Review on the Structure and Mechanical Properties of Mollusk Shells – Perspectives on Synthetic Biomimetic Materials

*Francois Barthelat · Jee E. Rim · Horacio D. Espinosa**

Key words: Mollusk shells, Biomaterials, Fracture, Microfabrication, MEMS

19.1
Introduction

Natural materials can exhibit remarkable combinations of stiffness, low weight, strength, and toughness which are in some cases unmatched by manmade materials. In the past two decades significant efforts were therefore undertaken in the materials research community to elucidate the microstructure and mechanisms behind these mechanical performances, in order to duplicate them in artificial materials [1, 2]. This approach to design, called biomimetics, has now started to yield materials with remarkable properties. The first step in this biomimetic approach is the identification of materials performances in natural materials, together with a fundamental understanding of the mechanisms behind these performances (which has been greatly accelerated by recent techniques such as scanning probe microscopy).

The mechanical performance of natural materials is illustrated in Fig. 19.1, a material properties map for a selection of natural ceramics, biopolymer, and their composites [3]. The upper left corner of the map shows soft and tough materials such as skin, with a mechanical behavior similar to elastomers. The lower right corner of the chart shows stiff but brittle minerals such as hydroxyapatite or calcite. Most hard biological materials incorporate minerals into soft matrices, mostly to achieve the stiffness required for structural support or armored protection [4]. These materials are seen in the upper right part of the map and show how natural materials achieve high stiffness by incorporating minerals while retaining an exceptional toughness. Alternatively, one can consider how natural materials turn brittle minerals into much tougher materials, in some cases only with a few percent additions of biopolymers. These materials have in general relatively complex structures organized over several length scales (hierarchical structures [1, 2]) with mechanisms operating over several length scales, down to the nanoscale [5, 6].

Mollusk shells, the topic of this chapter, are an excellent example of such high-performance natural materials. Mollusks are composed to at least 95% of minerals

---

* Corresponding author

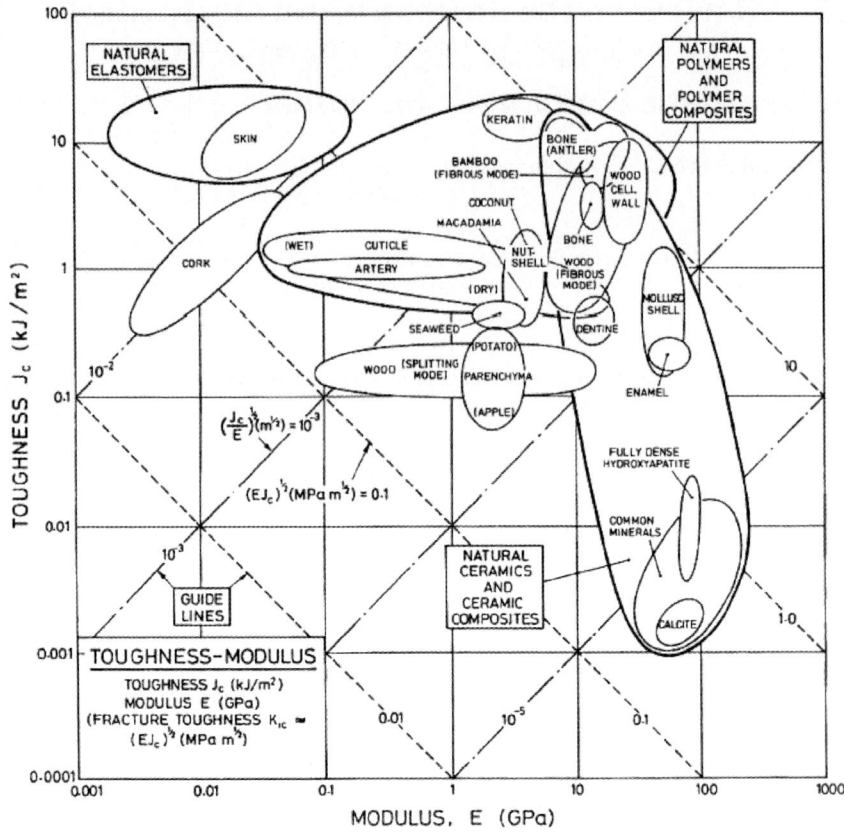

Fig. 19.1. Materials property map for a variety of ceramics, soft natural tissues, and their composites. The vertical axis (toughness) is a measure of the ability of the material to resist cracking, while the horizontal axis (modulus) is a measure of the stiffness of the material. Reproduced with permission from [3]

such as calcium carbonate (calcite, aragonite), yet by comparison with these brittle materials, mollusk shells are about 1,000-times tougher (see Fig. 19.1), at the expense of a small reduction in stiffness. How is such performance achieved? Can it be duplicated in artificial materials? This chapter gives first an overview of mollusk evolution and general characteristics. The next two parts then focus on the detailed structure and mechanics of two of the materials found in shells: the calcitic Pink Conch shell and Nacre from Red Abalone. Finally, the intricate structures and mechanisms of these materials have already inspired artificial materials, which are discussed in the last part of the chapter.

19.1.1
Mollusk Shells: Overview

Mollusks appeared 545 million years ago, and comprise about 60,000 species [7]. They have a very soft body (*mollis* means soft in Latin) and most of them grow a hard shell for protection. The earlier mollusks were small (2–5 mm) with shell structures very similar to the modern forms. The size and the diversity of the mollusk family increased dramatically 440–500 millions years ago, with the apparition of various classes. Currently, the class that includes the largest number of species is the Gastropoda, with about 35,000 living species. These include mostly marine species (Conch shell, top shell, abalone), but also land species (land snails). The second largest class, the Bivalvia, counts about 10,000 species and includes clams, oysters, and freshwater mussels.

The shell of mollusks is grown by the mantle, a soft tissue that covers the inside of the shell. A great variety of shell structures has emerged from this process. They include prismatic, foliated and cross lamellar structure, columnar and sheet nacre (Fig. 19.2). All of these structures use either calcite or aragonite, with a small amount of organic material that never exceeds 5% of the composition in weight. In order to provide an efficient protection, the shell must be both stiff and strong. Mechanical tests on about 20 different species of seashells by Currey and Taylor [8] revealed an elastic modulus ranging from 40 to 70 GPa, and a strength in the 20–120-MPa range. By comparison, human femoral bone is softer ($E = 20$ GPa) but stronger (150–200-MPa strength). Amongst all the structures found in shells, nacreous structures appear to be the strongest: The strength of nacre can reach 120 MPa for the shell Turbo, as opposed to a maximum of 60 MPa for other non-nacreous structures.

The shells of mollusks offer a perfect example of a lightweight, tough armor system, that now serve as models for new armor designs. The structure and mechanical properties of the materials that compose these shells are of particular interest, and they are the focus of numerous studies.

19.2
Cross-Laminar Shells: The Pink or Queen Conch (*Strombus gigas*)

19.2.1
Structure

The giant pink conch or *Strombus gigas* is part of the *conus* family of shells. The conch shell has a logarithmic spiral shape, and exhibits the highest level of structural organization among mollusk shells. The shell has a particularly high ceramic content of 99 wt.%, composed of lath-like aragonite crystals arranged in a crossed-lamellar or ceramic "plywood" structure. The crossed-lamellar structure is the most common structure in mollusk shells, represented in ∼90% of gastropods and ∼60% of bivalves. While nacre with the brick and mortar microstructure exhibits the highest tensile and compressive strengths among the various mollusk shell micro-architectures, the crossed-lamellar structure is associated with the highest fracture toughness.

The conch shell is arranged in a laminated micro-architecture over five different length scales: the macroscopic layers, the first, second, and third-order lamellae, and internal twins within each third-order lamella. The three macroscopic layers are termed the inner (I) (closest to the animal), middle (M), and outer (O) layers. Each

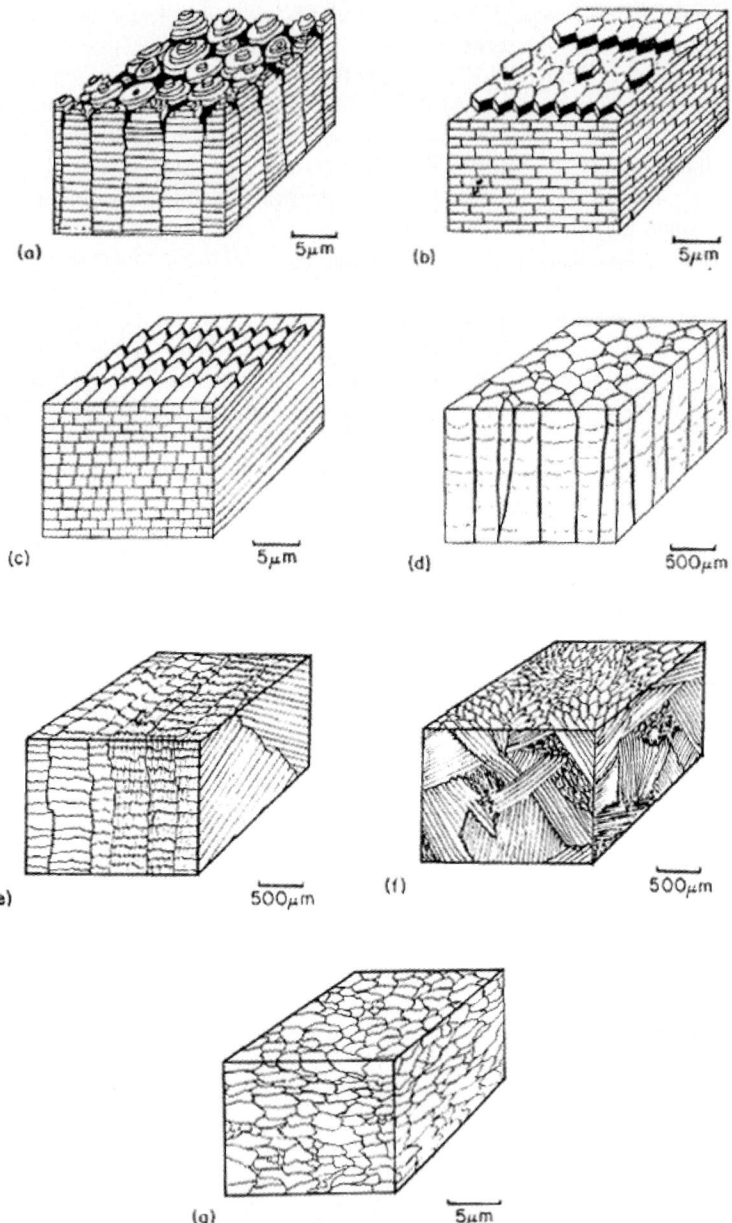

Fig. 19.2. Mineral structures found in seashells. (**a**): Columnar nacre. (**b**): Sheet nacre. (**c**): Foliated. (**d**): Prismatic. (**e**): Cross-Lamellar. (**f**): Complex crossed-lamellar. (**g**): Homogeneous. Reproduced with permission from [8]

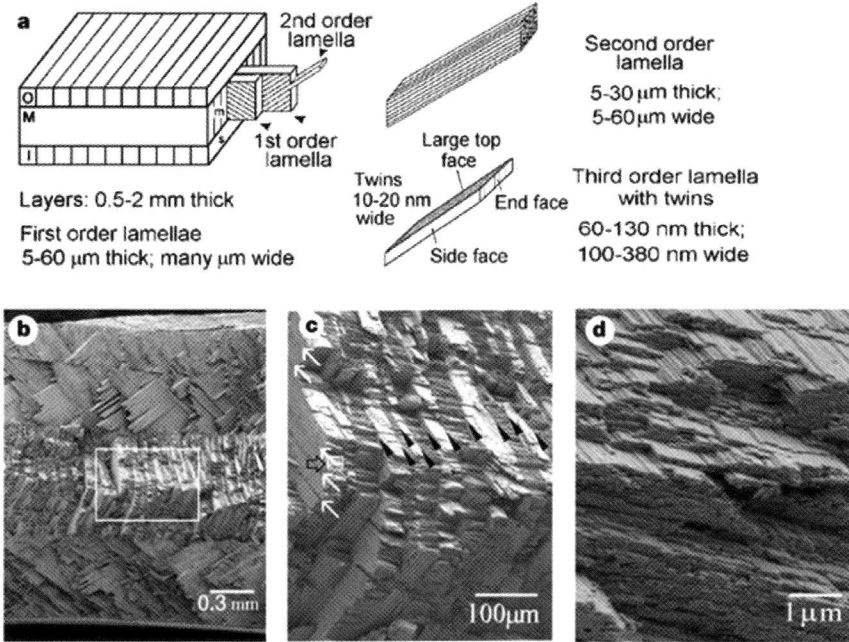

Fig. 19.3. Micro-architecture of the shell of *Strombus gigas*. (**a**) A schematic drawing of the crossed-lamellar structure, with characteristic dimensions of the three lamellar orders. (**b**), (**c**), and (**d**) SEM images of the fracture surface of a bend specimen at increasing magnification. Reproduced with permission from [9]

macroscopic layer is composed of parallel rows of first-order lamellae, and the first-order lamellae in the middle layer are oriented ∼90° to the first-order lamellae in the inner and outer layers. Each first-order lamella in turn is composed of parallel rows of second-order lamellae, which are oriented ∼45° to the first-order lamellae. The second-order lamellae are further subdivided into third-order lamellae. The basic building blocks are therefore the third-order lath-shaped aragonite crystals with internal twins. In particular, in the middle layer, the second-order lamellae in alternating first-order lamellae are rotated by ∼90°. The hierarchy of structural features and their characteristic dimensions are shown in Fig. 19.3(a). Each first, second, and third-order lamellae are enveloped in a thin organic matrix that composes only ∼1 wt. % of the shell [9–11].

Scanning electron microscope (SEM) images of fracture surfaces of bend test specimens are shown in Fig. 19.3(b)–(d), at increasing magnifications. They show clearly the three macroscopic layers [Fig. 19.3(b)] and the first, second and third-order lamellae and their relative orientations [Fig. 19.3(c)–(d)] [9].

19.2.2
Mechanisms of Toughening

The crossed-lamellar shells of *Strombus gigas* have an elastic modulus of 50 GPa, high bending strengths of 100 MPa, and an extremely high work of fracture up to $13 \times 10^3 \, \mathrm{J/m^2}$ [11, 13]. The work of fracture is defined as the area under the load-displacement curve divided by the fracture surface area. The large work of fracture is achieved through two energy dissipating mechanisms – multiple tunnel cracks along the first-order interfaces in the inner (or outer) layer, and crack bridging of the first-order lamellae once the cracks start to grow through the middle layer.

Figure 19.4 shows the visual appearance of a bending test specimen under progressively larger loads [12]. During bending deformation, multiple tunnel cracks develop at the interfaces between the first-order lamellae in the inner or outer layers (the layer experiencing tension). These cracks are arrested at the interface with the middle layer, and are due to the existence of weak proteinaceous interfaces between the first-order lamellae. As the load increases, one or more of these cracks progress into the middle layer along the organic interfaces, but the crack propagation is resisted by bridging forces due to the first-order lamellae with second-order interfaces perpendicular to the crack surfaces. This large-scale bridging is the dominant energy dissipation mechanism.

Multiple crack formation in the weaker inner (or outer) layer during fracture leads to enhanced energy dissipation. This has been modeled using a two-layer, elastically homogeneous structure with fracture toughnesses of K_c^m for the middle layer, and K_c^i for the inner layer [14]. K_c^i represents the fracture toughness of the proteinaceous interfaces between first-order aragonitic lamellae, while K_c^m represents an effective toughness associated with the extension of tunnel cracks into the middle layer. It was shown using energetic arguments that for $K_c^m > 2K_c^i$, multiple cracks will develop in the inner layer under uniform tension, preventing the first crack from causing failure of the specimen. The experimentally observed values for K_c^m/K_c^i are in the

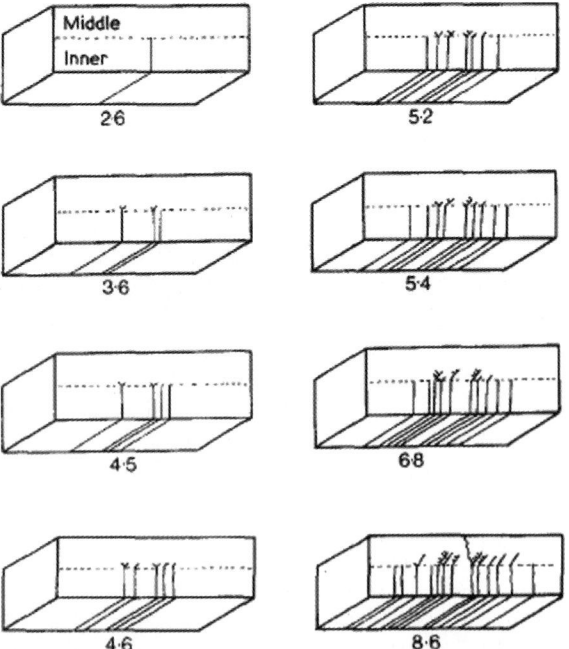

Fig. 19.4. Diagram of visual appearance of a test specimen under progressively greater loads. The loads are indicated in the figure in kg load. The lower surface is under tension. The cracks travel immediately to the junction of the inner and middle layers shown by the *dotted line*, but progress into the middle layer only with difficulty. Reproduced with permission from [12]

range 2.5–3.0 [9]. The interaction between the multiple closely spaced cracks leads to mutual shielding and thus a lower stress intensity factor at each crack tip compared to a specimen with a single crack. This in turn leads to a higher failure stress and strain, increasing the work of fracture.

However, multiple cracking accounts for a relatively small fraction of the toughness; a larger portion of the energy dissipated during fracture is associated with the crack bridging and microcracking in the tougher middle layer. The enhanced toughening is due to the alternating ±45° orientation of the second-order lamellae in the middle layer, which forces the crack to bifurcate at the interface between the inner and middle layers as seen in Fig. 19.5. The tunnel cracks start to propagate through the middle layer along the weak interfaces between the second-order lamellae, but they are retarded by the bridging action of the first-order lamellae with second-order interfaces perpendicular to the crack surfaces. Figure 19.6 depicts the crack bridging analogous to fiber bridging in fibrous composites with frictional sliding at fiber–matrix interfaces [15].

Fig. 19.5. Schematic of the geometry of crack growth at the interface between the inner and middle layers. Reproduced with permission from [12]

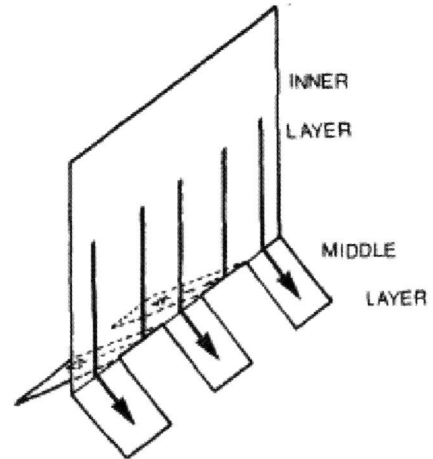

Fig. 19.6. Bridging with frictional sliding along debonded interfaces between fractured and bridging first-order lamellae. The low toughness second-order interfaces in the bridging lamellae are oriented perpendicular to the direction of crack propagation in the fractured lamellae. Reproduced with permission from [16]

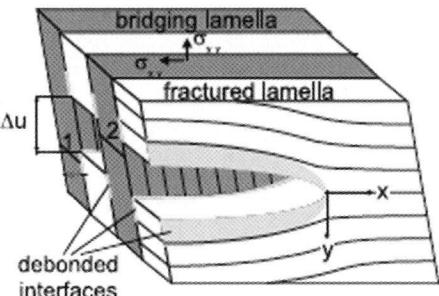

Although the details of this mechanism have not been completely characterized, an approximate crack-bridging model was developed by Kamat et al. [9, 16]. The cohesive law of the crack bridging according to their micromechanical model is given by

$$\sigma(\Delta u) = \beta \Delta u^2, \tag{19.1}$$

where σ is the traction on the crack surfaces and Δu is the crack opening displacement. β is an effective parameter that incorporates all possible energy-dissipating mechanisms. The additional energy release rate associated with crack bridging is then given as

$$J_b = \int_0^{\Delta u_{cr}} \sigma(\Delta u) d\Delta u = \frac{2}{3}\beta \Delta u_{cr}^{3/2}, \tag{19.2}$$

where Δu_{cr} is the critical crack opening displacement. Experiments yielded parameter values of the model as $\beta = 630\,\mathrm{M/mm}^{5/2}$, $\Delta u_{cr} = 5\,\mu\mathrm{m}$, and $J_b = 148\,\mathrm{N/m}$ [16].

The crack bridging results in a work of fracture two orders of magnitude larger than that of monolithic aragonite [16]. Kamat et al. also demonstrated that the intrinsic material length scales of the shell design is such that it enables the shell to approach the favorable Aveston–Cooper–Kelly (ACK) limit, where the crack-bridging fibers remain intact as cracks propagate across the specimen [17].

The organic material at the interfaces play a large role in determining the toughness of the shell, which was found by varying the ductility of the organic phase [16]. At lower temperatures (–120 °C), the fracture behavior was brittle with relatively smooth fracture surfaces, indicating a small amount of fiber pullout and bridging, while at higher temperatures (80 °C), significant pullout and associated ductility of the specimen was observed. This demonstrates that the ductility of the proteinaceous phase at the interfaces is critical in achieving the high toughness of the shell. However, the exact mechanism by which this is achieved has not been fully quantified.

In summary, the mechanical advantage of the highly organized crossed-lamellar structure is an increased fracture resistance. The relatively weak and ductile interfaces together with a hierarchical laminated microstructure at several length scales make this possible through a combination of energy-dissipation mechanisms such as multiple cracking and crack bridging.

19.3
Nacreous Shells

Nacre from seashell is another example showing how evolution can lead to a high-performance material made out of relatively weak constituents. Nacre can be found inside many species of seashells from the gastropod and bivalve groups [Fig. 19.7(a)]. Nacre is mostly made of a mineral (aragonite $CaCO_3$, 95% vol.), arranged together with a small amount (5% vol.) of softer organic biopolymers [1].

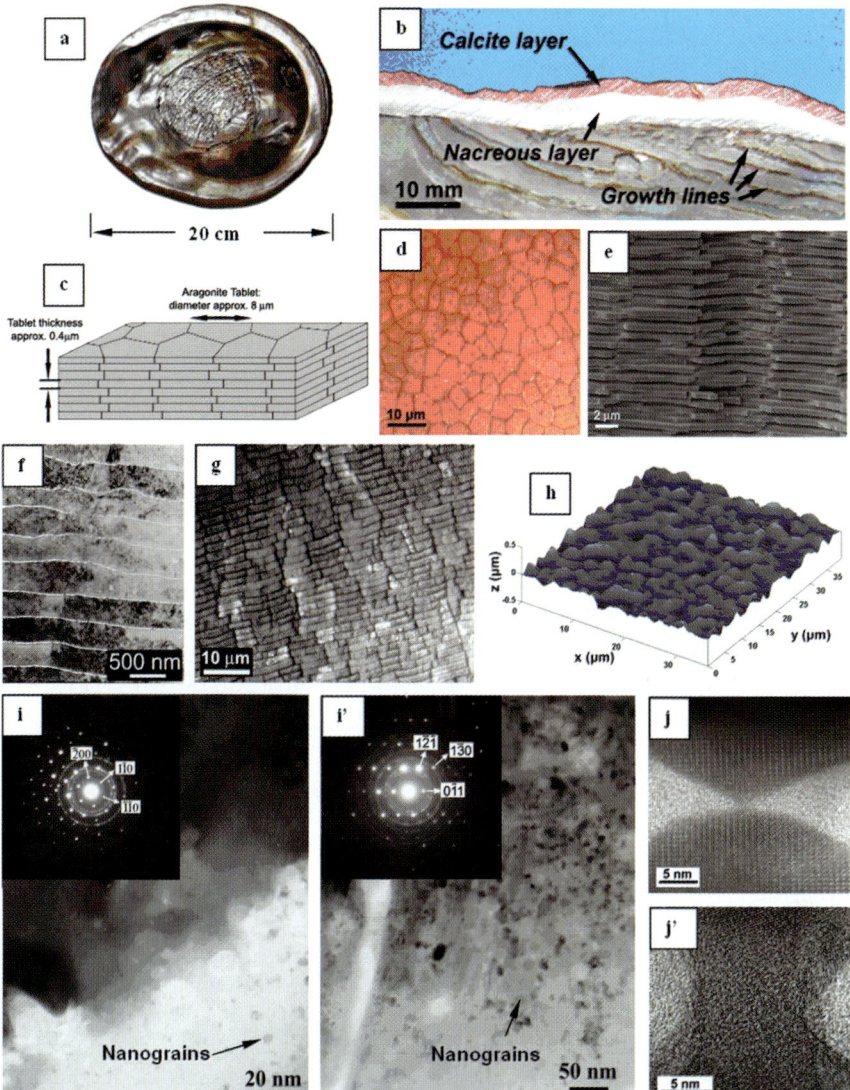

Fig. 19.7. The multiscale structure of nacre (all images from red abalone except **g**. (**a**) Inside view of the shell. (**b**) Cross section of a red abalone shell. (**c**) Schematic of the brick wall-like microstructure. (**d**) Optical micrograph showing the tiling of the tablets. (**e**) SEM of a fracture surface. (**f**) TEM showing tablet waviness (red abalone). (**g**) Optical micrograph of nacre from fresh water mussel (*Lampsilis cardium*). (**h**) Topology of the tablet surface from laser profilometry. (**i, i′**) TEM images showing a single aragonite crystal with some nanograins (*rings on the SAD*). (**j, j′**) HRTEM of aragonite asperity and bridge

While mostly made of aragonite, nacre is 3,000-times tougher than that material [3]. The structures and mechanisms behind this remarkable performance are examined in this section.

19.3.1
Overview of Nacre

Nacre is a highly complex biocomposite which, although made of a brittle mineral, is remarkably tough. Numerous mechanical experiments and models were therefore used to pinpoint which microstructural features are behind this performance, in order to duplicate them in artificial materials. It is now widely recognized that tablet sliding is a key mechanism in the toughness of nacre [18–21]. Because this mechanism is controlled by the interface between the tablets, many efforts have focused on investigating nanoscale mechanisms between the tablets [19, 22–25]. More recently, it was shown that these nanoscale mechanisms, while necessary, are not sufficient to explain the behavior of nacre at the macroscale [21]. Another key mechanism is actually found at the microscale, where the waviness of tablets generates progressive locking, hardening, and spreading of non-linear deformation around cracks and defects. The associated viscoplastic energy dissipation at the interfaces between tablets greatly enhances the toughness of nacre, arresting cracks before they become a serious threat to the shell and to the life of the animal. Nacre is therefore a perfect example of a natural material which developed a highly sophisticated microstructure for optimal performance, over millions of years of evolution. The structure and mechanisms of this remarkable material are now inspiring the design of the next generation of synthetic composites material.

19.3.2
Structure

Like the Pink Conch and many other biological materials, nacre has a hierarchical structure, meaning that specific structural features can be found at distinct length scales. At the millimeter scale the shell consists of a two-layer armor system, with a hard outer layer (large calcite crystals) and a softer but more ductile inner layer [nacre, Fig. 19.7(a), (b)]. Under external mechanical aggressions the hard calcite layer is difficult to penetrate, but is prone to brittle failure. Nacre, on the other hand, is relatively ductile and can maintain the integrity of the shell even if the outer layer is cracked, which is critical to protect the soft tissues of the animal. This design of hard ceramic used in conjunction with a softer backing plate is believed to be an optimal armor system [1]. Furthermore, within the nacreous layer itself there are a few sub-layers of weaker material, the so-called growth lines [26] which may act as crack deflectors [27].

The microscale architecture of nacre resembles a three-dimensional brick and mortar wall, where the bricks are densely packed layers of microscopic aragonite polygonal tablets (about 5–8 μm in diameter for a thickness of about 0.5 μm) held together by 20–30-nm thick layers of organic materials [Fig. 19.7(c)–(e)]. The tablets in nacre from abalone shell and other gastropods are arranged in columns (*columnar nacre*), while the tablets in nacre from bivalve such as mussels or oyster are arranged in a more random fashion (*sheet nacre*). Remarkably, the arrangement and size of the tablets in nacre is highly uniform throughout the nacreous layer. Optical microscopy on a cleaved nacre surface reveals Voronoi-like contours [Fig. 19.7(d)], with no particular orientation within the plane of the layer.

While the tablets are generally described and modeled as flat at the microscale [20, 22], they actually exhibit a significant waviness [28]. This feature could be observed using optical microscopy, scanning probe microscopy, and scanning and transmission electron microscopy. Tablet waviness is not unique to nacre from red abalone [Fig. 19.7(f)]; it was also observed on another gastropod species [top shell *Trochus niloticus*, Fig. 19.7(h)], and in a bivalve [freshwater mussel *Lampsilis Cardium*, Fig. 19.7(g)]. The waviness of the tablets can be also observed for many other species in the existing literature [1, 29–33]. For the case of red abalone laser profilometry was used to measure a roughness (RMS) of 85 nm, for an average peak-to-peak distance of 3 μm [21]. The roughness can reach amplitudes of 200 nm, which is a significant fraction of the tablet thickness [450 nm, Fig. 19.7(f)–(h)] [21]. The waviness of the tablets is highly conformal so that the tablets of adjacent layers fit perfectly together.

Nacre exhibits structural features down to the nanoscale. While transmission electron microscopy suggest that the tablets are made of large aragonite grains with a few inclusions of nanograins [Fig. 19.7(i), (i′)] [1, 34], recent scanning probe microscopy observations suggests that the tablet are nanostructured, with grains in the 30-nm range [35, 36]. These nanograins all have the same texture and they are delimited by a fine network of organic material [36]. At the 20–30-nm interfaces between the tablets, nanoscale features can also be found. The organic material that fills this space and bonds the tablets together is actually composed of several layers of various proteins and chitin [23, 37]. These sheets of organic layers contain pores with a 20–100-nm spacing, leaving space for two types of aragonite structures: nanoasperities [Fig. 19.7(j)] and direct aragonite connections across the interfaces [mineral "bridges" connecting tablets, Fig. 19.7(j′)]. These nanoscale features were observed using scanning probe microscopy [33, 34], scanning electron microscopy [19, 38], and transmission electron microscopy [24, 34, 38]. The height and width of these features varies from 10 to 30 nm while their spacing is in the order of 100–200 nm [34] The density, size, and shape of these asperities can vary significantly from one area to another (Fig. 19.8).

19.3.3
The Deformation of Nacre

The deformation behavior of nacre has been studied experimentally using a variety of configurations including uniaxial tension [21, 39, 40], uniaxial compression [34, 41], three- and four-point bending [18, 19], and simple shear [21]. The behavior of nacre at high strain rates was also explored [41]. Most of these tests were performed on millimeter-sized specimens. Smaller scale experiments were also used to determine the mechanical response of the individual components of nacre, including

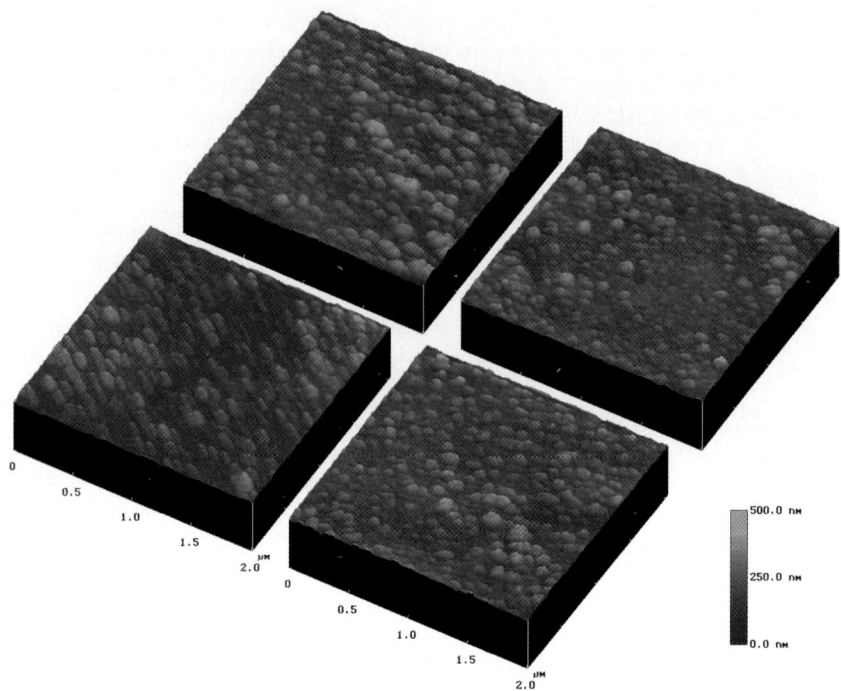

Fig. 19.8. 2×2-μm AFM scans from different areas of a cleaved specimen of nacre, showing the surface of the tablets. Asperities of various densities, heights, and shapes could be observed. Reproduced with permission from [28]

nanoindentation on single nacre tablets [33,34,42], and load-extension curves on single molecules of organic materials [23]. At the macroscale, the most striking mode of deformation is in uniaxial tension along the directions of the tablets (it is also the most relevant mode of deformation for nacre within the shell). Figure 19.9(a) shows the tensile behavior of nacre, showing some ductility at the macroscale [Fig. 19.9(a), (b)]. The stress-strain curve shows relatively large deformations, accompanied by hardening up to failure at a microscopic strain of almost 1%. Full strain field measurement revealed local strain values of 2% [40]. The transition from elastic to inelastic behavior is progressive (rounding of the curve), which probably results from the statistics of the microstructure. Unloading paths show a decrease in modulus, which indicates progressive accumulation of damage. The tensile behavior of aragonite is also shown on that graph: linear elastic deformation followed by sudden, brittle failure at small strains. Nacre, although made of 95% of that mineral, exhibit a ductile-like behavior with relatively large strain at failure.

This remarkable behavior is achieved by the following microscopic mechanism: At a tensile stress of about 60 MPa the interfaces start to yield in shear and the tablets slide on one another, generating local deformation. This phenomenon spreads over large volumes throughout the specimen, which translates into relatively large strains at the macroscale. Once the potential sliding sites are exhausted, the specimen fails

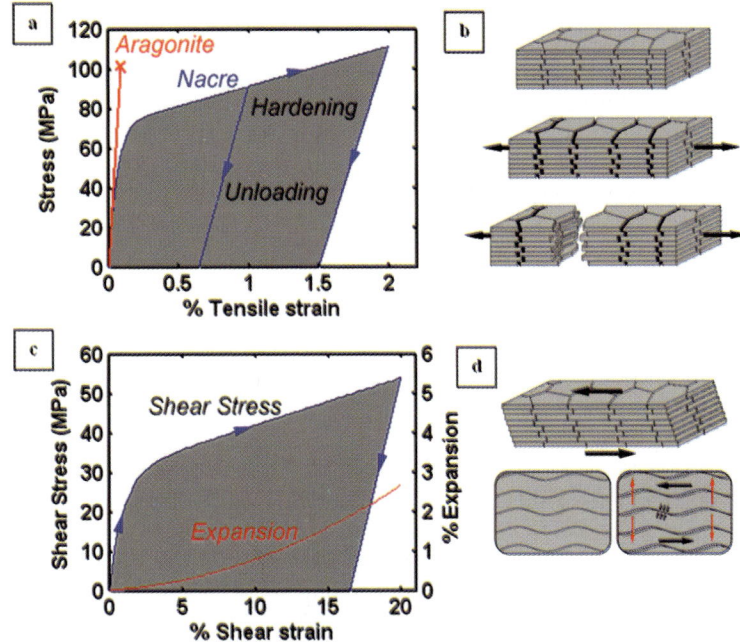

Fig. 19.9. (**a**) Experimental tensile stress-strain curve for nacre and (**b**) associated deformation modes. (**c**) Experimental shear stress-strain curve for nacre and (**d**) deformation mechanism. Tablet waviness generates resistance to sliding, accompanied by lateral expansion (*red arrows*)

by pullout of the tablets [see fracture surface, Fig. 19.7(e)], which occurs after local sliding distances of 100–200 nm. This type of micromechanism is unique to nacre, and it is the main source of its superior mechanical properties. For this reason numerous models were developed to capture this behavior [18, 20, 21, 43].

In order to achieve such behavior, however, some requirements must be met. First, the interface must be weaker than the tablets; otherwise, the tablets would fail in tension before any sliding could occur, which would lead to a brittle type of failure. Strong tablets are important in this regard, and it was shown that their small size confer them with increased tensile strength compared to bulk aragonite [5, 39]. It has also been suggested that the presence of nanograins provides some ductility to the tablets [44]. This would increase the tensile strength of the tablets, but would not significantly affect the deformation mode of nacre, which is dominated by tablet sliding. In addition, the aspect ratio of the tablets must be high enough to maximize sliding areas and produce strong cohesion within the material [45]. However, the aspect ratio is bounded by the fact that too thin tablets would lead to premature tablet failure and brittle behavior. Another fundamental requirement is that some hardening mechanism must take place at the local scale in order to spread sliding throughout the material. As tablets start to slide, higher stresses are required to slide them further so that it is more favorable for the material to initiate new sliding sites, thus spreading deformation over large volumes. Since the tablets remain essentially elastic in this process, the hardening mechanism has to take place at the interfaces. The best approach to interrogate the behavior of the interface directly is a simple shear test along the layers [21]. The shear stress-strain curve reveals a very strong hardening and failure that occurred at shear strains in excess of 15% [Fig. 19.9(c)]. The full strain field, measured by image correlation techniques, also captured a significant *expansion* across the layers. This important observation suggested that the tablets have to climb some obstacles in order to slide on one another. Either in tension or shear, strain hardening is the key to large deformation and is essential to the mechanical performance of nacre.

From this observation it is clear that the performance of nacre is controlled by mechanisms at the interfaces between these tablets. In particular, it is important to identify which mechanisms generate resistance to shearing and hardening. Several nanoscale mechanisms were proposed (Fig. 19.10) and are discussed next.

First, the tough organic material at the interfaces [Fig. 19.10(a)] has an extremely strong adhesion to the tablets [18, 23]. Some of the molecules it contains include modules that can unfold sequentially under tensile load, enabling large extensions [23] and maintaining cohesion between tablets over long sliding distances. The load-extension curve of a single of these molecules shows a saw-tooth pattern, where each drop in load corresponds to the sequential unfolding of the molecule [23]. When a bundle of these molecules is considered, however, the unfolding processes would operate more or less at constant load. Only when all the modules have unfolded does the chain stiffen significantly (at least upon a 100-nm extension [23]). If this type of molecule is attached to adjacent tablets and ensures their cohesion, little hardening should be expected from them, at least in the first 100 nm or so of tensile or shear deformation of the interface. This type of extension is on the order of the sliding distance observed in nacre tensile specimens, and therefore no significant hardening could be generated by the polymer during the tensile deformation of nacre. In the simple shear test, however, the shear strains at the interface are much higher and the polymer may contribute significantly to the hardening observed at the macroscale.

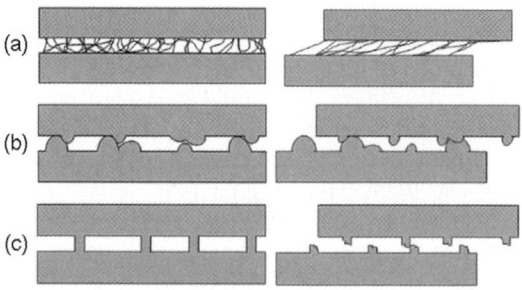

Fig. 19.10. Nanoscale mechanisms controlling the shearing of the tablet's interfaces: **(a)**: Biopolymer stretching; **(b)**: aragonite asperities contact; **(c)** aragonite bridges initially intact (*left*), and then showing some potential relocking after shearing (*right*). Adapted with permission from [25]

Another nanoscale mechanism is controlled by the nanoasperities on the surface of the opposed tablets, which may enter into contact and interact as the interface is sheared [Fig. 19.10(b)]. This mechanism was proposed as a source of strength and hardening at the interfaces [19, 22]. The strength of aragonite is sufficient for the nanoasperities to withstand contact stresses with very little plasticity [34]. However, the small size of these asperities restricts the range of sliding over which they provide hardening to about 15–20 nm [22], which is much smaller than the sliding of 100–200 nm observed experimentally. Beyond sliding distances of 15–20 nm, one must therefore assume that the resistance the nanoasperities provide remains constant (as shown in [22]).

The third nanoscale mechanism at the interface is associated with the aragonite bridges [Fig. 19.10(c)]. These bridges probably act as reinforcements for the interfaces, and probably influence the overall behavior of nacre [24]. However, given the brittleness of aragonite, they could not generate much resistance to tablet sliding after failure, which probably occurs at small interface shearing strains. After some sliding distance another mechanism was suggested, where broken bridges re-enter in contact, thereby generating re-locking of the tablets (Fig. 19.10(c) [25]). This mechanism has, however, not been demonstrated.

While the three nanoscale mechanisms described above contribute to the shearing resistance of the tablet interfaces, they cannot generate the level of hardening required for the spreading of non-linear deformations observed at the macroscale. In addition, none of these mechanisms could generate the transverse expansion associated with shearing of the layers [Fig. 19.9(c)].

A fourth mechanism was recently proposed where the hardening mechanism is generated by the waviness of the surfaces [21]. As the layers slide on one another in the simple shear tests, the tablets must climb each others waviness, which generates progressive tablet interlocking and an increasing resistance to sliding. In addition, such mechanisms could generate the observed transverse expansion, while the organic glue maintains the tablets together. While this mechanism can easily be envisioned in simple shear [Fig. 19.9(d)], it is less obvious in tension, because tablet sliding only occurs in the tablet overlap areas. Close examination actually reveals that waviness also generated locking in tension. Figure 19.11(a) shows an actual image

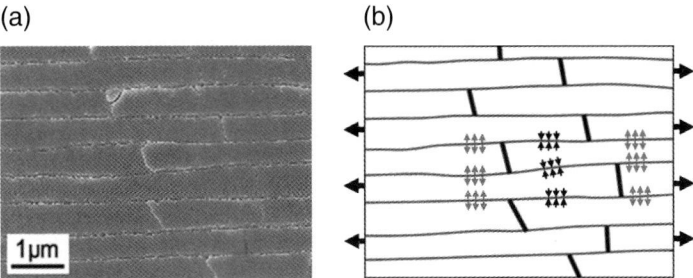

(a)   (b)

Fig. 19.11. (a) Scanning electron micrographs of a few dovetail-like features at the periphery of the tablets. (b) Outline of the tablets contours, showing some of the stresses involved when nacre is stretched along the tablets. In addition to shear the interface is subjected to normal compression (*black arrows*) which generates resistance to tablet pullout. Equilibrium of forces at the interfaces requires tensile tractions at the core of the tablets. Reproduced with permission from [40]

of the structure of nacre. Tablet waviness is evident, and it can be seen that it generates dovetail-like features at the end of some tablets. Such structure, loaded in tension, will generate progressive locking and hardening at large scales [Fig. 19.11(b)]. Microstructure-based three-dimensional finite element models have actually demonstrated that waviness was indeed the key feature that generated hardening in nacre [21]. Even though some of the nanoscale mechanisms of Fig. 19.10 are required to maintain cohesive strength between tablets, waviness is required for hardening. Such a hardening mechanism has a significant impact on the toughness of nacre, as described in the next section.

19.3.4
The Fracture of Nacre

Many flaws are present within nacre, for instance, porosity, Fig. 19.12(a), and defective growth, Fig. 19.12(b). These flaws are potential crack starters that can eventually lead to catastrophic failure under tensile loading [46]. For a material like nacre defects cannot be eliminated so the only alternative is to incorporate robustness in the material design such that cracks that might emanate from them are resisted. The resistance to cracking can be assessed with fracture testing [1, 18]. However, it was only recently that the full crack-resistance curve (toughness as a function of crack advance) was determined for nacre from red abalone [40] [Fig. 19.13(c)].

As the far field stress is increased on the fracture specimen, a white region appears and progressively increases in size [Fig. 19.13(a), (b)]. This whitening is an indication of tablet sliding and inelastic deformations, with the voids left by tablet separation scattering light (this phenomenon is similar to "stress whitening" associated with crazing in polymers). In the literature dealing with fracture mechanics, such an inelastic region is referred to as the *process zone* [46]. The process zone reaches about 1 mm in width [Fig. 19.13(b)] when the crack started to propagate at a J-integral value of $J_0 = 0.3\,\mathrm{kJ/m^2}$, which is already 30-times higher than

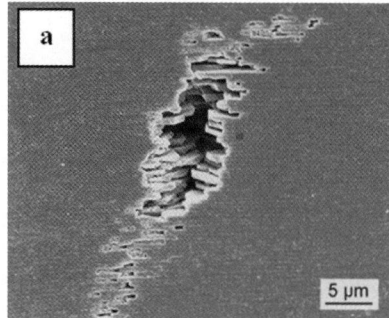
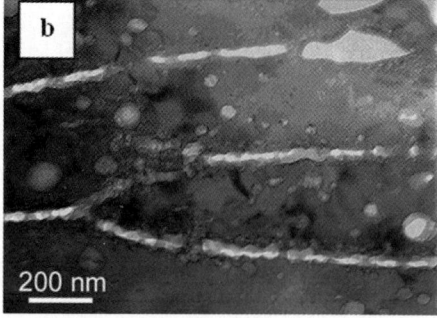

Fig. 19.12. (**a**) Large defect inside the nacreous layer (SEM). (**b**) "Stacking fault" in the tablet layers (TEM)

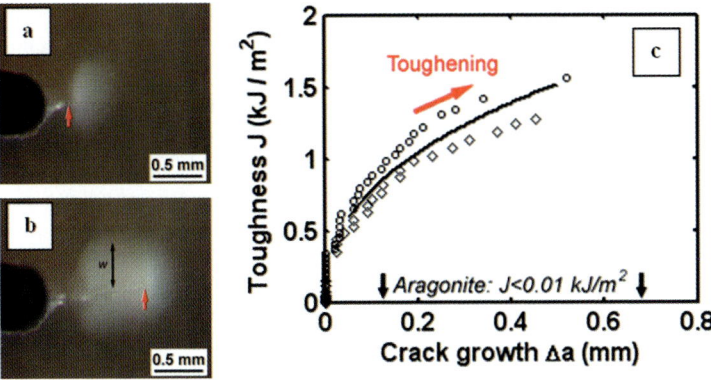

Fig. 19.13. (**a**) Prior to crack advance a tablet sliding zone develops ahead of the crack tip. (**b**) As the crack advances it leaves a wake of inelastically deformed material (**a** and **b**: optical images; *red arrow* shows location of crack tip at the onset of crack propagation and the steady state regime). (**c**) Crack resistance (J_R) curves for nacre from two experiments

the toughness of pure aragonite (about $0.01\,\mathrm{kJ/m^2}$). During the fracture test the crack propagated slowly in a very stable fashion that resembled tearing rather than the fast, catastrophic crack propagation typical to ceramics. On the crack-resistance curve the toughness increases significantly with crack advance Δa [Fig. 19.13(c)]. Such rising crack resistance curves have also been observed in dentin enamel [47] and bone [48]. Cracks in such materials tend to be very stable; upon propagation they tend to slow down and can even be arrested. The rising crack resistance curve and high toughness of nacre could be associated with the formation of the large, whitened region of inelastic deformation around the crack using fracture mechanics models [40].

By considering the energy dissipated upon an incremental crack extension or by straight use of the J-integral definition one can show that in the steady-state the initial (intrinsic) toughness J_0 is augmented by an energy dissipation term [46]:

$$J_{ss} = J_0 + 2\int_0^W U(y)dy \tag{19.3}$$

where w is the process zone width and $U(y)$ is the energy density, i.e., the mechanical energy (including dissipated and stored energies) per unit area per unit thickness in the z-direction, behind in the wake as $x \to \infty$ [49]. The analysis shares similarities with toughening in rubber-toughened polymers [50, 51] and toughening in transforming materials [49]. The exact calculation of $U(y)$ from strain fields requires accurate knowledge of the material constitutive law under multiaxial loading, including hysteretic unloading. Currently such constitutive description is only partially available for nacre. An estimate of the increase in J, in the steady-state, can be obtained, however, if one assumes that (1) the inelastic deformation associated with tablet sliding is the prominent energy dissipation and toughening mechanism (i.e., the effects of shear and transverse expansion in the wake, as well as elastic energies trapped outside of the wake are neglected); (2) the stress σ_{yy} around the tip can be predicted from the uniaxial tensile response [Fig. 19.9(a)]; and (3) the residual strain ε_{yy} in the wake decreases linearly from the crack face to the edge of the wake, as suggested by experimental observations [Fig. 19.13(c)]. Then, the total J can be written as,

$$J_{ss} \approx J_0 + 2 \int_0^W \int_0^{\varepsilon_{yy}(y)} \sigma(\varepsilon)d\varepsilon dy \qquad (19.4)$$

The upper bound of the inner integral, $\varepsilon_{yy}(y)$, is the residual strain across the direction of the crack at $x \rightarrow -\infty$ (which decreases linearly with y). The integration of the loading-unloading histories across the width of the wake [$2w = 1\,\text{mm}$, Fig. 19.13(b)] yields an increase of toughness of $0.75\,\text{kJ/m}^2$. Combined with the initial toughness, the predicted steady state toughness is therefore $1.05\,\text{kJ/m}^2$. This value is lower than the experimental maximum toughness of about $1.6\,\text{kJ/m}^2$ [Fig. 19.13(c)], but shows that dissipative energies associated to dilation and elastic deformation of the material surrounding the wake are significant contributors to the toughening of nacre. In comparison, the contributions of other mechanisms such as crack deflection or crack bridging [52, 53] are negligible. Note that this type of toughening is made possible by (1) spreading of non-linear deformations and (2) its associated energy dissipation.

19.4
Artificial Shell Materials

As described in Sects. 19.2 and 19.3, the attractive mechanical properties of mollusk shells such as nacre and conch shells have inspired a large class of biomimetic materials and organic–inorganic composites. The creation of artificial shell materials with their intricate microstructure is a challenge that requires both the design of optimum microstructures and the development of fabrication procedures to implement these designs. In the following section, we describe some of the efforts at mimicking the architecture of shell materials with different fabrication methods.

19.4.1
Large-Scale "Model Materials"

The challenge of trying to mimic the shell architecture at the μm level led to the development of model systems on the macroscopic scale [54, 55]. An important toughening mechanism of nacre is the crack deflection due to the presence of weak interfaces between the brittle aragonite tiles. Larger-scale segmented layered composites with ceramic tablets make use of this toughening mechanism to overcome the brittleness of ceramics.

Clegg et al. used thin square tiles ($50\,\text{mm} \times 50\,\text{mm} \times 200\,\mu\text{m}$) of SiC doped with boron [54]. The tiles were coated with graphite to retain a weak interface after

(a)

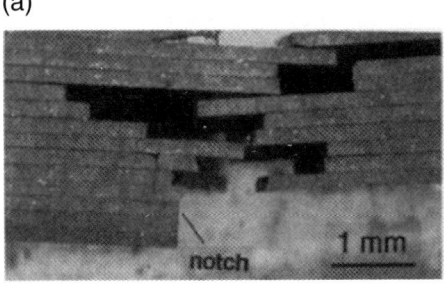

(b)

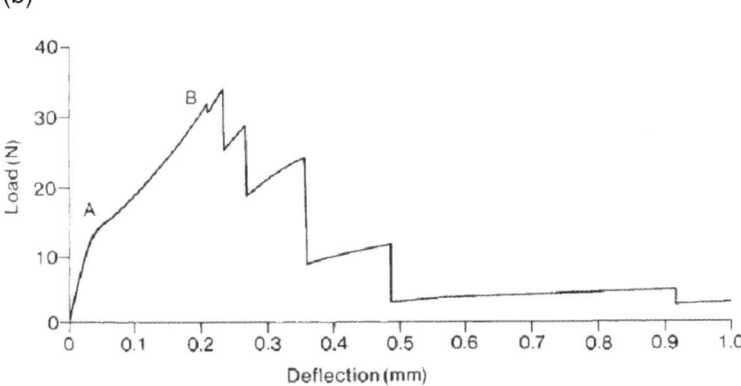

Fig. 19.14. (**a**) The fracture surface of a laminate composite specimen under three-point bending. The role of the interface in deflecting cracks can be clearly seen. (**b**) The load-deflection curve of the specimen. Crack growth begins at *A*, and is followed by a rising load for further deflection till *B* when crack growth becomes more rapid. Reproduced with permission from [54]

sintering. Under a three-point bending test, the crack is deflected along the weak interfaces, preventing catastrophic failure [Fig. 19.14(a)]. The load-deflection curve in Fig. 19.14(b) shows the load continuing to rise after crack growth starts. The laminated composite exhibited a toughness and work of fracture increase by factors of 5 and 100 over monolithic SiC, respectively.

Another larger-scale composite system used alumina tablets of about 50–76 mm long and 1 mm in thickness for the ceramic phase [55]. The plates were bonded with thin adhesive transfer tapes at the interfaces so that the composites achieved 70–90% volume fraction of the ceramic phase. The adhesive exhibited good resilience and extensive ligament formation [Fig. 19.15(a)], contributing to the toughness of the composite. Figure 19.15(b) shows the load-deflection curves of three sets of composite beams from four-point bending tests. The composites with continuous layers and segmented layers with 82 vol.% ceramic showed limited deflections before failure. However, the composite beam with 89 vol.% ceramic exhibited an extensive

(a)

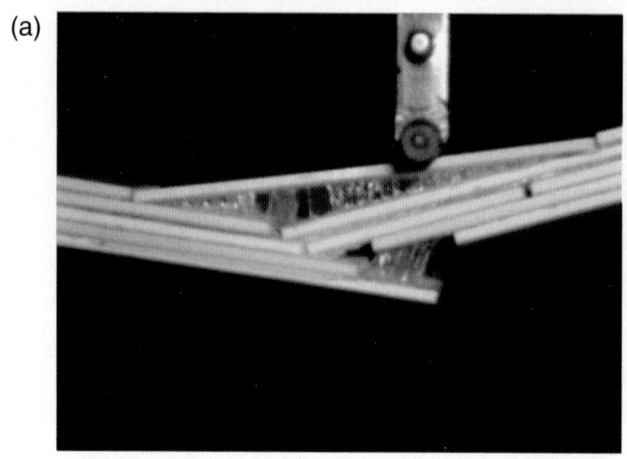

(b)

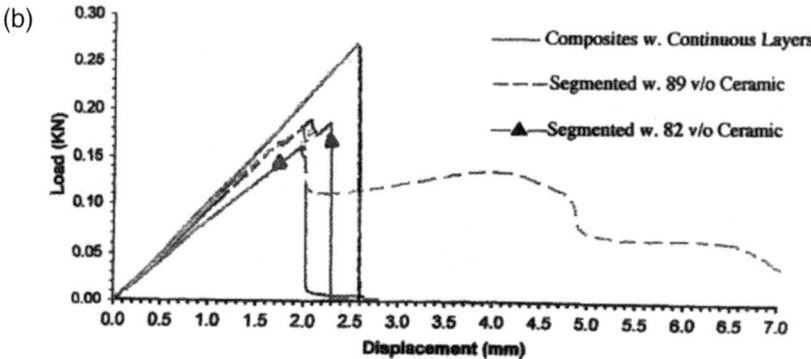

Fig. 19.15. (**a**) A beam being deflected in bending, showing tenacious ligament formation in the adhesive between the platelets. (**b**) Load-displacement curve from four-point bending tests of laminated composites. Reproduced with permission from [55]

deflection with a toughness six-times that of a monolithic alumina beam. The study indicates that a resilient, highly extensible interphase together with a segmented layered microstructure with an optimal ceramic volume fraction is required to achieve maximal toughening.

19.4.2
Ice Templation

Deville et al. utilized the physics of ice formation to develop layered-hybrid materials [56]. By freezing concentrated suspensions containing ceramic particles using precisely controlled freezing kinetics, a homogenous, layered, porous scaffold could be built. The porous scaffolds were then filled with a second phase, either organic or inorganic, to fabricate dense layered composites.

The freeze-casting process involves the controlled unidirectional freezing of ceramic suspensions. While the ceramic slurry is freezing, the growing ice crystals expel the ceramic particles, causing them to concentrate in the space between the ice crystals (Fig. 19.16). The ice is then sublimated by freeze drying, leaving a ceramic scaffold. The layers of the ice-templated (IT) scaffold are parallel to each other and very homogeneous throughout the entire sample [Fig. 19.17(a)]. A finer microstructure can be obtained by increasing the freezing rate, and a layer thickness of 1 μm could be achieved. Particles trapped in between the ice dendrites lead to a surface roughness of the layers as seen in nacre, while some of the rough asperities span the channels between the lamellae, mimicking the mineral bridges of nacre.

These porous scaffolds are filled with a selected second phase, either organic (epoxy) or inorganic (metal), to fabricate dense composites. The layered composite structure exhibited extensive crack deflection at the interface between layers, leading to increased toughness. The reported toughness of Alumina/Al-Si composite is about $5.5\,\mathrm{MPa\,m}^{1/2}$. Compared with the fracture toughness of aluminum oxide usually in

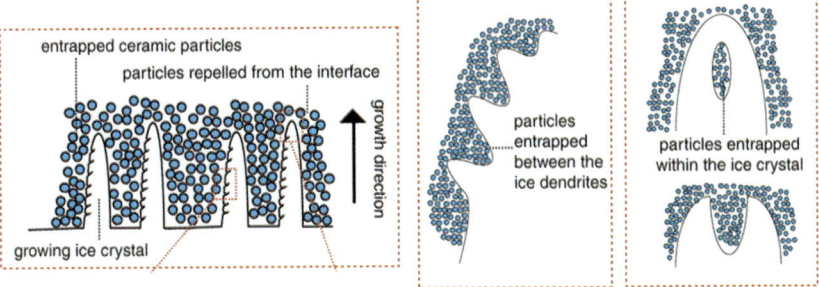

Fig. 19.16. Processing principles for ice templation. While the ceramic slurry is freezing, the growing ice crystals expel the ceramic particles, creating a lamellar microstructure oriented parallel to the direction of the freezing front. A small fraction of particles are entrapped within the ice crystals by tip splitting, leading to the formation of inorganic bridges and roughness on the walls. Reproduced with permission from [56]

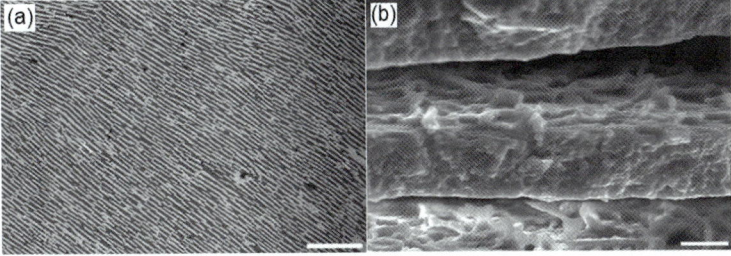

Fig. 19.17. (a) The layered microstructure of the IT dense composite (alumina-Al-Si composite). (b) The particles entrapped between the ice dendrites generate a characteristic roughness on the lamellar surface that mimics that of nacre. Reproduced with permission from [56]

the range of 3–5 MPa m$^{1/2}$, the increase in toughness is quite modest. The interface chemistry could be modified by adding 0.5 wt.% Ti to the aluminum alloy, which increased the fracture toughness further to 10 MPa m$^{1/2}$. Therefore, the IT process allows a measure of control over the morphology of the inorganic layers and the chemistry of the interface, as well as the ability to build mesostructural features and gradients into the structure, but the improvement of the fracture toughness seems to be limited.

19.4.3
Layer-by-Layer Deposition

A more conventional approach to making artificial nacre is by sequential deposition of organic and inorganic layers. Tang and coworkers used montmorillonite clay tablets (C) and PDDA polyelectrolytes (P), layering them by sequential adsorption of organic and inorganic dispersions in a method called layer-by-layer assembly [57]. The process consists of P-adsorption, rinsing, C-adsorption, and rinsing; repeating this process n times results in the $(P/C)_n$ multilayers. The thickness of each clay platelet is 0.9 nm, and a multilayer with $n = 100$ has a thickness of 2.4 μm, three orders of magnitude smaller than the characteristic dimensions of nacre.

Strong attractive electrostatic and van der Waals interactions exist between the negatively charged clay tablets and positive polyelectrolytes. During the adsorption step, the clay platelets tend to orient parallel to the surface in order to maximize the attractive energy. Irregular weakly adsorbed platelets are removed during the rinsing step. This results in a high degree of ordering of the microstructure, with the clay sheets strongly overlapping each other, as seen in Fig. 19.18.

The deformation of the multilayers under tension was homogenous, with no dilation bands between the tablets as observed in nacre. This difference was attributed to the extensive overlap of the clay sheets as well as the nanoscale nature of the $(P/C)_n$. The layered composite displayed an abrupt hardening after an initial plastic deformation as shown in Fig. 19.19(a). This was attributed to the stretching of the PDDA molecules from their initial tightly coiled configuration [Fig. 19.19(b)], and the subsequent breaking of P–C ionic bonds as the clay platelets slide over each other.

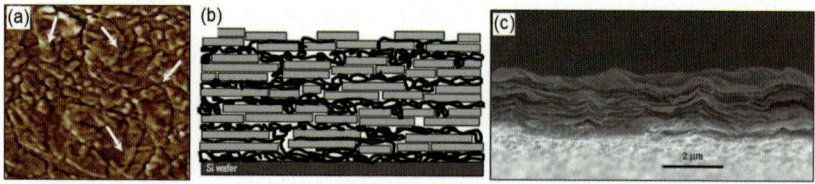

Fig. 19.18. The microscopic structure of the $(P/C)_n$ multilayers. (**a**) Phase-contrast AFM image of a $(P/C)_1$ film. The *arrows* mark the overlap of clay platelets. (**b**) A schematic of the $(P/C)_n$ structure. Note that n describes the number of deposition cycles rather than the number of layers, since several C–P layers may be deposited in each cycle. The thickness of each clay platelet is 0.9 nm. (**c**) SEM image of an edge of a $(P/C)_{100}$ film. Reproduced with permission from [57]

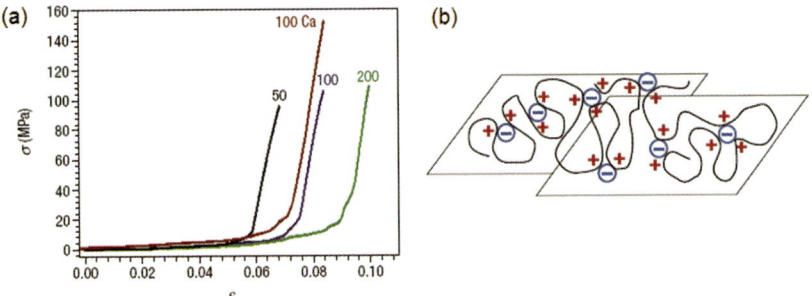

Fig. 19.19. (a) Stress-strain curves of free standing films (P/C)$_{50}$, (P/C)$_{100}$, (P/C)$_{200}$, and (P/C)$_{100}$ ion exchanged with Ca^{2+} ions. (b) Polyelectrolyte folding and P–C ion pair formation. Reproduced with permission from [57]

Although (P/C)$_n$ the multilayers system exhibited high ultimate tensile properties of $\sigma_u = 100$ MPa and $\varepsilon_u = 0.08$, and the segmented layered composite structure is characteristic of that in nacre, the deformation behavior is quite different from that observed in nacre as inferred from the stress-strain behavior shown in Fig. 19.19(a). In addition, the reliance on the strong attraction between the constituents limits the application of the method to a narrow range of materials.

19.4.4
Thin Film Deposition: MEMS-Based Structure

While a lot of effort has been put into reproducing the layered structure of nacre, an attempt was made recently for the first time to mimic the crossed-lamellar microstructure of the Queen conch (*Strombus gigas*), using MEMS (microelectro-mechanical systems) technology [11]. Polysilicon and photoresist were chosen as substitutes for aragonite and the organic matrix, and the microstructure was fabricated as a stack of three consecutively deposited films, the designs of which are shown in Fig. 19.20.

Each film is an approximation of the inner (or outer) and middle layers of the *Strombus gigas* shell, covering two length scales. The bottom half represents the inner layer with vertical interfaces, while the top half represents the middle layer with the interfaces oriented at $\pm45°$ to the horizontal. The films are deposited so that the angled interfaces are oriented in alternating directions designed to make the middle layer tougher than the inner layer.

The fabrication makes use of standard MEMS technology, repeating the deposition of a thin silicon film ($\sim2\,\mu$m) on which trenches for the interfaces are etched out with RIE (reactive ion etching) which are in turn filled with photoresist. Mechanical tests were performed as shown in Fig. 19.21. The results revealed that the micro-composite displayed significant ductility and toughness compared to monolithic silicon, with an estimated increase in energy dissipation of 36-times that for silicon. However, the energy-dissipation mechanisms were slightly different from those of the mollusk shell. Rather than the multiple tunnel cracking seen in the *Strombus gigas* shells, the micro-composite showed extensive delamination between the three

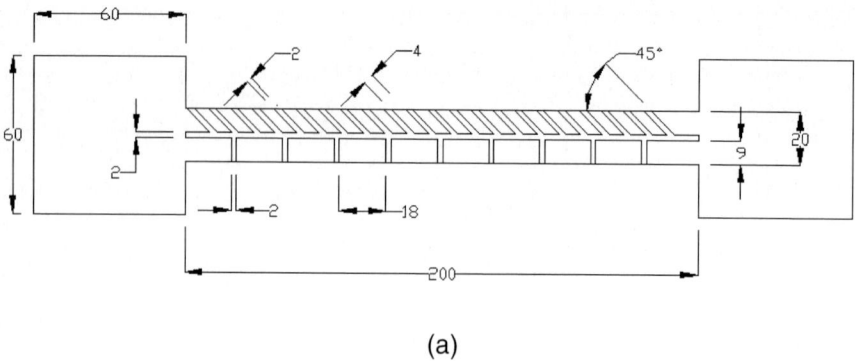

(a)

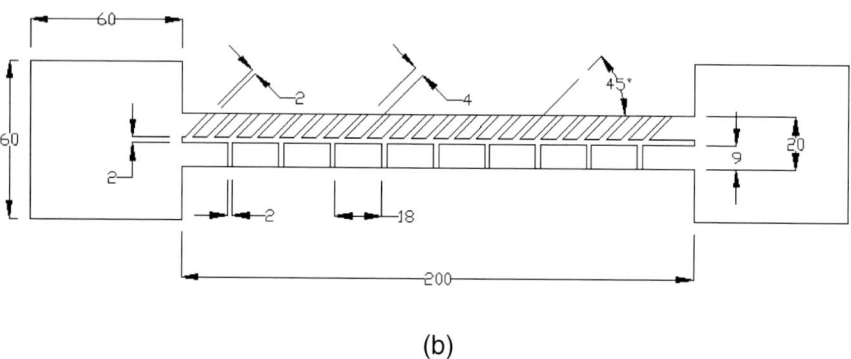

(b)

Fig. 19.20. Top view of designed structural geometry: (**a**) first and third film and (**b**) second film in the three-film stack. Dimensions are in μm. Reproduced with permission from [11]

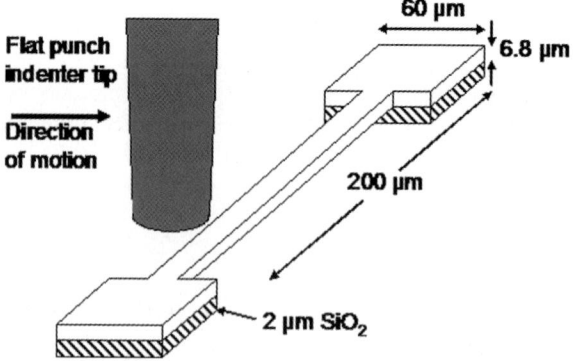

Fig. 19.21. Schematic illustration of a mechanical test. The load was applied in the lateral direction with a punch with a flat tip equipped with a lateral displacement transducer. Reproduced with permission from [11]

stacks. This could be due to the fact that the ratio of thickness of the interface ($2\,\mu m$) to thickness of the lamellae ($4\,\mu m$) is much larger than what is found in the structure of the conch shell. However, the bridged cracks along the $\pm 45°$ interfaces in the middle layer were similar to those seen in the Queen conch shells, and the microcomposite demonstrated a significant increase in strength and work of fracture.

19.5
Conclusions

Much progress has been made in characterizing the structure and mechanical properties of mollusk shells. However, a number of issues remain unresolved. In particular, the specific roles of nanograins, mineral bridges, and nanoasperities in nacre need further investigation. Likewise, general constitutive laws for these materials along various loading paths are currently unavailable.

Advances in in-situ microscopy experiments along with detailed multiscale modeling will facilitate and enhance our understanding of the relationship between material microstructure and mechanical properties.

In the manufacturing of artificial shell materials, creative and innovative methods are emerging. However, much work remains in order to obtain the morphological and chemical control of the interface needed to achieve the performance exhibited by mollusk shells. One lesson from studying these materials is that the desirable inelastic deformation mechanisms are the result of complex synergies between the constituents and the hierarchical structural features. Therefore, biomimicking of mollusk shells requires the optimum selection of constituent materials and manufacturing approaches to materialize the mechanistic synergies. It is anticipated that experiments and modeling will continue to be vital in addressing this challenge.

References

1. Sarikaya M, Aksay IA, (eds) (1995) Biomimetics, Design and Processing of Materials. Woodbury, NY.
2. Mayer G (2005) Rigid biological systems as models for synthetic composites. Science, 310(5751):1144–1147.
3. Wegst UGK, Ashby MF (2004) The mechanical efficiency of natural materials. Philos Mag 84(21):2167–2181.
4. Currey JD (1999) The design of mineralised hard tissues for their mechanical functions. J Exp Biol 202(23):3285–3294.
5. Gao HJ, Ji BH, Jager IL, Arzt E, Fratzl P (2003) Materials become insensitive to flaws at nanoscale: lessons from nature. Proc Natl Acad Sci USA 100(10):5597–5600.
6. Ballarini R, Kayacan R, Ulm FJ, Belytschko T, Heuer AH (2005) Biological structures mitigate catastrophic fracture through various strategies. Int J Fract 135(1–4):187–197.
7. Kohn AJ (2002) Encyclopedia of evolution: Mollusks. Oxford Unversity Press.
8. Currey JD, Taylor JD (1974) The mechanical behavior of some Molluskan hard tissues. J Zool (London), 173(3):395–406.
9. Kamat S, Su X, Ballarini, R, Heuer AH (2000) Structural basis for the fracture toughness of the shell of the conch *Strombus gigas*. Nature 405(6790):1036–1040.

10. Su XW, Zhang DM, Heuer AH (2004) Tissue regeneration in the shell of the giant queen conch, *Strombus gigas*. Chem Mater 16(4):581–593.
11. Chen L, Ballarini R, Kahn H, Heuer AH (2007) Bioinspired micro-composite structure. J Mater Res 22(1):124–131.
12. Currey JD, Kohn AJ (1976) Fracture in crossed-lamellar structure of Conus Shells. J Mater Sci 11(9):1615–1623.
13. KuhnSpearing LT, Kessler H, Chateau E, Ballarini R, Heuer AH, Spearing SM (1996) Fracture mechanisms of the *Strombus gigas* conch shell: implications for the design of brittle laminates. J Mater Sci 31(24):6583–6594.
14. Kessler H, Ballarini R, Mullen RL, Kuhn LT, Heuer AH (1996) A biomimetic example of brittle toughening .1. Steady state multiple cracking. Comput Mater Sci 5(1–3):157–166.
15. Cox BN, Marshall DB (1994) Overview no 111 – concepts for bridged cracks in fracture and fatigue. Acta Metall Mater 42(2):341–363.
16. Kamat S, Kessler H, Ballarini R, Nassirou M, Heuer AH (2004) Fracture mechanisms of the *Strombus gigas* conch shell: II – Micromechanics analyses of multiple cracking and large-scale crack bridging. Acta Mater 52(8):2395–2406.
17. Aveston J, Cooper GA, Kell A (1971) Properties of fiber composites. Conference Proceedings 15, National Physical Laboratory, IPC Science and Technology Press.
18. Jackson AP, Vincent JFV, Turner RM (1988) The mechanical design of nacre. Proc R Soc London 234(1277):415–440.
19. Wang RZ, Suo Z, Evans AG, Yao N, Aksay IA (2001) Deformation mechanisms in nacre. J Mater Res 16:2485–2493.
20. Kotha SP, Li Y, Guzelsu N (2001) Micromechanical model of nacre tested in tension. J Mater Sci 36(8):2001–2007.
21. Barthelat F, Tang H, Zavattieri PD, Li CM, Espinosa HD (2007) On the mechanics of mother-of-pearl: a key feature in the material hierarchical structure. J Mech Phys Solids 55(2):225–444.
22. Evans AG, Suo Z, Wang RZ, Aksay IA, He MY, Hutchinson JW (2001) Model for the robust mechanical behavior of nacre. J Mater Res 16(9):2475–2484.
23. Smith BL, Schaeffer TE, Viani M, Thompson JB, Frederick NA, Kindt J, Belcher A, Stucky GD, Morse DE, Hansma PK (1999) Molecular mechanistic origin of the toughness of natural adhesives, fibres and composites. Nature (London), 399(6738):761–763.
24. Song F, Bai YL (2003) Effects of nanostructures on the fracture strength of the interfaces in nacre. J Mater Res 18:1741–1744.
25. Meyers MA, Lin AYM, Chen PY, Muyco J (2008) Mechanical strength of abalone nacre: role of the soft organic layer. J Mech Behaiv Biomed Mater 1(1):76–85.
26. Su XW, Belcher AM, Zaremba CM, Morse DE, Stucky GD, Heuer AH (2002) Structural and microstructural characterization of the growth lines and prismatic microarchitecture in red abalone shell and the microstructures of abalone "flat pearls". Chem Mater 14(7):3106–3117.
27. Lin A, Meyers MA (2005) Growth and structure in abalone shell. Mater Sci Eng A Struct Mater 390(1–2):27–41.
28. Barthelat F, Tang H, Zavattieri PD, Li, CM, Espinosa HD (2007) On the mechanics of mother-of-pearl: a key feature in the material hierarchical structure. J Mech Phys of Solids 55(2):306–337.
29. Blank S, Arnoldi M, Khoshnavaz S, Treccani L, Kuntz M, Mann K, Grathwohl G, Fritz M (2003) The nacre protein perlucin nucleates growth of calcium carbonate crystals. J Microsc Oxford 212:280–291.
30. Feng QL, Cui FZ, Pu G, Wang RZ, Li HD (2000) Crystal orientation, toughening mechanisms and a mimic of nacre. Mater Sci Eng C Biomimetic Supramol Syst 11(1):19–25.

31. Manne S, Zaremba CM, Giles R, Huggins L, Walters DA, Belcher A, Morse DE, Stucky GD, Didymus JM, Mann S, Hansma, PK (1994) Atomic-force microscopy of the nacreous layer in mollusk shells. Proc R Soc London Ser B-Biol Sci 256(1345):17–23.
32. Song F, Zhang XH, Bai YL (2002) Microstructure in a biointerface. J Mater Sci Lett 21:639–641.
33. Bruet BJF, Qi HJ, Boyce MC, Panas R, Tai K, Frick L, Ortiz C (2005) Nanoscale morphology and indentation of individual nacre tablets from the gastropod mollusk *Trochus niloticus*. J Mater Res 20(9):2400–2419.
34. Barthelat F, Li CM, Comi C, Espinosa HD (2006) Mechanical properties of nacre constituents and their impact on mechanical performance. J Mat Res 21(8):1977–1986.
35. Li XD, Chang WC, Chao YJ, Wang RZ, Chang M (2004) Nanoscale structural and mechanical characterization of a natural nanocomposite material: the shell of red abalone. Nano Lett 4(4):613–617.
36. Rousseau M, Lopez E, Stempfle P, Brendle M, Franke L, Guette A, Naslain R, Bourrat X (2005) Multiscale structure of sheet nacre. Biomaterials 26(31):6254–6262.
37. Schaeffer TE, IonescuZanetti C, Proksch R, Fritz M, Walters DA, Almqvist N, Zaremba CM, Belcher AM, Smith BL, Stucky GD, Morse DE, Hansma PK (1997) Does abalone nacre form by heteroepitaxial nucleation or by growth through mineral bridges? Chem Mater 9(8):1731–1740.
38. Lin AYM, Chen PY, Meyers MA (2008) The growth of nacre in the abalone shell. Acta Biomater, 4:131–138.
39. Currey JD (1977) Mechanical properties of mother of pearl in tension. Proc R Soc London 196(1125):443–463.
40. Barthelat F, Espinosa HD (2007) An experimental investigation of deformation and fracture of nacre-mother of pearl. Exp Mech 47(3):311–324.
41. Menig R, Meyers MH, Meyers MA, Vecchio KS (2000) Quasi-static and dynamic mechanical response of *Haliotis rufescens* (abalone) shells. Acta Mater 48.
42. Barthelat F, Espinosa HD (2005) Mechanical Properties of Nacre Constituents: an inverse method approach. MRS 2004 Fall Meeting Boston.
43. Katti DR, Katti KS (2001) Modeling microarchitecture and mechanical behavior of nacre using 3D finite element techniques. Part 1. Elastic properties. J Mater Sci 36:1411–1417.
44. Li XD, Xu ZH, Wang RZ (2006) In situ observation of nanograin rotation and deformation in nacre. Nano Lett 6(10):2301–2304.
45. Ji BH, Gao HJ (2004) Mechanical properties of nanostructure of biological materials. J Mech Phys Solids 52(9):1963–1990.
46. Lawn BR (1993) Fracture of Brittle Solids. Cambridge University Press, New York.
47. Kruzic J, Nalla RK, Kinney JH, Ritchie RO (2003) Crack blunting, crack bridging and resistance-curve fracture mechanics in dentin: effect of hydration. Biomaterials 24(28):5209–5221.
48. Nalla RK, Kruzic JJ, Kinney JH, Ritchie RO (2005) Mechanistic aspects of fracture and R-curve behavior in human cortical bone. Biomaterials 26(2):217–231.
49. Budiansky B, Hutchinson JW, Lambropoulos JC (1983) Continuum theory of dilatant transformation toughening in ceramics. Int J Solids Struct 19(4):337–355.
50. Evans AG, Ahmad ZB, Gilbert DG, Beaumont PWR (1986) Mechanisms of toughening in rubber toughened polymers. Acta Metall 34(1):79–87.
51. Du J, Thouless MD, Yee AF (1998) Development of a process zone in rubber-modified epoxy polymers. International J Fract 92(3):271–285.
52. Evans AG, Hutchinson JW (1989) Effects of non-planarity on the mixed-mode fracture-resistance of bimaterial interfaces. Acta Metall 37(3):909–916.
53. Evans AG (1990) Perspective on the development of high-toughness ceramics. J Am Ceram Soc 73(2):187–206.

54. Clegg WJ, Kendall K, Alford NM, Button TW, Birchall JD (1990) A simple way to make tough ceramics. Nature 347(6292):455–457.
55. Mayer G (2006) New classes of tough composite materials – lessons from natural rigid biological systems. Mater Sci Eng C Biomimetic Supramol Syst 26(8):1261–1268.
56. Deville S, Saiz E, Nalla RK, Tomsia AP (2006) Freezing as a path to build complex composites. Science 311(5760):515–518.
57. Tang ZY, Kotov NA, Magonov S, Ozturk B (2003) Nanostructured artificial nacre. Nat Mater 2(6):413–U418.

20 Electro-Oxidative Lithography and Self-Assembly Concepts for Bottom-Up Nanofabrication

Stephanie Hoeppener · Ulrich S. Schubert

Key words: Scanning probe lithography, Self-assembly monolayers, Nanofabrication, Chemical surface reactions, Nanostructures, Chemically active surface templates

20.1
Introduction

"There is plenty of space at the bottom" [1]. This sentence is still one of the most well-known statements about nanotechnology and inspires many researchers in different fields of science to look closer at confined nanometric systems. Richard P. Feynman introduced this claim in a lecture in 1959, a time when nanometer dimensions were still an "undiscovered" land. In the meantime powerful new techniques [2, 3] have been developed that allow the investigation and even the manipulation of material at the bottom. Atoms and molecules are used as building blocks for electronic devices [4] and the precise positioning of individual building blocks is possible. A well-known example is the work of Eigler et al. [5] who placed single atoms into a corral structure and they were able to observe concentric standing wave patterns in the local density of states by means of scanning tunneling microscopy (STM); an effect, that is caused by the quantum-mechanical interference patterns formed by the scattering of the two-dimensional electron gas on the atoms forming the corral. However, such structures are frequently only stable under low-temperature conditions. The stabilization of nanomaterials is therefore regarded as an important issue to integrate nanomaterials into useful and applicable device structures. This is a difficult task, as common CMOS (complementary metal–oxide–semiconductor) techniques are not very compatible with nanomaterials. Looking back to Feynman's lecture a practical solution for this problem is given. He claimed ". . .we can arrange the atoms the way we want. . .within reason, of course; you can't put them so that they are chemically unstable. . ." [1]. Therefore, the role of chemical interactions can be regarded as an important key that will help to integrate nanometric (molecular) building blocks into devices. Nature has optimized this approach, and many examples are found that use chemical interactions to form well-defined structures, i.e. binding pockets in proteins, that identify other molecules by using the lock-and-key principle [6, 7].

The integration of such chemical binding motifs into device structures might provide a versatile means to obtain stable structures consisting of individual nano-objects. These binding schemes include electrostatic interactions, covalent binding of molecules, hydrogen bonding, complexation and supramolecular approaches, etc. Chemical synthesis provides a toolbox of reaction schemes that can be applied to functional molecular building blocks, and tailor-made molecules with well-designed physico-chemical properties can be linked together. Introducing these additional possibilities to tailor the properties of surfaces will contribute to a powerful mechanism to anchor nanometric building blocks into device frameworks of higher complexity.

Common structuring techniques usually do not take advantage of chemical binding schemes and the integration of nano-objects remains challenging. This chapter will highlight recent developments in chemical lithographic approaches, which utilize chemical binding motifs to guide the assembly of nano-objects into predefined structures, whereby special emphasis is placed on electro-chemical approaches to generate addressable surface templates.

20.2
Chemically Active Surface Templates

Chemically active surface templates play an important role for nanofabrication as they provide structures that can guide the assembly of desired nano-objects into predesigned configurations utilizing self-assembly strategies. The implementation of self-assembly functionalization schemes offers here important advantages, such as their relatively easy processability, stability, and the fast preparation process that allows a rapid fabrication of nanostructures. The templates offer moreover the possibility to tailor the interaction motifs and properties of the surface structures by means of chemical surface reactions. Tailor-made surface templates have the potential to reliably guide the integration of nano-objects into defined positions on a substrate due to the high selectivity of such interactions. In the last years key technologies have been developed that allow the reliable fabrication of chemically active surface templates. Thereby two major strategies are followed. While dip-pen and soft lithographical approaches introduce chemical functionality by the locally restricted formation of monolayers on substrates, a second class of patterning techniques concentrates on the active modification of self-assembled monolayers, by degradation or activation protocols.

20.2.1
Locally Confined Self-Assembled Monolayers

Dip-pen lithography was introduced in 1999 and many examples of possible functional structures were demonstrated [8] (Fig. 20.1a). Thereby the tip of a scanning force microscope (SFM) is used as a reservoir for different inks that are transferred to the substrate if the tip stays in contact with the surface. Functional features with nanometer resolution could be produced by patterning of alkylthiols (e.g. octadecylthiol or thiohexadecanoic acid) on gold substrates [9–13] or using silazanes to

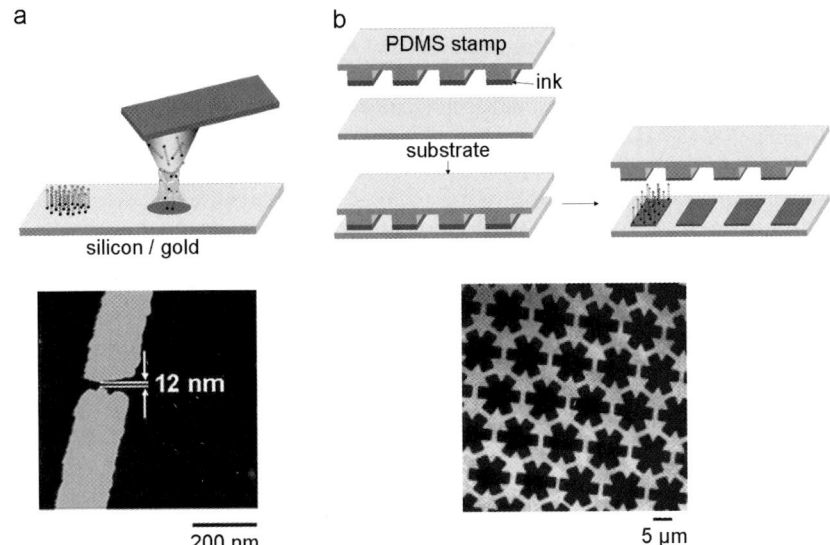

Fig. 20.1. Lithographic techniques employing the local deposition of self-assembled monolayers. (**a**) Schematic representation of dip-pen lithography. The transfer of ink molecules by a SFM tip allows the fabrication of a plethora of features with nanometer resolution. For example, metallic gap features can be developed [28]. (**b**) Patterned surface structures can be fabricated by using μ-contact printing. The ink transfer is established by forming conformal contact between an elastomeric stamp and the substrate [29]

pattern the surface such as silicon oxide or GaAs [14, 15]. Alkoxysilanes were used as inks on silicon oxide, whereby the control of the humidity is an important issue in the patterning process [15, 16]. Viscous solutions of colloidal particles [17, 18], DNA [19], fluorescent dyes [20] etc., have been transferred by the tip onto silicon substrates. Device structures include protein patterns [10, 21–23]. The large variety of different structures is an impressive example of surface templates that potentially find wide applications in different fields of research, i.e. biology, material science, and nanoelectronics. Examples of conjugated polymer structures [24, 25], combinatorial polymer brushes [26], and the fabrication of redox-active nanostructures [27] have been demonstrated. Small metallic gap features of 12 nm could be fabricated [28] (Fig. 20.1a, bottom).

The inherently slow patterning speeds that can be realized by scanning probe-based techniques is regarded as a major drawback and would substantially limit the applicability of these approaches with respect to their integration into high-throughput fabrication processes. Therefore, important developments in the parallel patterning with an array of 55,000 cantilevers have been made. It could be demonstrated that large area phospholipid arrays with nanometer resolution could be fabricated in reasonable time scales [30]. Advantage is taken here of the fact that the ink transfer process is independent of the applied force. Thus, the tip array can be operated without active feedback control of each individual tip. Alternatively, soft lithography approaches have been established using the μ-contact printing approach introduced by Whitesides et al. [29,31]. In this approach an elastomeric stamp is used to transfer reactive molecules onto surfaces in a parallel fashion (Fig. 20.1b). Critical parameters are here the proper choice of stamp material to form a conformal contact between the stamp structures and the surfaces. Moreover, the diffusion characteristic of the inks has to be taken into account to allow the patterning of large surface areas with high-resolution structures [32]. Optimization of the printing conditions and the transfer of the process into a roll-to-roll approach fuels high expectations in regard to this technique for high-throughput fabrication.

Numerous reviews highlight the developments in this field of research and are therefore only shortly acknowledged in the present overview. Detailed information about both techniques can be found in a number of recent reviews [33–37].

20.2.2
Electrical Structuring Techniques of SAMs

In contrast to monolayer deposition related approaches, electro-chemical patterning techniques have been developed that allow the local chemical modification of self-assembled monolayers. Thereby the monolayer can be used as a sensitive resist material or can be used to locally activate the monolayer to generate chemically active surface groups. Both approaches can be used and attractive strategies for the bottom-up fabrication of nanostructures emerge from both techniques.

20.2.2.1
Destructive Mode

The local modification of surfaces by their local oxidation was discovered in an early stage of the era of scanning probe microscopy. Dagata et al. modified a hydrogen-terminated silicon surface by voltage pulses applied to the substrate via a STM tip [38] and observed the formation of protrusions. It could be demonstrated that modified areas contained silicon oxide. With this work he fuelled a rapid development in the use of scanning probe techniques to locally modify surfaces by means of an electrical bias voltage.

These experiments were expanded and in 1993 the first approach to perform the local surface oxidation by means of a SFM tip was reported [41] (Fig. 20.2a). Besides the use of other substrate materials a broad diversity of applications evolved from this process and a large variety of different structures could be produced. While early examples of local probe oxidation focused on the use of Si(111) and polycrystalline tantalum as a substrate, a number of patternable substrates widely emerged including metallic substrates [42,43], III–V semiconductors [44], silicon carbide, titanium [45], and silicon nitride films [46].

The local oxidation process allows the fabrication of dielectric barrier layers, and templates and can be furthermore used for selective etching processes, whereby the oxidation pattern serves as a mask [47]. Villarroya et al. integrated local probe oxidation into the fabrication process of a cantilever-based sensor device to create nanometric gap structures [48]. They used an approach which includes classical fabrication routines for MEMS (microelectromechanical systems) devices, i.e. optical

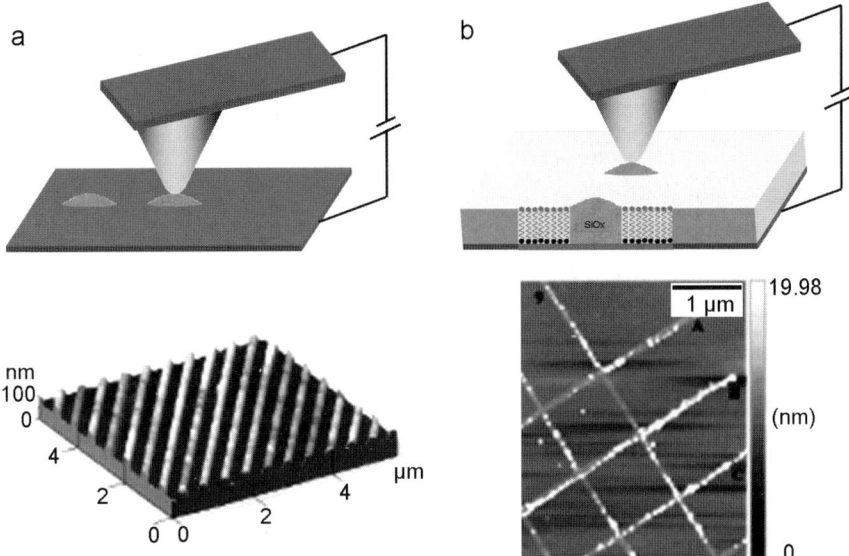

Fig. 20.2. Local probe oxidation can be performed either directly on the substrate (**a**) or through a self-assembled monolayer (**b**). In both cases silicon oxide structures can be produced. (**a**) Stripe features produced at ambient conditions via the water meniscus induces anodization of silicon [39]. (**b**) Oxidation of the silicon substrate through a self-assembled monolayer of OTS allows the further functionalization of the silicon oxide structures, e.g. by self-assembly of APTES and DNA molecules [40]

lithography. Desired nanometric finger-structures, which are an essential component of the sensor device, had to be integrated into these structures, demanding a gap between the electrodes of less than 100 nm. This part of the device was implemented by means of SFM-based nano-oxidation approaches [49].

Moreover, local probe oxidation has been used to fabricate (permanent) data storage memories with high storage capacities [50, 51], conducting wires [52], side-gated field-effect [47] and single electron transistors [53], metal oxide devices [54], superconducting quantum interference devices, quantum points and rings, Josephson junctions [55] etc. Especially interesting is here the approach used by Ensslin et al. who integrated the local probe oxidation patterning for the fabrication of quantum mechanical devices, i.e. quantum rings or quantum dots, on AlGaAs/GaAs semiconductor heterostructures containing a two-dimensional electron gas [56]. The local oxidation of the substrates influences the two-dimensional electron gas underneath the modified areas, and oxidized areas are rendered insulating. However, the electro-oxidation of silicon remains one of the most widely used processes, even if here the produced structures do not directly influence the electronic properties of the substrates.

Intensive investigations were performed to study the kinetics [57, 58] of the oxidation process and a self-limiting growth of silicon oxide is regarded as the underlying mechanism of the anodization reaction [59]. In these investigations the oxidation of silicon oxide terminated and H-passivated silicon served as a well-suited model system and investigations of the required electrical currents during the oxidation process could demonstrate that the local oxidation proceeds in a rather effective way. Typical currents involved in the anodization reaction depend on the final size of the oxidized spot, but are typically in the femto-ampere regime [60].

Severe improvements in the obtainable line-width of the structures was reported by the use of dynamic SFM modes, and the use of carbon nanotube tips further improved the line width down to 15 nm [61]. Cavallini et al. demonstrated that this oxidation process is scaleable and can be integrated into a parallel oxidation process. They used the nanometer size features of a DVD stamp that could be used to induce a silicon oxide feature replica of the stamp. They predict an obtainable resolution better than 10 nm [62].

Sugimura et al. applied local probe oxidation to structure self-assembled monolayers consisting of trimethylsilyl and octadecyltrimethoxysilanes on a silicon substrate [39, 63, 64]. The use of an additional monolayer provides here significant advantages as it can be used, for example, as a passivation layer or to control subsequent etching processes [65]. During the patterning process the monolayer is locally destroyed and the anodization of the released silicon surface proceeds upon prolonged application of the bias voltage on the tip.

n-Octadecyltrichlorosilane (OTS) has developed into a widely used monolayer for such investigations (Fig. 20.2b) and allowed the fabrication of chemically active templates by the site-selective self-assembly of reactive monolayers onto the silicon oxide patterns. This process is very compatible with common silanization routines as the anodization process creates silicon oxide structures that are possible binding sites for additional molecules. Examples utilizing this approach include, for example, the site-selective placement of colloidal nanoparticles [66]. Thereby an additional functionalization is obtained by introducing amino-propyl triethoxy silane (APTES) molecules on the silicon oxide structures via a subsequent self-assembly process. Exposing the structures to the colloidal nanoparticle solution results in the electrostatic interaction of the negatively charged gold colloids and the positive charges provided by the amino termination of the introduced self-assembly layer. High fidelity in terms of the selective binding could be achieved. A similar approach was used by Shin et al. to align λ-DNA onto predefined linear grid structures [40]. Also here advantage is taken of the electrostatic interaction potential. To assemble the DNA strands in a grid-like fashion the preparation was optimized to avoid curling, loop formation, or "branching" of the DNA by a controlled slow withdrawal of the substrate from the DNA solution. This directed assembly resulted in the reliable organization of the strands in one direction. A "perpendicular" preparation cycle could be applied to completely fill the grid-like pattern. Thereby the chemical stabilization of the DNA strand, prepared in the first step, was proven to be successful.

The use of other SAMs than the non-reactive OTS, which is used as a reliable passivation layer, was recently introduced. The use of different functional groups varies the surface properties and has a significant influence on the oxidation process. Not only the meniscus formation can be significantly altered in the presence of different functional groups, due to different water contact angles, but also the charge injection into the monolayer/silicon is modified. Because electrons are transferred from a tip to a substrate via the resist molecules this effect is of special interest as the functional groups could be optimized to ensure a reliable, fast oxidation process.

Kim et al. investigated this effect by the use of monolayer films of metal phosphates [67]. Divalent and tetravalent metal ions were attached to phosphoric acid-modified substrates. The comparison of Zr^{4+} and Ca^{2+} ions, resulting in a positive surface charge or neutral surfaces respectively, allowed them to conclude a beneficial influence on the electron transfer via positively charged surfaces. It could be concluded that a localized positive charge on the surface of a resist enhances the electron transfer in AFM anodization lithography to accomplish better and faster patterning.

Kim et al. investigated also the effect of the length of the alkyl chain on the anodization process [68]. The length of the alkyl chain affects the widths of line patterns, and the diameters of dot patterns. Hexyl-, octyl-, dodecy- and octadecyl-trichlorosilane monolayers were investigated with respect to their oxidation characteristics. Higher oxidation voltages resulted in the formation of broader features on each of the monolayers. At the same applied voltage longer alkyl chains could be patterned with narrower lines, and the rate of the line variation decreased as the alkyl chain length of the monolayer increased. Even though, the wetting properties of the monolayers were comparable, as elucidated by contact angle measurements, the differences in the resulting line width were evident. It was suggested that the alkyl chain length of the monolayer has an influence on the water column area and its resistivity against degradation. Significant variations in the required threshold voltages, which have to be applied to initiate the silicon anodization of monolayer-coated substrates, were observed. The threshold voltage increased with increasing the alkyl chain length of the SAM. This was assigned to the required electric energies which are needed to degrade the monolayers, and the organization of the monolayer itself, which depends on the alkyl chain length of the precursor molecules.

Recently, Martínez et al. introduced a different approach where they replaced the OTS monolayer by a monolayer formed of aminopropyl triethoxy silane (APTES) [69]. They demonstrated that the local probe oxidation can also be applied in the anodization mechanism and the APTES locally degrades upon the application of a sufficient bias voltage. In contrast to other approaches they utilized the electrostatic repulsion between positively charged $[Mn_{12}O_{12}(bet)_{16}(EtOH)_4]^{14+}$ single-molecule magnets and the aminie terminated, unmodified APTES areas, and guided by this means the assembly of the magnets on the oxide structures. They point out that the trapped charges within the oxide structures play an important role in the self-assembly step of the magnets. This could be verified by the preparation of neutral single-magnets on such structures, which demonstrated a lower affinity to the silicon oxide patterns and showed moreover a strong tendency to unspecifically bind on the APTES monolayer due to the absence of repulsive forces. A similar strategy was followed by Yoshinobu et al. who used local probe oxidation to fabricate protein patterns [70]. It was demonstrated that positive and negative tone protein patterns of ferritin could be obtained. The approach to form positive structures, follows the previously introduced approach to structure OTS as a protein-inhibiting SAM, and introducing proteinphilic binding sites via the functionalization of the oxide structures by means of APTES. Interesting are the approaches that lead to the formation of negative-tone images and two possible strategies were introduced. The first approach takes advantage of the fact that proteins adhere to a bare silicon surface, whereas they are washed off from silicon oxide-terminated surfaces. By electro-oxidation silicon oxides were produced that protrude several nanometers over the silicon wafer level. In its natural state these silicon wafers are coated with a native oxide layer, thus a thicker oxide is expected to have formed upon the anodization process on the inscribed structures. By a short dip of the patterned substrate in HF the silicon oxide is partially etched away. By careful control of the etching conditions it was possible to remove the thin native oxide layer without completely removing the oxide patterns [71]. Protein assembly and rinse of the substrate resulted in the site-selective functionalization of the ferritin on the bare silicon, whereas the silicon oxide patterns remained free of proteins. A similar result could be obtained also by the electro-oxidative patterning of an APTES monolayer, which resulted in the local degradation of the amine-terminated surface and the formation of silicon oxide structures. The affinity of the proteins to APTES was used to bind ferritin proteins to the monolayer whereas silicon oxide structures did not permit their binding. The possibility to locally guide the position of proteins on a surface is of special interest for the realization of bio-electronic devices, including biosensors and enzyme field effect transistors [72] but also for applications in proteomics [73].

Haensch et al. used the local oxidation patterning to create inverted templates for the surface functionalization [74]. In contrast to the approach by Martínez et al. local oxidation was performed on a bromine-terminated alkyl silane layer. This choice was inspired by the fact that amine-terminated SAMs do not permit the subsequent treatment of the silicon oxide structures with additional silane molecules, as they tend to form monolayers also on the amino-terminated, unstructured parts of the substrate. However, bromine-terminated SAMs can be employed in a variety of additional chemical surface reactions to generate a large diversity of functional groups by nucleophilic substitution reactions [75]. It could be demonstrated that the silicon oxide structures, embedded into the bromine-terminated monolayer, can be selectively functionalized with fluorine-terminated precursor molecules, which can be used as an ultrahydrophobic passivation layer, without degrading the integrity of the surrounding, chemically active layer. Subsequent conversion of the bromine to azide and reduction of the azide to amine allowed to circumvent the multilayer formation and the high fidelity assembly of silica particles on the amine-terminated surrounding could be demonstrated (Fig. 20.3). This approach offers promising options for the fabrication of gap structures, nanoelectronic circuits, or to separate for example biological systems, such as cells, into predefined arrangements. Furthermore it highlights the requirement to carefully design the fabrication process if surface structures with more than one functional group are to be realized. It demonstrates moreover clearly the advantages of introducing chemical functionality into nanometric device features, as it allows the site-selective placement of nano-objects with high precision and reliability.

20.2.2.2
Constructive Mode

In 1999 Sagiv et al. introduced an alternative approach that utilizes the local oxidation of a self-assembled monolayer of nonadecyltrichlorosilane (NTS) [76], a process that was also used to locally functionalize a chemically inert *n*-octadecyltrichlorosilane [77]. In contrast to the previously introduced approach to locally destroy the monolayer it could be demonstrated that the application of

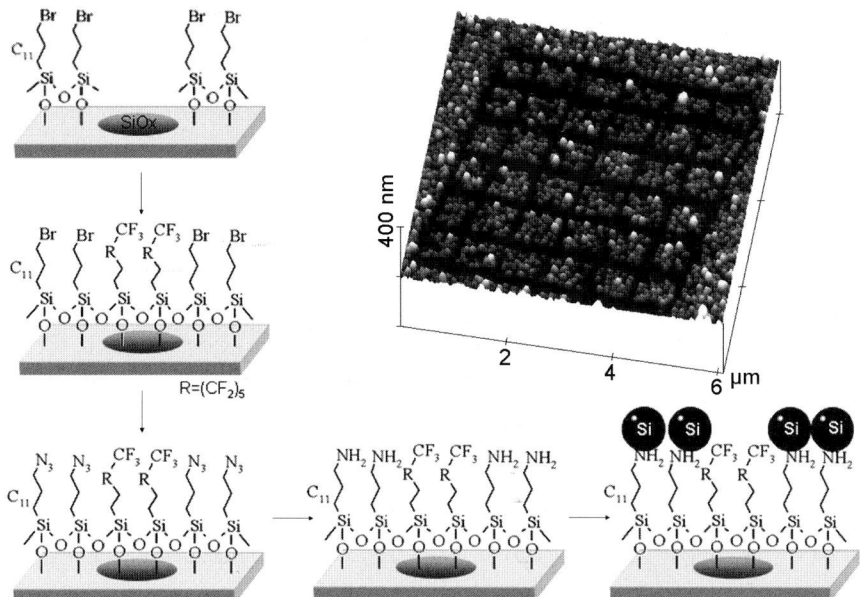

Fig. 20.3. Fabrication of negative-tone images of silica nanoparticles by patterning of reactive monolayers of 11-bromo undecyltrichlorosilane. Post self-assembly of a fluor-terminated passivation layer creates inactive areas on the silicon oxide structures. Subsequent chemical surface reactions convert the bromine-terminated monolayer into amine, which allows the electrostatically driven self-assembly of negatively charged silica nanoparticles

milder oxidation conditions result in the chemical conversion of the surface terminated methyl groups of the OTS monolayer into acid functions (Fig. 20.4a). The process was investigated by means of scanning force microscopy and by macroscale simulation experiments by Fourier Transform Infrared Spectroscopy (FT-IR), which confirmed the presence of carboxylic acid groups on the surface after the application of the bias voltage and confirmed moreover, that the quality of the monolayer was not significantly decreased. The underlying chemical processes are up to now not precisely known but it was observed that neither the application of positive tip bias voltages, nor the oxidation in dry atmosphere resulted in a detectable activation of the monolayer. It was concluded that the formation of a reliable water meniscus is essential to perform the chemical oxidation of the OTS monolayer.

In a moderate regime of oxidation conditions the monolayer is preserved and chemical conversion of the top functional groups of a OTS monolayer is performed. Surface terminated $-CH_3$ groups are locally converted into $-COOH$ functions. The inscription process can be guided by (a) a conductive SFM tip or (b) by conductive stamps. Chemically active surface patterns with a resolution from 10 nm to several micrometers can be obtained.

Pignataro et al. further investigated the oxidation patterns by means of Time-of-Flight Secondary Ion Mass Spectrometry (TOF-SIMS) experiments [78]. The method was employed to perform chemical imaging and analysis of the obtained

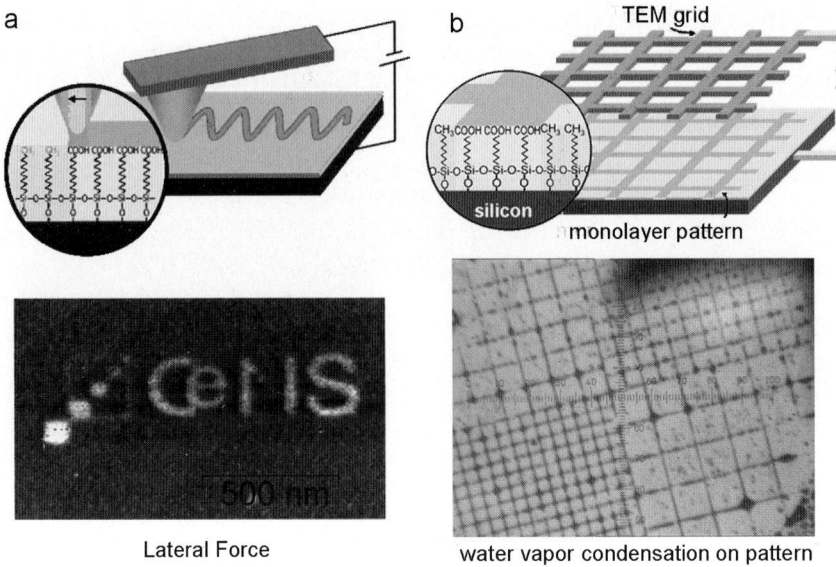

Fig. 20.4. Constructive electro-oxidative patterning modes

spectra to gain information about the chemical nature of the generated surface patterns with a high lateral resolution of 100–200 nm [79]. Experiments were conducted on alkyl monolayers self-assembled on hydrogenated silicon surfaces. Surface structures with a size of several micrometers were inscribed on the surface within different bias voltages. Interestingly, a modification of the surfaces could be obtained with this approach also by the application of positive tip bias voltages, an effect that might be related to the quality of the formed layers and/or by the different linkage of the molecules by Si−C bond formation, in contrast to the formation of Si−O linkages formed during the self-assembly of OTS. However, also here a stronger effect of negative tip bias voltages suggests that the reaction is preferably initiated by an electron-injection process from the tip to the substrate. Positive TOF-SIMS spectra from the modified regions inscribed with negative tip bias voltages showed the presence of CxHyO and CxHyN-type peaks that increase with the tip bias, whereas an intensity decrease of the SiCxHy signals with respect to the unmodified regions was observed. The presence of the CxHyO and CxHyN signals was explained by the formation of polar organic moieties due to the nanoelectro-chemical modification. In contrast to results obtained by FT-IR [77] investigations no significant generation of −COOH groups was observed. However, the formation of CHO moieties is suggested due to a degradation of the alkyl chain by C−C bond cleavage.

Further characterization of nanometer-sized surface templates remains difficult, as the small amount of modified molecules places limitations on the investigations. The sensitivity of characterization techniques and/or the resolution of the investigation tools are frequently not sufficient. A powerful, however indirect, technique is provided by friction force microscopy that can be used to monitor the change of the surface properties. Upon oxidation of the monolayer a strong hydrophilic/hydrophobic contrast is created, which results in a clearly visible contrast due to the absorption of a water layer on the structures in ambient conditions. The scanning tip is subjected to a higher mechanical resistance in these areas, which results in a strong corresponding friction signal. Only minor changes in the simultaneously recorded height images are observed. Small depletions/protrusions in the topography images in the oxidized areas could be identified as a scanning artifact that vanishes if the resonant lateral force imaging mode is used [80]. The high friction force can be used to monitor the result of the oxidation process directly after the inscription is performed and can be furthermore used to determine the parameter space for the oxidation [80]. The result is a diagram that correlates the applied voltage and the duration of the oxidation process with the obtained surface patterns. Wouters et al. found a relatively small corridor where the chemical activation of the OTS monolayer takes place. The application of higher oxidation voltages, and/or oxidation times results in the destruction of the monolayer and the anodization of the underlying silicon substrate. Scanning force spectroscopic investigations performed with a biased tip, allowed Hoeppener et al. to investigate the changes of the surface properties during the oxidation process itself [81]. The analysis of the snap-in and snap-out events allows the interpretation of the oxidation kinetics and it was found that especially the transition between the monolayer oxidation mode and the degradation of the monolayer proceeds very fast.

In this intermediate bias voltage regime Maoz et al. reported a third patterning process that is characterized by a faint contrast visible only in friction mode [77]. These structures vanished upon the exposure of the substrate to OTS. This was attributed to a surface/solution exchange of OTS molecules, where de-bonded monolayer molecules are replaced by OTS molecules from the solution. This effect might be caused by the tip-induced cleavage of the siloxane bonds during the patterning of the monolayer, with no significant oxidation of the top methyl groups. This exchange was confirmed by exposing a surface pattern to a solution of a functional nonadecyl-trichlorosilane (NTS) and subsequent wet-chemical conversion of the vinyl groups by $KMnO_4$, to induce a hydrophilic/hydrophobic contrast that can be created exclusively if the proposed surface/solution exchange takes place. The inscribed structures reappeared in the friction force images and the replacement of OTS molecules by NTS was confirmed. However, the control of this process remains difficult and this writing mode is not used commonly for patterning.

It was found that the oxidation conditions moreover critically depend on the tip material and the quality of the tip and a screening of the proper conditions to obtain the desired oxidation result is required for each individual tip. A clear correlation between the applied voltage and the inscription time was found. Shorter oxidation times require higher oxidation voltages to work in the desired patterning mode. The applied bias voltage has also important implications on the resolution of the patterning process. It is observed that higher oxidation voltages result in the formation of thicker line patterns, an effect which has to be taken into account if high-resolution surface templates are to be formed. Moreover, the line-width depends essentially on the quality of the tip and therefore on the size of the water meniscus that is formed. The optimization of these conditions is an essential step and improvement of the pattern resolution is obtained by using highly doped, non-coated SFM tips. Line patterns as thin as 6 nm could be obtained [82].

Critical issues for use of this electro-chemical oxidation process in high-throughput fabrication processes are here also the scalability and the possibility to speed up the pattern inscription process. A straightforward approach is here to automate the inscription process by using, for example, a completely software-driven SFM set-up, that controls the approach and oxidation process, before the tip is automatically removed and the sample is transferred by a motorized θy-stage to a different area chosen for oxidation [83]. Automatic landing of the tip ensures comparable oxidation conditions for all oxidation cycles.

Large surface patterns have been produced by Wouters et al. and the structures could be investigated by means of XPS (X-ray photoelectron spectroscopy). One-thousand structures could be inscribed on a surface while maintaining a functional tip. However, the automatization does not improve the patterning speed. First feasibility studies have been performed to implement the electro-chemical oxidation into a parallel process utilizing multiple tip arrays [84]. In contrast to the relative straight-forward adaptation of dip-pen lithography into this parallel fabrication concept, important issues, influencing the electro-chemical oxidation process, have to be taken into account. Especially the force applied to the individual tips has important effects on the inscribed structures. Higher forces exerted by the scanning force microscopy tips result in a decrease of the efficiency of the oxidation process and significantly longer oxidation times are required, which are 10–20-times longer than for a tip that was approached with normal force values used also for imaging. An active control of each individual tip would be beneficial; however common SFM set-ups do not permit the feedback control of tip arrays. An elaborated horizontal alignment of the tip arrays is therefore required. With this approach it was demonstrated that reliable patterning of the surface could be obtained by operating a four-cantilever array [84]. An alternative approach replaces the SFM tip by a conductive stamp [85] (Fig. 20.4b). This approach represents an easy and precise technique to obtain chemically active surface templates by a single, parallel inscription process. Surprisingly, it is observed that the pattern transfer proceeds even if relatively rough, commercially available TEM grids are used. The observed corrugation of these rigid stamps allows the faithful pattern transfer due to the formation of water menisci which level out the roughness of the stamp. Therefore, high humidity, resulting in the condensation of water vapor, is applied to the stamp structure prior to contact formation. It is observed that the pattern transfer process required oxidation times of approximately 30 s, and higher oxidation voltages, in the range of 30–35 V, were necessary compared to the inscription process which is initiated with a conductive SFM tip. The larger distance of the TEM grid and the thicker water layer, which is required to obtain the conformal contact, are supposed to be major reasons for the more demanding oxidation conditions. The presence of carboxyl groups were identified in a recent paper by Andruzzi et al. by XPS investigations [86].

Instead of using a conductive stamp pattern it was demonstrated that an oxidation pattern can be directly transferred from one substrate to another. Such surface templates have no topography and it was concluded that the pattern transfer is established exclusively by the formation of water bridges between the planar surfaces and the important role of the water meniscus for the patterning process was confirmed [87]. A mirror image of the hydrophilic/hydrophobic stamp is inscribed on the OTS-covered substrate. These findings would allow a simple and straightforward means to reproduce structures from a single master stamp in practically unlimited numbers, which makes the whole patterning process cost effective.

The strong hydrophilic/hydrophobic contrast between the oxidized structures and the surrounding substrate can be directly used to assemble liquid materials on surface templates. Checco et al. used this effect to study the wetting behavior of ethanol and octane on patterned line features [88]. This allows the precise control and stabilization of liquid objects in desired confinements, enables the study of wetting phenomena, and allows the determination of liquid profile shapes with sub-100-nm resolution. Measurements of the liquids were preformed by Amplitude Modulation SFM to avoid deformations of the liquid due to interaction with the tip. Experiments were performed in a closed chamber that allowed investigation of the liquid adsorption under equilibrium conditions. Variation of the substrate temperature enabled the controlled absorption of the liquids on the pattern structures and nanoscale morphological wetting transitions could be investigated, i.e. the bridging of lines with a relatively narrow distance, which tend to merged together if the amount of liquid increases on the structures. Quantitative determination of the weak line tension of the condensed liquids could be determined, which were in agreement with the estimated values [89]. Further applications for this method are seen for the investigation of biological phenomena, nanofluidic morphological wetting transitions, e.g. where neighboring lines merge together to form a single larger channel, and the investigation of the relative role of the line tension (compared to the surface tension) by defining the profile of liquid nanodrops. A major advantage of the use of chemical nanopatterns is seen in the fact that wetting phenomena are not influenced by topographical features, which are well known to have a strong influence on the wettability and might affect the values of the measured line tension.

Recently Cai [94] reported on the formation of iodine patterns on surface templates, produced by electro-chemical oxidation. Iodine was used here as a model tracing and visualization agent for studying the liquid behavior at the nanometer level and might help to understand the evaporation dynamics of liquid solvents in nanometric confined surfaces areas. It was observed that the height of the patterns depends critically on the initial concentration of the iodine solution. The shape of the iodine films is strongly affected by the evaporation speed of the used solvent and the size of the pattern. A qualitative model was developed to explain the resulting evaporation patterns. In general, during the evaporation process, both the "coffee ring" (i.e. aggregation of material at the borders of the structures) [95] and the "Mexican-Hat" effect (preferential deposition of the material in the middle) [96–98] will occur. In the fast evaporation case, the Mexican-Hat effect plays a dominant role, while the coffee-ring effect dominates during slow evaporation. The coffee-ring formation could moreover be suppressed by the reduction of the pattern size. Interesting is moreover the fact that iodine-covered patterns exhibited an unusually low vapor pressure compared with the bulk crystalline iodine, which undergoes sublimation in an open environment or under heating. However, iodine patterns on the carboxylic acid templates were stable even after the incubation at 80 °C for 48 h, and only a minor decrease in the pattern height was observed. This effect was observed to be dependent on the layer thickness of the iodine and the interaction between the carboxylic chemical pattern and the iodine adsorbate was identified as the reason for the reduced vapor pressure of the iodine on the pattern structure.

Hoeppener et al. introduced an approach that allows the in-situ generation of magnetic nanoparticles on carboxylic acid-terminated surface patterns [90]. The particle formation is initiated by applying an aqueous solution of Fe^{II} ions to the surface followed by a reduction step by hydrazine vapor. After the reduction process the structures are covered with a dense layer of uniform particles (Fig. 20.5a). Remarkable is here the uniform size of the nanoparticles, which was found to be 6–7 nm, as determined by the height and roughness of the formed particle layer. By reducing

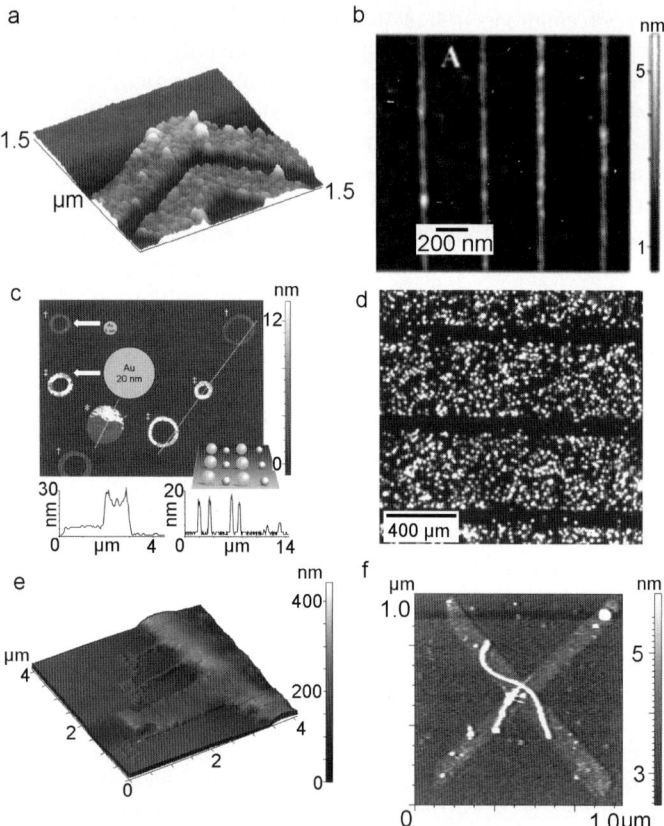

Fig. 20.5. Modification schemes for chemically active surface templates fabricated in the constructive patterning mode on OTS monolayers. (**a**) In-situ generation of uniform structures covered with magnetic particles (particle diameter 6–7 nm) [90]. (**b**) Wetting driven self-assembly of eicosene after a subsequent conversion step into reactive species that can be employed for the site-selective generation of elemental silver [82]. (**c**) Site-selective placement of different nanoparticles in a sequential patterning approach [91]. (**d**) Epithelial cells cultured on a polyethyleneglycol (PEG)-functionalized OTS pattern. *Dark stripes* represent the pegylated areas, which inhibit the adhesion of cells [86]. (**e**) Conductive wire structures fabricated in close vicinity to a macroscopic electrode [92]. (**f**) Carbon nanotubes site-selectively attached to an aminopropyltrimethoxy silane (APTMS)-coated cross-structure fabricated by tip-mediated monolayer oxidation [93]

the template size individual nanoparticles could be arranged in a grid-like fashion with very small next neighbor spacing of 50 nm. The magnetic properties of individual particles were studied by means of Magnetic Force Microscopy that revealed the superparamagnetic character of the small iron nanoparticles. The integration of magnetic nanoparticles into well-defined array structures is regarded as an important step that allows the fabrication of magnetic structures with decreasing device dimensions. Additional properties of such structures, i.e. single domain magnetization, will become accessible and offer new possibilities for the effective construction of new devices [99, 100]. With respect to fundamental research interesting possibilities arise from this approach to study flux closure and magnetic orientation effects in designed particle configurations.

Recently, Sagiv et al. introduced the wetting-driven self-assembly concept (WDSA) [82]. Advantage is taken here of the fact that materials with low melting points and suitable surface tensions can be prepared from the melt onto lyophilic/lyophobic surface patterns and solidified upon withdrawal of the substrate from the liquid melt phase into ambient temperature conditions. Optimization of the withdrawal speed allowed the fabrication of nanostructures consisting of different test materials. Desired chemical functionalities of the surface patterns are introduced by using readily available, cheap functional alkanes. A feasibility study to use this approach to create metal structures by loading functionalized surface patterns with silver and gold ions and further post-assembly chemical treatment to form metallic nanoparticles resulted in the formation of metallic features (Fig. 20.5b). However, spill-over effects of the liquid in the interface region of lyophilic/lyophobic areas decrease the fidelity and obtainable resolution of the assembly process. Potential applications of this approach are seen in the in-situ chemical synthesis in "nanoreactors" confined at predefined areas on a surface. Moreover, surface templates might be utilized for repeated use as a base monolayer pattern in different post-patterning assembly processes. A critical issue of this approach remains the stability of the structures formed by the wetting-driven self-assembly process as no covalent linkage of the material is obtained and the restrictions imposed by this fact have to be investigated.

Besides the fact that structures can be generated by wetting-driven effects additional possibilities emerge from the direct use of the generated functional groups by chemical means. Thereby especially electrostatic interaction provides a straightforward means to assemble charged nano-objects, such as nanoparticles onto the predefined patterns. The $-COOH$-terminated surface patterns are negatively charged and can be directly used for the site-selective assembly of positively charged nanoparticles. Different kinds of particles have been tested, i.e. positively charged magnetite nanoparticles, which have been self-assembled on concentric circular surface templates by Hoeppener et al. [101]. However, the use of relatively weak electrostatic interactions poses limitations on the stability of the formed arrays, especially if the structures are subjected to further preparation processes. Wouters et al. demonstrated that the chemical fixation of amine-terminated CdSe/ZnS core-shell nanoparticles is an essential step if different nanoparticles are placed on sequentially inscribed nanopatterns [91]. Simple self-assembly approaches resulted in the reliable coverage of $-COOH$-terminated surface templates and a corresponding height increase of 20 nm could be observed, reflecting the diameter of the used

CdSe/ZnS nanoparticles. In a further patterning step additional surface templates have been oxidized on the surrounding OTS monolayer, making use of a closed-loop system that allows the faithful control of the patterning area by eliminating the drift usually present in common SFM set-ups. The straightforward use of these newly inscribed areas to assemble 5-nm cationic Au nanoparticles, coated with a poly-L-lysine, resulted in a complete exchange of the CdSe/ZnS nanoparticles by gold nanoparticles due to competing interactions. An intermediate fixation step of the CdSe/ZnS particles is required to prevent their displacement during the second self-assembly step, which was performed by annealing the CdSe/ZnS particle arrays at 90 °C for 6 h. This process leads to a stabilization of the nanoparticles on the −COOH patterns and allows the exposure of the substrate to a solution containing different nanoparticles (Fig. 20.5c), leaving the stabilized particle arrays, produced in the first lithographic step, unaffected. The possibility to combine different materials in close vicinity to each other implies attractive solutions for the fabrication of complex structures. The demand to integrate tailor-made binding groups into the fabrication process is therefore evident.

Andruzzi et al. introduced the synthetic approach to integrate the carbodiimide N-hydroxysuccinimide (NHS) group amplification into the functionalization scheme [86]. This synthesis protocol is known as a very useful tool in organic chemistry for producing chemical bonds under mild reaction conditions [102]. The broad scope of the reaction includes the coupling of, for example, semiconductors to bioactive molecules [103], the preparation of N-terminus functionalized proteins [104], and nanocrystal conjugates [105]. It was demonstrated that bioselective surfaces could be fabricated consisting of micrometer-sized features of polyethyleneglycol (PEG) and OTS. The approach uses the previously introduced electroprinting approach [85] by means of a conductive stamp and subsequent functionalization with amino-terminated PEG via NHS chemistry. The functionalization was investigated by means of XPS and patterned substrates were checked for their applicability as bioselective surfaces. Preliminary tests demonstrated that PEG-functionalized areas of the patterns efficiently suppressed the adsorption of bovine serum albumin (BSA), as investigated by means of fluorescence microscopy. Cell-adhesion experiments were carried out on PEG-patterned surfaces (Fig. 20.5d), which were treated with adhesion-promoting fibronectin prior to the cell preparation to support epithelial cell adhesion to non-PEGylated regions. The cell adhesion was investigated by fluorescence microscopy tracing the Enhances Green Fluorescent Protein (EGFP) expressed by the epithelial cells. High selectivity for the cell adhesion was found in the fibronectin-coated regions. Cells showed healthy growth and maintained gene expression, while no cytotoxicity effects were observed. The NHS approach holds promise for the rapid and simple fabrication of bio-selective and bio-functional surfaces. Further possibilities are seen in the fabrication of biofunctional surfaces using amino-terminated molecules, e.g., amino-terminated PEG, DNA, peptides, proteins, and sugars.

Willner et al. extended the possible reaction scheme of the formed −COOH patterns by using enzymes [106]. Tyramine was covalently linked to the −COOH-terminated templates resulting in lines terminated with hydroxyphenyl moieties. The exposure of the structure to tyrosinase induced the conversion of the hydroxyphenyl end-groups to a catechol functionality. The successful conversion was indirectly proven by the site-selective assembly of Au nanoparticles, which carried boronic acid residues that interact with the catechol functions. Alternatively, magnetic particles were used that linked to the structures through surface Fe^{3+} ions. Additional EDX (Energy dispersive X-ray spectroscopy) investigation of the surface-bound particles revealed the iron content of the nanoparticles linked to the surface structure. The use of enzyme encoded and activated structures open new possibilities for the fabrication of nano-biosensors, but might have also implications for the fabrication of nanocircuits. Because of the known high selectivity of catalytic units the fabrication of an encoded pattern consisting of several different substrates for various enzymes might be possible. This would allow full control on the orthogonal fabrication of nanostructures avoiding chemical cross-talk during the fabrication sequence and highly selective binding.

To further increase the spectrum of surface chemistry functional monolayers can be site-selectively assembled on the inscribed patterns as introduced by Maoz et al. [77]. Advantage is taken of the fact, that the chemical patterning creates hydroxylated surface areas, which are embedded into a matrix of methylene-terminated OTS monolayer. The application of suitable trichlorosilane molecules resulted in the site-selective formation of an overlayer structure on the inscribed patterns either by hydrogen bonding or covalent binding of the overlayer to the patterned base monolayer, depending on the preparation conditions [107]. This approach allows the template-guided hierarchical self-assembly of chemically designed templates. A very versatile precursor molecule that is capable of forming reliable and well-ordered multilayer structures is nonadecyl trichlorosilane that provides an ethylenic surface termination of the inscribed surface templates [77]. A sequence of chemical surface reactions, involving the photochemical conversion of the ethylenic groups, has been introduced to generate thiol and amine functionalities. The thiolation of bilayer structures is e.g. established by photoreaction of the surface-bound NTS molecules with H_2S in the gas phase [77, 108].

Maoz et al. used such templates for the in-situ generation of nanoparticles, i.e. silver particles, by treatment of the thiolated templates with an aqueous solution of $HAuCl_4$. Subsequent use of a silver enhancer solution resulted, according to new insight into the process, in the amplification of the particle size by catalytic metal deposition on tiny gold nanoclusters produced upon the spontaneous reduction [77, 82].

Another approach allows the site-selective formation of CdSe nanoparticles on such substrates [77, 92]. Thereby a chemical reduction is performed on a surface template, which is loaded with Cd acetate solution. Larger surface structures functionalized in this way showed the preferential formation of particles at the structure boundaries, which might be influenced by wetting and/or diffusion processes [77].

Moreover, thiol functions can be directly used to bind gold and silver nanoparticles and the functionalization scheme was used to fabricate well-defined arrays and structures of these nanometric building blocks. The site-selective assembly of $Au_{55}(Ph_2PC_6H_4SO_3Na)_{12}Cl_6$ clusters with a diameter of 1.4 nm was demonstrated by Liu et al. [109]. Au_{55} clusters have been demonstrated to show single-electron tunneling effects at room temperature [110] and are promising candidates for single-electron devices. Electro-oxidative patterning provided a powerful means to control the position of the particles in a faithful fashion by a simple self-assembly preparation process. Thereby a ligand-exchange process is used to anchor the Au_{55} clusters to the thiol-terminated surface templates. Previous attempts to integrate individual clusters into device frameworks were difficult to control or required substantial preparation [111]. Circular dots with a lateral size of 30–50 nm were generated and it was found that a small number (2–5) of clusters were localized in these areas. On line features with a template width of 30 nm two densely packed parallel rows of clusters were formed, indicating the preferential assembly of nanoclusters at the hydrophilic/hydrophobic boundary. The prototyping of cluster layouts, resembling the shape of more complex circuits with wires and contact pads and isolated metal clusters localized within a gap structure, demonstrated the possibility to also manipulate the precise control of individual particles due to an improved resolution within the tip inscription process. Towards the all-chemical fabrication of nanoelectronic devices the thiol functionalization was used by Hoeppener et al. to fabricate conductive wire structures with nanometer dimensions that are connected to a macroscopic electrode [92] (Fig. 20.5e). Also for the macroscopic electrode a chemical approach was used. Starting from a chemically inert OTS covered monolayer a droplet of liquid gallium was applied to one part of the substrate. The gallium solidified and represented an easily replaceable barrier, which was utilized in the subsequent irradiation process of the uncovered OTS part with UV light, as a protective mask. During this process the OTS monolayer was degraded and the bare silicon oxide surface was obtained. Replacement of the gallium protection layer and self-assembly of an NTS monolayer on the clean part of the substrates generated a platform to create stabilizing thiol functions by the previously described photochemical process and metallization of the layer by metal. The OTS monolayer provided the possibility to inscribe −COOH-functionalized wire structures close to the macroscopic electrode and even to establish a well-defined contact at the interface of the electrode. Metallization of the inscribed structures by loading of the acid functions with silver ions and subsequent reduction resulted in the formation of elemental silver nanoparticles. During this process the −COOH functions are regenerated and the process can be repeated until a complete coverage of the structures is obtained. Silver enhancer solution can be used to fill gaps between the individual particles. Finally, the conductivity of such structures was tested by means of lateral conductivity mapping performed by applying a voltage between the conductive SFM tip and a simultaneous recording of the current that flows through the tip–sample contact area. The macroscopic electrode area was used as the counter electrode. It was demonstrated that the samples show a moderate conductivity, however the determination of the resistance of the wire structures was not possible due to the unknown contact resistance of the SFM tip. Structures were stable enough to withstand contact-mode measurements, which were applied due to the more reliable contact that is established by the conductive tip to the structures. This is regarded as the first example that functional structures can be fabricated by means of an electro-chemical patterning process.

The conversion of the ethylenic groups of the NTS-covered templates to thiol is only one example of a number of possible modification schemes. Alternatively, amine functions can be generated as demonstrated by Liu et al. [112]. Thereby, a photochemical reaction is performed to treat the vinyl groups of the NTS precursor with formamide. This process adds a terminal amide group to the alkyl chain of the NTS and is converted into a terminal amine function upon the reduction with $BH_3 \cdot THF$. The reduction step resulted also in a conversion of the interfacial $-COOH$ functions of the inscribed monolayer pattern to $-CH_2OH$, however, the conversion does not seem to have implications on the stability of the bilayer structure. Amino-terminated surface templates are important as the protonation of the amino groups results in a positive charging of the template. Thus, negatively charged nano-objects, i.e. nanoparticles, can be stabilized by means of electrostatic interactions [113–118]. It was demonstrated that citrate-stabilized colloidal gold nanoparticles could be anchored to even complex amino-terminated templates in a reliable fashion. Thereby the colloidal gold particles acquire a negative charge due to the adsorption of citrate and chloride anions present in the colloidal suspension.

Even though the use of NTS provides a versatile strategy to generate a large variety of different tailor-made bilayer structures, it is not commercially available and the screening for more readily available precursors is desired. Thereby, long-chain precursor molecules, which form highly stable monolayers of good quality, are rare and compromises have to be made. The general tendency of shorter alkyl chain precursors to form more disordered monolayers poses drawbacks in the stability of these systems. On the other hand the number of available functional groups is acknowledged as an important issue that has to be taken into account to obtain reliable template functionalization. Especially in confined surface areas the number of available groups is essential to obtain a sufficient quantity of potential binding sites; a requirement that is more difficult to achieve in disordered monolayer systems.

However, there are applications where a densely packed monolayer is of minor importance. This is especially valid for the binding of one-dimensional nanosystems, i.e. carbon nanotubes. Carbon nanotubes have been used in a number of prototyping experiments and their useful electronic properties, for example the formation of functional transistors, has been reported [119–121]. A serious problem remains their controllable integration into suitable device frameworks. Smalley et al. reported on the use of aminopropyltrimethoxy silane (APTMS) to promote the adhesion of carbon nanotubes. This approach could be combined by Hoeppener et al. with the patterning of the OTS monolayer and subsequent self-assembly of APTMS via a vapor-phase self-assembly process [93]. Even though the quality of the formed bilayer is rather low, subsequent assembly of carbon nanotubes was limited to the APTMS-functionalized surface structures, for example cross-structures, which lead to the formation of a basic transistor layout (Fig. 20.5f). The major advantage of the implementation of the electro-chemical oxidation approach is the sequential character of the inscription process, which will permit the controlled step-by-step assembly of such transistor layouts, even with the possibility to hierarchically assemble preselected carbon nanotubes (diameter, electronic properties, length) or allow combination of tubes consisting of different materials.

A very versatile functional group in chemical synthesis is bromine. 11-Bromo undecyltrichlorosilane is commercially available from different suppliers. Balachander et al. [75] demonstrated the versatile use of bromine-terminated monolayers for further functionalization by nucleophilic substitution reactions. They demonstrated the possibilities to generate thiol, amine, azide, cyanate, thiocyanate, and thioacetate-terminated surfaces. A careful comparison of the quality of the 11-bromo undecyltrichlorosilane with OTS monolayers reveals a less-ordered formation of the shorter 11-bromo undecyltrichlorosilane monolayer. However, sufficient stability of the formed layers was found.

In recent approaches these readily available precursor molecules were integrated into the electro-chemical patterning approach. The functionalization scheme was used by Haensch et al. to perform the 1,3-dipolar cycloaddition of acetylene-functionalized building blocks onto azide-terminated monolayers, via intermediate formation of azide groups [122], and on nanometric bromine-functionalized surface structures. This reaction scheme, known as click-chemistry [123, 124], represents a very versatile route in organic synthesis. The fact that reactions can be performed under mild conditions and the general high yield of this reaction make it a very useful protocol for the modification of surfaces. As a small test molecule propargyl alcohol was used and nanostructured line features with a line width of 50 nm could be functionalized with good reliability [125]. The propargyl alcohol is only one member of a whole class of acetylene functionalized molecules, i.e. dendrimers [126], complexes [127], and nanoparticles [128], that are either commercially available or can be synthesized in a straightforward fashion and can be used as tailor-made building blocks. Therefore, the integration of click-chemistry is regarded as an important step to enlarge the possibilities to fabricate functional nanostructures. The stability of the nanostructures within this sequence of functionalization steps demonstrates moreover the general possibility to use multi-step chemical modification schemes (Fig. 20.6).

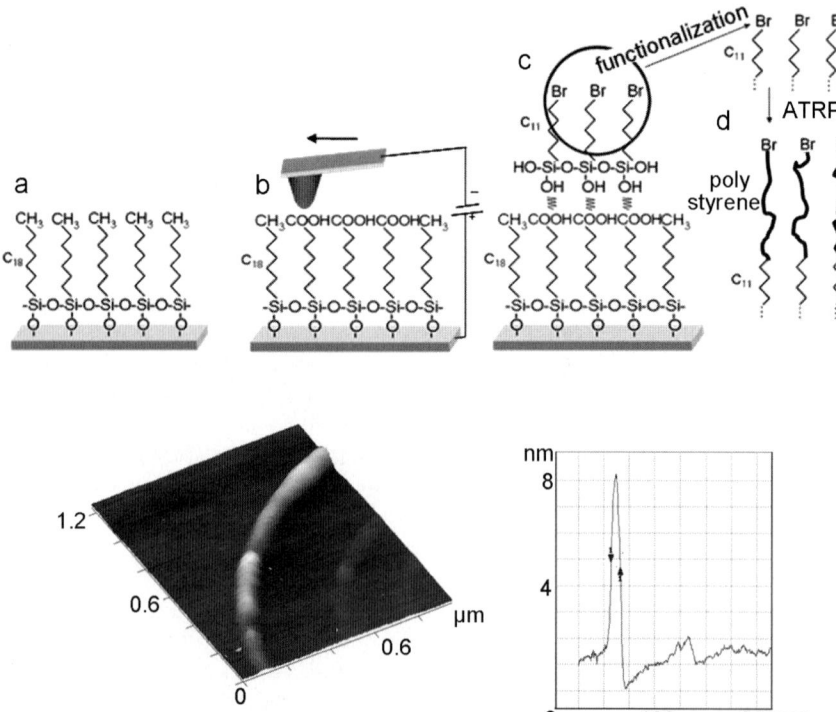

Fig. 20.6. Surface-initiated grafting of polymer brushes by ATRP polymerization

The site-selective assembly of 11-bromo undecyltrichlorosilane on top of an inscribed surface pattern provides suitable initiator structures for the controlled polymerization of polystyrene (PS) onto the structures. Well-defined polymer brushes could be formed that faithfully follow the inscribed pattern.

Another approach, reported by Becer et al. uses the surface terminal bromine functions as the initiator for attenuated transfer radical polymerization (ATRP) of polystyrene [129]. The surface-initiated "grafting from" polymerization process is regarded as a favorable technique to obtain well-defined polymer brush systems, as the steric hindrance of macromolecular species, i.e. predefined polymer chains, is avoided, a well-known effect in "grafting to" approaches; the bulkiness of polymer chains and entropic factors prevent frequently the formation of dense polymer brushes. The surface-initiated polymerization process resulted in the formation of dense polymer brushes with a height of several tens of nm. The polymerization process proceeded linearly with the reaction time, a fact that gives evidence for a living polymerization reaction. As the bromine function is transferred to the end of the polymer chain after each addition of a monomer unit, the polymer brushes remain active and can be employed also for the formation of block copolymer brush systems, which could be demonstrated by the addition of *tert*-butyl acrylate (TBA) to continue the polymerization. Further increase in height of the structures was observed, indicating that the PS polymer brush exhibited at least partially bromine termination. The growth of polymer brushes faithfully follows the inscribed surface patterns, which were in this case inscribed by the electroprinting approach and polymer brush patterns with a line width of 7 μm could be obtained. Recent results demonstrate the formation of polymer brushes on 75-nm features, demonstrating a perfect selectivity of the polymerization process, which proceeded exclusively on the patterned bromine-terminated surface templates. This approach allows to tailor the chemical, mechanical, electrical, and energetic properties of surfaces in a versatile fashion by designing the block copolymers. The introduction of a topographic contrast has interesting implications for biomedical applications of such surfaces. Responsive brush systems, that change their configuration upon external stimuli, can be obtained by introducing polymers with lower critical solution temperatures (LCST) or pH responsive polymers.

Frequently the time consumed for the chemical post-assembly functionalization is not regarded as an important issue. However, there is a potential to reduce nanofabrication times by combining the patterning and the chemical functionalization step. Cai et al. introduced the electro-pen lithography approach [130]. In fact this approach represents the combination of electro-chemical lithography and dip-pen lithography. In a single step an ink-coated, biased conducting SFM tip induces the oxidation of the OTS monolayer to form $-$COOH-functionalized areas. Simultaneously, ink molecules are transferred directly to the regions of higher surface energy. The resolution obtained in this approach was reported to be 50 nm and fast writing times of 10 μm/s have been reported. Resolution limiting is in this process the line width of the $-$COOH-terminated surface templates rather than the diffusion properties of the inks. Two different inks have been tested, including quaternary ammonium salts, a material that was previously used by Wouters et al. to demonstrate the site-selective assembly on patterned OTS monolayers [131], and trialkoxysilanes. Thereby the careful tuning of the inks was performed and advantage was taken of the slow reactivity of alkoxysilanes with water. The site-selective in-situ assembly of the molecules is expected here to be catalyzed by the acidic properties of the surface patterns and generated cross-linked, stable bilayer structures on the inscribed patterns. The successful ink transfer of mercaptopropyltrimethoxysilane was tested by the local binding of 2-nm maleimide mercapto crosslinker-modified gold nanoparticles. The particles appeared to be stable and withstand even the investigation by contact mode SFM.

As a special feature the fabrication of multilayer structures by using this approach is reported. The step-wise growth of patterns can be used to assemble three-dimensional architectures and might find application in multibit data storage.

An important issue that is rarely addressed in the available literature is the task to induce reversible chemical modifications of a monolayer, thus to implement a writing/erasing process. Sugimura et al. addressed this subject and stressed the importance that both, oxidation and reduction, reactions must be performed on the same monolayer to achieve a reversible patterning [132]. Electrochemical patterning in the ambient water meniscus of a SFM tip was used to locally initiate the reduction and oxidation processes. The experiments were conducted on a p-aminophenyltrimethoxysilane monolayer, prepared by a vapor-phase self-assembly process. Local application of bias voltages was found to manipulate the chemical state of the monolayer. With positive substrate bias voltages greater than $+2$ V, applied during the scanning process, the surface was oxidized and nitroso functions were generated at the interface. With a negative bias scanning of less than -3 V an amino-termination of the monolayer could be generated. The oxidation/reduction cycle could be repeated a few times, demonstrating the writing/erasing of the inscribed structures. Conformation for the underlying chemical redox-cycle is difficult and surface potential measurements based on Kelvin Probe Microscopy were used to investigate the changes induced in the modified areas. However, the investigation remained indirect and additional tests were performed to confirm the presence of the amino groups within the patterns by the self-assembly of $-$COOH-functionalized latex beads. The preparation solution was adjusted to pH $= 4$, which causes the protonation of the amine to NH_3^+, while the $-$COOH groups of the latex beads were deprotonated to COO^-, thus a electrostatic driving force was created that allows the site-selective decoration of the reduced regions of the monolayer. Besides the difficulties to determine the created chemical species it was observed that the precursor monolayer does not show a well-defined electrochemical state after the preparation process and some of the precursor molecules were already in the oxidized state. However, this study stimulates further research in this direction and might trigger the development of new concepts that allow the reversible manipulation of molecular monolayers or molecular architectures. The introduction of supramolecular binding schemes is here another approach that might open new possibilities.

20.3
Conclusions

The combination of lithographic approaches to pattern self-assembled monolayers, resulting in a chemically selective structures, has been established by different methodologies. Electro-chemical oxidation approaches are compatible techniques, that provide advantages especially if chemical binding units can be introduced. The introduction of tailor-made functional groups results in a plethora of fabrication schemes with very attractive possibilities for the smart and reliable design of functional nanostructures. The tailor-made formation of binding sites allows the highly selective, cost-effective integration of self-assembly strategies to integrate nano-materials into device frameworks of higher complexity. Chemically active, addressable surface templates offer in this respect beneficial advantages over common lithography techniques. Even though convincing feasibility studies have been performed there are still severe challenges to be addressed to finally be able to make full use of these attractive templating approaches.

Some of the most important targets for future developments are for example the fabrication of three-dimensional surface architectures, the development of orthogonal binding mechanisms that will allow the site-selective assembly of different nano-objects within a reduced number of assembly steps, and the reversible functionalization of surface structures. Also the implementation of parallelization approaches is up to now only at a very early stage of development and significant progress has to be made. However, probe-based oxidation techniques provide a useful means ready to be implemented in prototyping.

Acknowledgments. The author is grateful for the financial support of the Nederlandse Wetenschappelijk Organizatie (NWO), who supported the research with a VICI grant awarded to Prof. Ulrich S. Schubert. Prof. Schubert is gratefully acknowledged for continued support. Post-Docs, Ph.D. and undergraduate students were actively involved in the investigations and are kindly acknowledged for their enthusiasm and motivation to participate in this field of research.

References

1. Feynman RP (1960) Engineering and Science p 20
2. Binnig G, Rohrer H, Gerber Ch, Weibel E (1982) Phys Rev Lett 50:120
3. Binnig G, Quate CF, Gerber Ch (1986) Phys Rev Lett 56:930
4. Heinrich AJ, Lutz CP, Gupta JA, Eigler DM (2002) Science 298:1381
5. Crommie MF, Lutz CP, Eigler DM (1993) Science 262:218
6. Behr JP (ed) (1995) The Lock and Key Principle, The State of the Art -100 Years on, Wiley
7. Eschenmoser A (1994) Angew Chem Int Ed 33:2363
8. Piner RD, Zhu J, Xu F, Hong SH, Mirkin CA (1999) Science 283:661
9. Zhang H, Chung SW, Mirkin CA (2003) Nano Lett 3:43
10. Hong SH, Mirkin CA (2000) Science 288:1808
11. Weeks BL, Noy A, Miller AE, De Yoreo JJ (2002) Phys Rev Lett 88:255505
12. Sheehan PE, Whitman LJ (2002) Phys Rev Lett 88:156104
13. Zhang M, Bullen D, Chung SW, Hong S, Ryu KS, Fan SF, Mirkin CA, Liu C (2002) Nanotechnology 13:212
14. Ivanisevic A, Mirkin CA (2001) J Am Chem Soc 123:7887
15. Pena DJ, Raphael MP, Byers MJ (2003) Langmuir 19:9028
16. Jung H, Kulkarni R, Collier CP (2003) J Am Chem Soc 125:2096
17. Ben Ali M, Ondarcuhu T, Brust M, Joachim C (2002) Langmuir 18:872
18. Garno JC, Yang YY, Amro NA, Cruchon-Dupeyrat S, Chen SW, Liu GY (2003) Nano Lett 3:389
19. Demers LM, Park SJ, Taton TA, Li Z, Mirkin CA (2001) Angew Chem Int Ed 40:3071

20. Noy A, Miller AE, Klare JE, Weeks BL, Woods BW, De Yoreo JJ (2002) Nano Lett 2:109
21. Lee KB, Park SJ, Mirkin CA, Smith JC, Mrksich M (2002) Science 295:1702
22. Lee KB, Lim JH, Mirkin CA (2003) J Am Chem Soc 125:5588
23. Zhang H, Lee KB, Li Z, Mirkin CA (2003) Nanotechnology 14:1113
24. Maynor BW, Filocamo SF, Grinstaff MW, Liu J (2002) J Am Chem Soc 124:522
25. Lim JH, Mirkin CA (2002) Adv Mater 14:1474
26. Liu XG, Liu SW, Mirkin CA (2003) Angew Chem Int Ed 42:4785
27. Ivanisevic A, Im JH, Lee KB, Park SJ, Demers LM, Watson KJ, Mirkin CA (2001) J Am Chem Soc 123:2424
28. Zhang H, Chung SW, Mirkin CA (2003) Nano Lett 3:43
29. Xia YN, Whitesides GM (1998) Angew Chem Int Ed 37:550
30. Lenhert S, Sun P, Wang Y, Fuchs H, Mirkin CA (2007) Small 3:71
31. Xia Y, Whitesides GM (1998) Annu Rev Mater Sci 28:153
32. Peterson EJ, Weeks BL, DeYoreo JJ, Schwartz PV (2004) J Phys Chem B 108:15206
33. Salaita KL, Wang YH, Mirkin CA (2007) Nature Nanotech 2:145
34. Ruiz SA, Chen CS (2007) Soft Matt 3:168
35. Huck TS (2007) Angew Chem Int Ed 46:2754
36. Wouters D, Schubert US (2004) Angew Chem Int Ed 43:2480
37. Krämer S, Fuierer RR, Gorman CB (2003) Chem Rev 103:4367
38. Dagata J, Schneir J, Harary HH, Evans CJ, Postek MT, Bennett J (1990) Appl Phys Lett 56:2001
39. Sugimura H, Okiguchi K, Nakagiri N, Miyashita M (1996) J Vac Sci Technol B 14:414
40. Shin M, Kwon C, Kim SK, Kim HJ, Roh Y, Hong B, Park JB, Lee H (2006) Nano Lett 6:1334
41. Day HC, Allee DR (1993) Appl Phys Lett 62:2691
42. Brandow SL, Calvert JM, Snow ES, Campbell PM (1997) J Vac Sci Technol A15:1455
43. Takano H, Fujihira M (1996) Thin Solid Films 273:312
44. Held R, Vancura T, Heinzel T, Ensslin K, Holland M, Wegscheider W (1998) Appl Phys Lett 73:262
45. Irmer B, Kehrle M, Lorenz H, Kotthaus JP (1997) Appl Phys Lett 71:1733
46. Pyle JL, Ruskell TG, Workman RK, Yao X, Sarid D (1997) J Vac Sci Technol B15:38
47. Campbell PM, Snow ES (1998) Mater Sci Eng B51:173
48. Davis ZF, Abadal G, Hansen O, Borrisé X, Barniol N, Pérez-Murano Boisen A (2003) Ultramicroscopy 97:467
49. Villarroya M, Pérez-Murano F, Martín C, Davis Z, Boisen A, Esteve J, Figueras E, Montserrat J, Barniol N (2004) Nanotechnology 15:771
50. Cooper EB, Manalis SR, Fang H, Dai H, Matsumoto K, Minne SC, Hunt T, Quate CF (1999) Appl Phys Lett 75:3566
51. Tello M, Garcia F, Garica R (2002) J Appl Phys 92:4075
52. Clement N, Tonneau D, Dallaporta H, Bouchiat V, Fraboulet D, Mariole D, Gautier J, Safarov V (2002) Physica E 13:999
53. Matsumoto K, Gotoh Y, Maeda T, Dagata JA, Harris JS (2000) 76:239
54. Snow ES, Campbell PM, Rendell RW, Buot FA, Park D, Marrian CRK, Magno R (1998) 72:3071
55. Garcia R, Martinez RV, Martinez J (2006) Chem Soc Rev 35:29
56. Fuhrer A, Luscher S, Ihn T, Heinzel T, Ensslin K, Wegscheider W, Bichler M (2001) Nature 413:822
57. Dagata J, Perez-Murano, Martin C, Karamochi H, Yokoama H (2004) J Appl Phys 96:2386
58. Avouris Ph, Hertel T, and Martel R (1997) Appl Phys Lett 71:285
59. Calleja M, Anguita J, García R, Birkelund K, Pérez-Murano F, Dagata JA (1999) Nanotechnology 10–34

60. Kuramochi H, Ando K, Tokizaki T, Yokoyama H (2004) Appl Phys Lett 84:4005
61. Kuramochi H, Ando K, Tokizaki T, Yasutake M, Pérez-Murano F, Dagata JA, Yokoyama H (2004) Surf Sci 566–568:343
62. Cavallini M, Mei P, Biscarini F, García R (2003) Appl Phys Lett 83:5286
63. Sugimura H, Okiguchi K, Nakagiri N (1996) J Vac Sci Technol B 14:4140
64. Sugimura H, Hanji T, Hayashi K, Takai O (2004) Adv Mater 14:524
65. Perkins FK, Dobisz EA, Brandow SL, Calvert JM, Kosakowski JE, Marrian CRK (1996) Appl Phys Lett 68:550
66. Zheng JW, Zhu Z, Chen HF, Liu ZF (2000) Langmuir 16:4409
67. Kim SM, Ahn SJ, Lee H, Kim ER, Lee H (2002) Ultramicroscopy 91:165
68. Kim B, Pyrgiotakis G, Sauers J, Sigmund W (2005) Colloids and Surfaces A 253:23
69. Martínez RV, García F, García R, Coronado E, Forment-Aliaga A, Romero FM, Tatay S (2007) Adv Mater 19:291
70. Yoshinobu T, Suzuki J, Kurooka H, Moon WC, Iwasaki H (2003) Electrochim Acta 48:3131
71. Moon WC, Yoshinobu T, Iwasaki H (1999) Jpn J Appl Phys 38:483
72. Caras S, Janata J (1980) Anal Chem 52:1935
73. Emili AQ, Cagney G (2000) Nat Biotechnol 18:393
74. Haensch C, Wouters D, Hoeppener S, Schubert US (2008) submitted
75. Balachander N, Sukenik CN (1990) Langmuir 6:1621
76. Maoz R, Cohen SR, Sagiv S (1999) Adv Mater 11:55
77. Maoz R, Frydman E, Cohen SR, Sagiv S (2000) Adv Mater 12:725
78. Pignataro B, Licciardello A, Cataldo S, Marletta G (2003) Mat Sci Eng C 23:7
79. Pignataro B, Panebianco S, Consalvo C, Licciardello A (1999) Surf Interface Anal 27:396
80. Wouters D, Willems R, Hoeppener S, Flipse CFJ, Schubert US (2005) Adv Funct Mater 15:938
81. Hoeppener S, van Schaik JHK, Schubert US (2006) Adv Funct Mater16:76
82. Chowdhury D, Maoz R, Sagiv S (2007) Nano Lett 7:1770
83. Wouters D, Alexeev A, Kozodaev D, Saunin S, Schubert US (2006) Mat Res Soc Symp Proc 894:111
84. Wouters D, Schubert US (2007) Nanotechnology 18:485306
85. Hoeppener S, Maoz R, Sagiv S (2003) Nano Lett 3:761
86. Andruzzi L, Nickel B, Schwake G, Rädler JO, Sohn KE, Mates TE, Kramer EJ (2007) Surf Sci 601:4984
87. Hoeppener S, Maoz R, Sagiv J (2006) Adv Mater 18:1286
88. Checco A, Cai Y, Gang O, Ocko BM (2006) J Am Chem Soc106:703
89. Checco A, Gang O, Ocko BM (2006) Phys Rev Lett 96:56104 [Erratum to document cited in CA144:136131] (2006) Phys Rev Lett, 97:039902/1
90. Hoeppener S, Schubert US (2005) Small 1:628
91. Wouters D, Schubert US (2005) J Mater Chem 15:2353
92. Hoeppener S, Maoz R, Cohen SR, Chi LF, Fuchs H, Sagiv J (2002) Adv Mater 14:1036
93. Hoeppener S, Wei G, Schubert US (2005) 13th International conference on scanning tunneling microscopy/spectroscopy and related techniques, JSAP catalogue number: 051224, 234
94. Cai Y (2008) Langmuir 24:337
95. Deegan RD, Bakajin O, Dupont TF, Huber G, Nagel SR, Witten TA (1997) Nature 389:827
96. Haw MD, Gillie M, Poon WCK (2002) Langmuir 18:626
97. Gorand Y, Pauchard L, Calligari G, Hulin JP, Allain C (2004) Langmuir 20:5138
98. Pauchard L, Parisse F, Allain C (1999) Phys Rev E 59:3737
99. White RL (2002) J Magn Magn Mater 242–245:21

100. Tsang C, Lin T, MacDonald S, Pinarbasi M, Robertson N, Santini H, Doerner M, Reith T, Lang V, Diola T, Arnett P (1997) IEEE Trans Magn 33:2866
101. Hoeppener S, Susha AS, Rogach AL, Feldmann J, Schubert US (2006) Curr Nanosci 2:135
102. Williams A, Ibrahim IT (1981) Chem Rev 81:589
103. Beamson G, Briggs D (eds) (2000) XPS of Polymers Database, Surface Spectra Ltd., Manchester UK
104. Patel N, Davies MC, Harthorne M, Heaton RJ, Roberts CJ, Tendler SJB, William PM (1997) Langmuir 13:6485
105. Shavel A, Gaponik N, Eychmueller A (2005) ChemPhysChem 6:449
106. Basnar B, Xu JP, Li D, Willner I (2007) Langmuir 23:2293
107. Maoz R, Sagiv S (1985) Thin Solid Films 132:135
108. Frydman E (1999) Ph.D. Thesis, Weizmann Institute of Science, Israel
109. Liu S, Maoz R, Schmid G, Sagiv J (2002) Nano Lett 2:1055
110. Chi LF, Hartig M, Drechsler T, Schwaak T, Seidel C, Fuchs H, Schmid G (1998) Appl Phys A 66:S187
111. Schmid G(ed) (2003) Nanoparticles – From Theory to Application,Wiley-VCH, Weinheim
112. Liu S, Maoz R, Sagiv J (2004) Nano Lett 4:845
113. Doron A, Katz E, Willner I (1995) Langmuir 11:1313
114. Freeman RG, Grabar KC, Allison KJ, Bright RM, Davis JA, Guthrie AP, Hommer MB, Jackson MA, Smith PC, Walter DG, Natan MJ (1995) Science 267:1629
115. Sato T, Brown D, Johnson BFG (1997) Chem Commun 1007
116. Zhu T, Fu X, Mu T, Wang J, Liu Z (1999) Langmuir 15:5197
117. Yonezawa T, Onoue S, Kunitake T (1998) Adv Mater 10:414
118. Liu S, Zhu T,Hu R, Liu Z (2002) Phys Chem Chem Phys 4:6059
119. Huang Y, Duan X, Cui Y, Lauhon LJ, Kim KH, Lieber CM (2001) Science 294:1313
120. Tans SJ, Verschueren ARM, Dekker C. (1998) Nature 393:49
121. Martel R, Schmidt T, Shea HR, Hertel T, Avouris Ph (1998) Appl Phys Lett 73:2447
122. Haensch C, Hoeppener S, Schubert US (2008) Nanotechnology 19:035703
123. Kolb HC, Finn MG, Sharpless KB (2002) Angew Chem Int Ed 40:2004
124. Huisgen R, Padwa A (ed) (1984) 1,3-Dipolar Cycloaddition Chem, Vol 1 pp. 1–176 Wiley, New York
125. Haensch C, Hoeppener S, Schubert US (2008) submitted
126. Fernandez-Megia E, Correa J, Riguera R (2006) Biomacromolecules 7:3104
127. Wang XY, Kimyonok A, Weck M (2006) Chem. Comm. 37:3933
128. Fleming DA, Thode CJ, Williams ME (2006) Chem Mater 18:2327
129. Becer CR, Haensch C, Hoeppener S, Schubert US (2007) Small 3:220
130. Cai Y, Ocko BM (2005) J Am Chem Soc 127:16287
131. Wouters D, Schubert US (2003) Langmuir 19:9033
132. Sugimura H, Lee SH, Saito N, Takai O (2004) J Vac Sci Technol B 22:L44

21 Application of SPM and Related Techniques to the Mechanical Properties of Biotool Materials

Thomas Schöberl · Ingomar L. Jäger · Helga C. Lichtenegger

Abstract. Soon after the introduction of scanning probe microscopes (SPMs) originally developed for imaging purposes, their potential for mechanical analyses at the smallest scales was recognized and soon the method was applied to a variety of materials, including some of biological origin. Experimental techniques range from phase imaging to indentation and scratch testing. This chapter focuses on the use of instrumented indentation and related techniques such as scanning wear testing on biological tool tissues, materials often characterized by a high abrasion resistance. A brief overview of structure and composition of biological materials is given, since these factors are crucial in determining the mechanical properties and a basic understanding of such correlations is indispensable for the interpretation of the results. Furthermore, the influence of sample storage, preparation, and environmental conditions on mechanical tests is discussed, and relevant evaluation methods described. Finally, examples from the literature illustrating the successful application of SPM techniques on biotool tissues (mainly teeth) are presented.

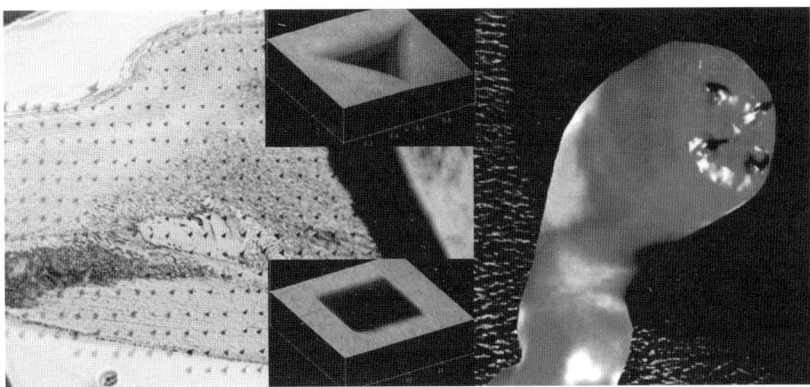

Key words: SPM, Indentation, Scratch test, Wear test, Biotool materials, Teeth, Jaws

Abbreviations

AFM	Atomic force microscope
O.P.-method	Oliver–Pharr method, a method for extracting hardness and elastic modulus from load-displacement curves

CSM	Continuous stiffness measurement. ... a dynamic indentation mode patented by MTS instruments, Oak Ridge, TN, USA
nanoDMA	A dynamic indentation mode patented by Hysitron Inc, Minneapolis, MN, USA
PTFE	Polytetrafluorethylen
SBR	Styrene butadiene rubber
HBSS	Hank's balanced salt solution, a salt solution containing phosphate and carbonate buffers
SEM	Secondary electron microscope
FTIR-RM	Fourier-transformed infrared spectroscopy in reflectance mode
SAXS	Small-angle X-ray scattering
qBEI	Quantitative backscattered electron imaging
DEJ	Dentin–enamel junction
TEM	Transmission electron microscopy
EDS	Energy-dispersive X-ray spectroscopy
EDTA	Ethylenediaminetetraacetic acid, used to sequester di- and trivalent metal ions
XPS	X-ray photoelectron spectroscopy
M	mol/l
His	Histidin

21.1
Introduction

Even small animals can cause huge detrimental effects: in the summer of 2004 coveys of desert locusts attacked eight countries in northern Africa, causing huge amounts of damage to the crop. Every day they ate 300 tons of forage. The largest covey observed had a number of 150 million distributed over an area of $15\,km^2$. Even though their number was huge, the eating activity of each single animal was enormous. For cutting and chopping up the forage, locusts possess mandibles which have to be stiff, hard and abrasion resistant, since their food often contains hard minerals. The material of the mandibles is, compared to common technical wear resistant materials, much softer and more compliant. Thus, at first sight, it looks like a mystery, how such a high abrasion resistance is achieved. Since many biological materials offer, apart from individual qualities, excellent and often specific mechanical properties, it is not surprising that they attract the attention of material scientists. One goal of this interest is to predict macroscopic material behaviour from the knowledge of chemistry, structures, and local mechanical properties on all smaller scales. As soon as at least part of this topic is understood the door is open to application of these concepts for the development of new synthetic materials. Even though the corresponding scientific field, biomimetics, is rather at its beginning, there already exist some successful applications, see [1–3].

The mechanical properties of biological materials span a wide range, depending not only on the specific demands, e.g. high elasticity, rigidity, or wear resistance, but also on the availability of chemical elements in the environment of the living organism.

A nice example is given by the formation of jaws and teeth of vertebrates: the chemical composition and the basic material structure on the nanometre level are more or less adopted from the skeleton (see below). For non-vertebrates such an already existing concept is at least not known, so that probably other strategies had to be found, taking into account the availability of chemical elements.

Before we delve into the details of biotools, let us shortly consider what we are talking about. In analogy to the Merriam-Webster Online Dictionary of a tool we define a biotool as "... something produced by an animal in a natural way to perform an operation otherwise impossible". In this strict sense the range of biotools is severely limited: only the tools for cutting and grinding food (the teeth of vertebrates and the mandibles of invertebrates) and the ovipostors of certain insects remain. But let us take a somewhat broader view; then the claws of a cat are biotools, as well as the hooves of ungulates, or the thick horny soles the camelids evolved for travelling over burning-hot sand or rugged terrain, because a cat without claws or a horse with a damaged hoof are severely handicapped. And since the competition for reproduction is a very important item for animals, maybe we should also include symbolic tools like, e.g. deer antler, which is regularly used for fighting for the dominance of a "harem", but seldom for real defence against predators. Or the bizarre "horns" some beetles carry (staghorn beetle!), which are quite useless for practical purposes, except, of course, to impress the females and for levering off rivals from a leaf. By the way, it is not that easy to draw the line because at least in some cases the fancy headgear is put to real use. The horns of a steer may easily be a symbol of male dominance, but a charging bull can use its horns with deadly precision, as any torero can tell you – if he survives. Finally, maybe a border case, the molluscan shells. They are certainly not "tools" to work something with, rather protective armour. But in a world full of hungry predators such armour can be life-saving for a stationary animal unable to flee or take cover. Therefore they should, at least, be mentioned.

For an insight into biological materials today a large number of experimental techniques exist. Apart from microscopy methods based on the interaction between electromagnetic waves or particles (visible light, X rays, electrons, ions, neutrons) with solid matter (optical methods in a more general meaning), scanning probe microscopy (SPM) and related techniques have become more and more useful and important. The former type of methods provides mainly structural and local chemical information, whereas the latter allows, apart from the imaging ability, the determination of mechanical properties on a very local scale, down to microns or even to the nanometre range. Apart from the classic SPM techniques, several refinements of mechanical measurements have been proposed, like dynamic mechanical analysis [4] or modulated lateral force microscopy [5]. However, the experimental limits of all devices can be clearly seen by the example of bone tissue. The basic building block is represented by mineral platelets embedded in a matrix of collagen [6]. The length and width of the platelets is some 10 nm, the thickness only a few nm. Most promising would be to measure directly on this scale, which is not practicable due to the small size. Systems on a larger, micron-scale, resembling the bone model do exist, e.g. the mineral rods in enamel or the structure of nacre. In both cases, however, the amount of the "hard" material is predominant. Even if a material with the basic bone structure would exist on larger scales, conclusions from such experiments on the material behaviour of bone tissue would be doubtful anyway, since its mechanical properties are intensely determined by the typical dimensions [7]. Since a straightforward experimental approach to "true" nanometre mechanical properties of solid-state matter does not yet exist (except tests on macromolecules), only sophisticated investigations on the submicron or micron scale can help to understand the effects which proceed on nanometre dimensions.

21.2
Typical Biotool Materials

21.2.1
Chemistry

In general, biological tool tissues consist of organic matter (mainly biopolymers), water and – as a further optional component – inorganic matter in the form of metal ions or mineral salt. Structural biopolymers can be grouped into proteins (chains of amino acids) and polysaccharides (chains of sugar molecules), both of which may occur in fibrous form or as a filler material. Fibrous biopolymers very commonly act as reinforcing elements in a fibre composite architecture, the main structural motif in biological tool tissues. In animal tissue, the fibrous component is usually proteinaceous (e.g. collagen in humans), an exception being the tissue of arthropods that consists of polysaccharide fibres (chitin) embedded in a protein matrix. Fibrillar structures with a moderate degree of cross-linking between the participating molecules can be very stiff in tension, but are notoriously weak in bending and compression. Therefore, they are used in nature to transmit tensile forces, if necessary around bends (Achilles tendon). Biotools such as teeth or claws, however, are usually stressed in various ways (compression, bending, shear), thus requiring additional stabilization strategies. One way towards greater hardness is to stiffen the organic matrix, by increasing the density of crosslinks. For example in keratin-containing biotools such as rhinoceros horn, bird beak, ungulate horn sheaths and hooves, finger nails and claws of predators [8], the hardness of the tissue is increased by disulphide bonds established between cysteines (sulphur-containing amino acid). Consequentially, high sulphur content is found predominantly in the hardest keratin-based tissues such as tortoise shell or biotools, in particular claws and beaks [9]. In chitinous arthropod tissue, the mechanical properties are controlled to a great extent by the degree of cross-linking of the protein matrix (sclerotization). Sclerotization is achieved by a process called tanning that usually also involves darker colouring and most importantly decreases the degree of hydration. The stiffness of fully tanned dry arthropod cuticle is about 20 GPa, while Young's moduli in the MPa range are found for wet and untanned cuticle [10]. Although chitin is mainly found in arthropods, it also occurs in the radula (a tongue-like rasping organ) of molluscs, the beak of squid and octopus, or in certain worms (e.g. grasping spines of arrow worms [11]). Alternatively or in addition to organic crosslinking, proteinaceous tissue may increase the density of crosslinks via coordination with metal ions. Such a mechanism was found for example in byssal threads of *Mytilus edulis* (holdfasts used by the marine mussel to stay attached to rocks), and also rather recently in the jaw of the marine worm *Nereis* that uses Zn ions to cross-link the protein matrix of its teeth [12]. A similar case may be present in *Glycera* worms (also belonging to the family of Annelida), albeit with Cu. Both mandibles contain considerable amounts of histidine, an amino acid well known for its affinity to transition metals. Remarkable amounts of transition metals (Zn and Mn) are also known to occur in arthropod tool tissue, in particular at the cutting edge where stresses are highest, e.g. locust and leaf cutter ant incisor, but the chemistry is not well known.

Very commonly, however, reinforcement of soft tissues in order to build structural elements and biotools is achieved by incorporation of mineral particles. In comparison to organic reinforcement strategies, mineral precipitation from solution is clearly a low-cost strategy: it saves metabolic energy needed to synthesize heavily cross-linked organic material and instead uses metals salts easily available from the respective environment as hard filler. From the chemical point of view, the assortment is rather limited. Vertebrates use bio-hydroxyapatite, a calcium phosphate mineral with the ideal stoichiometric formula $Ca_5(PO_4)_3OH$, but in vivo with up to 5% CO_3 groups substituting either the phosphate or the OH group. In vertebrate bone tissue, the mineral content is typically around 40–50% vol, embedded in a (hydrated) organic matrix consisting of mainly collagen (plus small amounts of proteoglycans and bone-specific proteins like osteopontin). In vertebrate tooth enamel, the mineral content reaches 90% mineral in an amelogenin matrix. By contrast, invertebrate animals favour a different calcium-containing mineral, namely calcium carbonate ($CaCO_3$). It is found in sea shells but also used by arthropods such as lobster (subphylum: Crustacea). In crustaceans the mineral is embedded in a protein–chitin matrix and used for the hard cuticle and their biotools, the claws [13]. Finally, there is a group of marine molluscs, the chitons, living in tidal waters and scraping algae from the rocks with a rasping tongue called radula. This radula carries many rows of tiny teeth and in order to prevent those teeth from being worn down too quickly by scraping calcite rocks they are quite literally armoured. Their cutting edge is impregnated with magnetite, whereas the rest of a tooth may contain Ca-minerals, silica (opal), and other Fe-minerals like limonite or lepidocrocite [14, 15]. Since iron minerals as structural elements are otherwise unknown in nature, this example is of special interest, the more so because for a long time it was believed that magnetite can only be synthesized using high temperatures.

Last, but not least: the marine worm *Glycera* makes use of both mechanisms to harden its teeth: reinforcement of a proteinaceous matrix by coordination with metal ions (Cu) and reinforcement by Cu mineral (atacamite) [16, 17], but the degree of mineralization ($\approx$10% vol) is much lower than that observed for other mineralized tissues. This rather unique combination is complemented by other peculiarities: *Glycera* worms are the only organism known to employ a Cu-based biomineral for structural purposes and their proteinaceous jaw matrix contains a considerable amount of melanin, not only providing the organic matrix with impressive hardness but also exceptional chemical stability.

21.2.2
Structures

As mentioned above, the most common structural concept for biotools is a fibre composite reinforced either by enhanced cross-linking or by incorporation of metal salts. In the following we shall concentrate on mineral particles as reinforcements and discuss a few principles of structuring biological materials using bone as an example. Bone is not only the most important structural material for vertebrates, but also a biotool: deer antler, for example, is slightly less mineralized bone, the sharp bill of sword fish and marlin consists of highly mineralized bone, and the horns of ungulates are bony protrusions, clad with a horn sheath. The structure of bone (like all biomaterials) is dominated by hierarchical levels. That is, certain principles and motifs can be found at various length scales, similar to a multi-fractal. Starting at the bottom, tropocollagen molecules self-assembly into subfibrils that are then mineralized. The mineral, bio-hydroxyapatite, usually forms platelets (although needle-like crystals are sometimes reported) within the fibrils as well as on their surface. Dimensions of the platelets vary considerably, as a rule of thumb they may be some tens of nanometres of length and width but, of course, not regularly shaped and only a few (3–5) nanometres thick. The volume fraction of the mineral component of bone and bone-like structures usually lies around 45%. The role of surface mineralization is not well understood, but within a fibril the platelets are arranged layer-by-layer, not in rank and file but in a staggered way, overlapping like the bricks in a well-laid wall. Jäger and Fratzl [6] succeeded in estimating the elastic modulus of such a structure using a simplified, two-dimensional model (Fig. 21.1), and were able to show that its longitudinal stiffness can be increased nearly arbitrarily simply by increasing the aspect ratio (the length-to-thickness ratio) of the platelets and therefore, implicitly, of the collagen layers in between. And – very important for biotools – this increase of stiffness does not cause a proportional decrease of fracture toughness. However, naturally, there is a price to pay. In this case it is a pronounced anisotropy. The model predicts very good mechanical properties along the direction of the longer axis of the platelets (which is in nature frequently oriented along the major stress direction), not quite as good along the minor axis and rather poor in the direction perpendicular to the plane of the particles. In certain cases, however, this is not necessarily a disadvantage. Trabeculae (the small struts forming the foam-like interior of bones, like the femur head and neck, but also vertebral discs and ribs) have been shown by SAXS (Small Angle X-Ray Scattering) [18] to have their mineral platelets oriented preferentially parallel to their long axes. It should be noted, however, that these trabeculae are arranged in such a way that they are almost exclusively stressed in tension or compression, but never in shear or bending. In cases where such a high degree of anisotropy would be less desirable the fibrils are arranged either in a crossed-plywood style, with the stiff axes alternating in their orientation with respect to the main stress, like in the secondary osteons in bone, or in an irregular, felt-like structure like in dentine.

Fig. 21.1. Staggered model of bone: a tensile load F causes shear forces in the matrix (*light grey* regions) between the mineral platelets (*dark grey* regions). The lengths D, L, d and δ strongly influence the resulting mechanical properties

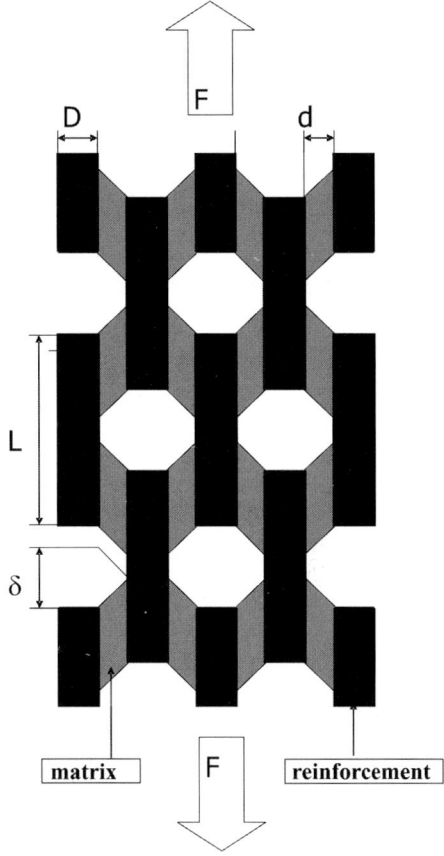

Another interesting aspect is the size of the particles. The staggered ("brick-and-mortar") model itself has no built-in length scale. But Gao et al. [7], using the Griffith criterion, showed that mineral particles below a certain critical length are insensitive to cracks and therefore display their (very high) theoretical strength, much in contrast to macroscopic ceramic bodies whose sensitivity to even small defects causes their brittleness. They estimated this critical length to be some tens of nanometres, which exactly corresponds to the lengths of the mineral platelets occurring in bone.

The existence of mineralized collagen fibrils, beautifully shown in SEM pictures of [19] and obviously connected by a very tough protein has led Gupta [20, 21] to propose a second hierarchical level of staggered arrangement consisting of stiff mineralized fibrils in a tough matrix of other proteins. But this idea remains controversial.

Other mineralized biomaterials follow similar concepts. The cuticle and claws of crustaceans, e.g. lobsters, are built from a chitin–protein fibre-composite, reinforced with calcite crystals and arranged in lamellae, in a crossed-plywood manner to avoid macroscopic anisotropy [13]. Enamel and nacre, the most heavily mineralized biomaterials known (above 90% vol mineral) strongly rely on very thin polymer layers between their mineral components in order to enhance fracture toughness. Enamel consists of "rods", bundles of thin but long hydroxyapatite crystals bound together by thin layers of the protein amelogenin and oriented more or less radially outward from the dentine–enamel junction perpendicular to the surface. Here the anisotropy mentioned above is fully exploited. Nevertheless enamel contains Hunter–Schreger bands, narrow regions of near-perfect crystal orientation that differ from one band to the next. Nacre is made from roughly hexagonal aragonite platelets, again bonded together by very thin layers of protein and arranged in a perfectly staggered manner. The small fraction of hydrated organic material (≈5% vol) is reported to increase the fracture toughness of the composite by a factor of 3,000 [22] with respect to pure aragonite.

Finally, in the Cu-mineralized jaws of the marine worm *Glycera*, the Cu-mineral atacamite is arranged into polycrystalline fibres embedded in a proteinaceous matrix and oriented along the outline of the needle-like tip, an arrangement quite different from the ones described above. It should be kept in mind, however, that the degree of mineralization in *Glycera* jaws is very low and mineralization might be regarded as a complementary strategy in addition to the reinforcement of the organic matrix via coordination with Cu-ions. According to a recent investigation by Moses et al. [23] employing, among other methods, wear tests, the outer layer of the *Glycera* teeth is reinforced exclusively by Cu ions, resulting in a very hard, water and wear resistant surface, whereas the atacamite-mineralized fibres are restricted to the layer underneath and are thought to convey enhanced bending strength [24]. Most remarkably this layer is less wear resistant in spite of the mineral reinforcement [23].

21.2.3
Mechanical Properties

Most of the first micro- and nanomechanical tests have been performed on metals, glass, and ceramics. The "simplest" class of materials is represented by glasses, being homogeneous, isotropic and, at room temperature, almost non-viscous. The results of mechanical measurements should, thus, neither depend on the choice of the investigated area, nor on the orientation of the sample or the velocity of the test. Biomaterials, however, are usually highly anisotropic, inhomogeneous and their deformation behaviour depends on the deformation rate and on the degree of hydration [25–31]. In the literature several models describing the viscous deformation of polymers and/or biological materials can be found. They usually imply various combinations of springs (describing the response of an elastic solid) and dashpots (describing the response of a viscous fluid). The simplest version is the combination of one spring and one dashpot either in parallel (Voigt system) or in series (Maxwell system). A refinement was made by a three-element combined model [32], where a Voigt system is arranged in series with an additional spring. At least partly physical models for time-dependent deformation behaviour have been published by Sasaki and Enyo [33] and by Jäger [34, 35]. Another, physics-based type of model has been proposed using poroelasticity [36–40], where the material is regarded as an elastic, porous body. With any deformation the fluid in the material must be "pressed" through the pores, causing a damped, viscous deformation response. Oyen et al. [40] performed a poroelastic analysis of nanoindentation data on bone with various fluid contents and compared the results with those already existing in the literature for poroelasticity at larger length scales. The resulting pore size was consistent with the scale of fundamental collagen–apatite interactions. However, such a model is somehow in contradiction with the fact that bone shows viscous behaviour even in a dry state. Models describing the mechanical behaviour of materials can never completely substitute experimental work. Therefore, for a full investigation of local mechanical properties it would be necessary to perform tests in various orientations, on a large number of spots in three dimensions ("mechanical tomography"), varying loading velocity and fluid content. In practice, however, such a vast amount of measurements is not feasible. It is, thus, necessary to reduce the total number of measurements by restricting the study to a few clearly defined questions. But even then, some fundamental constraints remain: the elastic modulus determined by indentation experiments is not identical with the classical, orientation-dependent Young's modulus [41, 42], but reflects some average over the various orientations. Moreover, the choice of a certain position in the sample is often restricted by the available preparation methods. Under such conditions, thorough test protocols are highly recommended.

21.3
Experimental Methods and Setups

21.3.1
SPM and Indentation

The basis of all experiments described here is a probe tip interacting with a sample surface. A detailed description of these interactions can be found for example in [43]. For the purpose of mechanical tests the elastic–plastic interaction is of predominant interest, even though forces like adhesion may play a role, especially for very soft materials. The "classic" suspension of probe tips is that on a flexible "bending arm" (cantilever) which is the common setup for an SPM. The interaction between sample and tip leads to a bending of the cantilever acting as a force sensor [44, 45]. Several techniques have been proposed for the spring force calibration of AFM cantilevers [46–60]. In addition to vertical probe tip–sample interactions, also lateral forces can be detected: by choosing the fast scan direction perpendicular to the cantilever axis, topographical extrusions as well as friction cause a torque on the cantilever with the consequence of tilting. The difference between the tilt angles from trace and retrace represents a measure of friction, which is usually only a semi-quantitative measure, mainly due to the fact that the resistance of a cantilever against torque is even harder to calibrate than the resistance against bending. Another, not fully quantitative mechanical measure is the damping of an oscillating cantilever hitting the sample surface, due to the interaction between tip and sample material, which affects the oscillation in frequency, amplitude and phase, where most commonly the resulting phase lag is utilized. The recorded "images" are maps of phase values with the high resolution of an AFM. However, the interpretation of such maps is not always straightforward. The most common application of this technique is

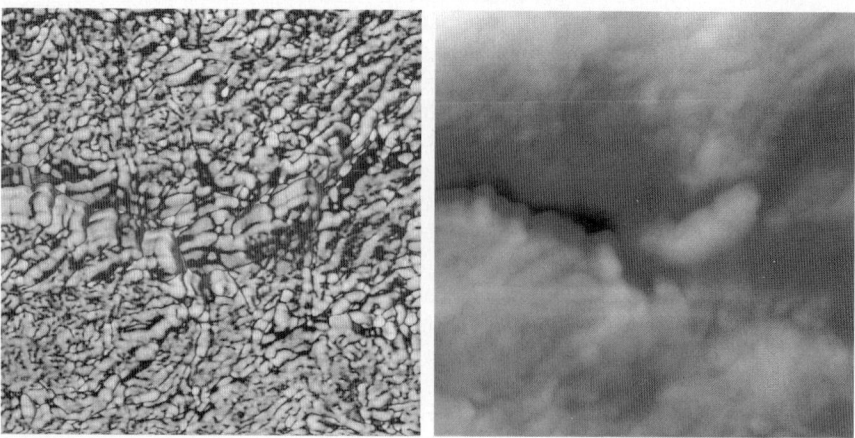

Fig. 21.2. AFM image of bone tissue. On the left the phase image, on the right the topographical image. Scan size $2 \times 2\,\mu$m

imaging of components with highly different mechanical properties without the need of special preparation techniques. A nice example is given in Fig. 21.2 showing the phase image of bone in cross-section with the hard hydroxyapatite minerals in the soft collagen matrix, in comparison with the topographical image, containing only random surface features.

Another mapping technique is force volume imaging [61–66]. Here, force-distance curves are recorded at every point corresponding to a pixel of an AFM image. Since the tip is not only scanned over the surface, but travels also perpendicularly, the denomination "force volume" imaging has been established. By post-processing the force-distance curves, two dimensional maps of tip–sample interaction can be obtained, giving information on physico-chemical properties and mechanical material response. The main problem of the method is the long acquisition time which can give rise to sample drift. This fact and the memory usage for storage of all data points limits the resolution of force images, problems that can partly be overcome by the pulsed mode force technique [67, 68], where recording of force curves is possible at rather high frequencies. To avoid problems with data acquisition, only selected points of the force curves are captured. Apart from the mapping methods described above, the measurement of force-distance curves also allows one to obtain information about single molecules and molecule pairs [69–73]. In the first case, tensile tests are performed where a large molecule is bound to a substrate at one end, the other end is "picked up" and bound to the AFM tip. In the second case, specific interactions between biological pairs are investigated. In this chapter, however, these experiments are not discussed further.

In the contact regime of the tip–sample interaction, both in approach and in retraction, the elastic deformation of the sample can be related to the elastic modulus. There are several theories describing the sample deformation. In the Hertz model [74] the adhesion forces are neglected, whereas in the Johnson–Kendall–Roberts (JKR) model [75] and the Derjaguin–Müller–Toporov model [76–78] adhesion between tip and sample is taken into account. Since materials forming biotools are rather stiff, adhesion effects can usually be neglected. Probing the mechanical properties by driving and retracting a hard tip vertically into a sample and recording load-displacement curves is known as nanoindentation or instrumented indentation. One of the advantages over the conventional hardness test is that there is no need to measure the area of the residual imprint, which is a rather difficult endeavour and usually accompanied with large errors, especially for very small imprints. The most common data extracted from load-displacement curves are hardness and elastic modulus, or more precisely, the reduced elastic modulus which takes into account the deformation of the indenter tip and the lateral deformation of the sample material via its Poisson's ratio [79]. Since for many biological materials the latter is not or only poorly known, it is rather convenient to give the results of indentation experiments by the reduced modulus. As long as there is no plastic deformation of the sample, the modulus can be obtained simply by a Hertzian contact. In case of a spherical contact geometry a simple relationship between applied force, contact depth and the reduced elastic modulus can be found, i.e. the force is proportional to the reduced elastic modulus, to the square root of the (deformed) tip radius and to $h^{1.5}$, h being the penetration depth. The radius can be obtained from experiments on a calibration material with well-known reduced modulus. At the onset of plasticity, often a sudden displacement burst appears which allows us to calculate the yielding pressure at this point. In biological materials, however, such a distinct onset of plastic deformation is usually not found. Hardness and elastic modulus are, commonly, extracted from the unloading part of load-displacement curves, assuming that during unloading only elastic deformation takes place. The most popular method of this type was introduced by Oliver and Pharr in 1992 [79]. In contrast to earlier methods [80], it is suitable also for experiments with a curvature of the unloading curve, i.e. if the contact area reduces during unloading which is commonly observed in biomaterials. The method takes into account the elastic deflection of the surface immediately adjacent to the contact, based on the general relationships derived by Sneddon between load, displacement, and contact area [81].

In [79] it is shown that the load-displacement relationship for many simple punch geometries can be written as $P = \alpha\, h^m$, P being the load, h the displacement, and α and m constants. Values of the exponent m are, for example, $m = 1$ for flat cylinders, $m = 2$ for cones, and $m = 1.5$ for spherical indenters. Most of the commercially available instruments work with a software package based on the Oliver–Pharr (O.P.) method. The user should, however, be well aware of some shortcomings: indents are frequently surrounded by raised material, so-called pile-ups or by permanent sink-ins (not to be confused with the elastic, non-permanent sink-ins). Since the contact area is determined via the displacement, these effects are not taken into account by the Oliver–Pharr method. Special attention has to be focused on the calibration of the tip shape as well as on the machine compliance as will be described later.

21.3.2
Scratch and Wear Tests

Mechanical testing on small scales can be classified by dimensions of lateral tip movement: from this point of view indentation has the dimension zero (0D), followed by scratch tests (1D) and scanning wear experiments (2D). Besides the basic research on the origin of friction and wear, scratch and scanning wear tests turned out to be a valuable tool for the investigation of ultra-thin coatings, lubricants on hard discs and magnetic recording heads [82–88]. The physical mechanisms of deformation can be expected to be different for the various types of experiments: while with indentation elastic and plastic deformation should appear mainly in compression, in scratch and wear tests shear forces and the effect of ploughing and/or cutting play a role. In a scratch test, a sharp tip is moved over the surface of a test material. The scratch depth at a given load or the load at which the material fails serves as a measure of scratch and wear resistance. Scratch tests can be performed in various modes: with constant or increasing load, as a single scratch or oscillating scratch. Using scanner-based instrumentation, scratching and imaging can be performed with the same tip, so that immediately after scratching the scratch depth can be determined. In addition, pile-ups and debris caused by scratching can be imaged. However, depending on the tip shape, one must be aware that the result of imaging represents a convolution of topography and the tip contour which may lead in scanning wear experiments to entirely different values of volume of the worn crater and the pile ups, respectively. Depending mainly on the tip shape and radius, a minimum vertical load is necessary for the onset of measurable wear which can be estimated by simple geometrical considerations in connection with a Hertzian contact. In [89] it was shown for a sphero-conical indenter with a tip radius of $1.8\,\mu$m that essentially two mechanisms occur. At loads below a threshold there is only ploughing in the material, no real material removal occurs. Eventually existing pile-ups are redistributed into the wear crater. At loads above the threshold material is removed from the beginning of the test. The transition from ploughing to immediate wear mainly depends on the lateral attack (critical contact) angle between tip and sample (including the pile ups in front of the tip). In [89] the attack angle was determined to be $19 \pm 2°$, independent of the material under investigation. This correlates to a transition from mainly compressive stresses to shear stresses. Similar thresholds have been observed for example by Degiampietro and Colaco on stainless steel [90] and Zhao and Bhushan on silicon [91].

21.3.3
Dynamic Modes

The techniques described involve "quasi-static" movement of the probe tip. An expansion of the utility of indentation devices has been achieved by introducing dynamical modes. The two most prominent variants from two different producers of indentation devices are described in the following, continuous stiffness (CSM) [79, 92, 93] measurements and nanoDMA. The CSM method is accomplished by superimposing a small, sinusoidally varying signal on top of the DC signal that drives the indenter. Data are obtained by means of a frequency specific amplifier (lock-in technique). At any point of the loading curve the contact stiffness can be measured, allowing the determination of the elastic behaviour at minimum penetration depths. Because of the short measurement time, due to frequencies above $40\,$Hz, this method is insensitive to thermal drifts and their fluctuations. CSM allows a fast examination of graded materials, layered structures as well as creep measurements at shallow penetration depths. Moreover, load cycles of sinusoidal shape allow the performance of fatigue tests on the nanoscale. Li and Bhushan described several applications of CSM [94]. Experiments on fused silica, PTFE and styrene butadiene rubber (SBR) revealed different results: for fused silica and PTFE, the contact stiffness increased linearly with the contact depth, elastic stiffness and hardness showed no change. For the SBR, however, the contact stiffness increased linearly with depth, while the measured elastic modulus decreased with depth, an effect not expected for this material. This result suggested that the contact depth calculated from the theory based on rigid plastic materials is not applicable to viscous materials, since time-dependent deformation is neglected. The dynamic method nanoDMA is also based on sinusoidal oscillation of the indenter tip. Various modes allow measurements with variable load, frequency or amplitude. Since this instrument is mounted on the scanner head of an AFM, it can be used, for example, for mapping the elastic modulus of a whole sample area within a rather short time.

21.3.4
Fracture Toughness Tests

In addition to hardness, elastic modulus, scratch and wear resistance fracture toughness is an important material parameter. Micro- and nanoindentation have been used to measure the fracture toughness of many brittle materials, mainly ceramics [95–108]. When brittle materials are indented with a sharp tip, radial cracks are generated during unloading. The length of these cracks can be correlated with the fracture toughness by various semi-empirical relationships. Rather simple relationships are given in [109–111], the fracture toughness being a function of elastic modulus, hardness and the crack length, all parameters which can be obtained immediately from indentation experiments. For details the reader is referred to [112]. However, a significant problem exists, namely that there are minimum loads for the occurrence of cracks, so-called cracking thresholds. Cracking thresholds have been measured, for example, by means of acoustic emission during indentation [113]. The cracking condition depends on the indenter geometry: the indentation thresholds can be reduced by using indenter tips with a small included tip angle, e.g. a Cube Corner tip instead of a Berkovic-type. Harding et al. showed [114] that the relationship between fracture toughness and crack length is the same for Berkovich and Cube Corner indenters, the only difference is a different geometric constant. The determination of fracture toughness is not totally restricted to pointed, pyramidal indenters on brittle materials: Yan et al. [115] reported on a simulation of internal cone cracks induced by conical indentation in brittle materials, and an attempt at estimating the fracture toughness of ductile materials by instrumented indentation was made by Lee and co-workers [116]. The applicability to biomaterials, however, has not yet been confirmed. Almost each of the experimental techniques above can be performed with a conventional AFM probe, however, with a few limitations, especially regarding indentation experiments and friction measurements. Calibrating the cantilever precisely is still not a simple task, especially a calibration of the torque resistance. Even the shortest and stiffest cantilevers have a spring constant too low to perform indentations in hard materials like enamel. Moreover, due to cantilever bending, the tip rotates during an indentation, thus the contact geometry is not known. From this point of view the introduction of a system where the cantilever plus its holder is replaced by a complete indentation setup (transducer) represents a considerable progress, combining indentation and scanning techniques [117].

21.4
Samples

21.4.1
Choice

Because of the inherent individual variations in biotool materials and the range of external factors (temperature, moisture, etc.) potentially influencing the mechanical properties, samples must be chosen with care and with regard to the questions to be answered. Bone, for example, occurs in a very large structural variety. Although composed of the same building blocks as bone on the nanometre scale, dentin exhibits a somewhat simpler structure. It contains rather densely packed channels (tubules) with a mutual distance on the order of 5–10 μm and a diameter of 1–3 μm, and consists of at least two types of material: intertubular dentin between the tubules and peritubular dentin surrounding the tubules, the two differing in the mineral content and, consequently, in hardness and stiffness. In order to elucidate the influence of the degree of mineralization, similar materials varying only in mineral content have to be found. Such an approach was undertaken, for example, by the investigation of turkey leg tendons [118], exhibiting large variations of mineral content with age: the mineral content starts from zero until an age of about 10 weeks and increases up to about 25% until an age of 23 weeks. In this study it could be shown that mechanical properties of bone-like material do not exclusively depend on the mineral content, they are influenced by variations of the collagen properties as well, a point of view important for clinic research on bone diseases. Clinical research usually requires statistical analysis making experiments on many samples from different individuals necessary. This approach, however, usually does not allow monitoring of the influence of external parameters and is therefore not feasible for basic research. In basic research usually a limited number of samples are thoroughly characterized under varying experimental and storage conditions, in order to extract valid property parameters. A study of the bone–cartilage interface of three human patellae [119] illustrates the latter strategy: the experimental approach was to perform indentation line scans across the interface and subsequently measure the local mineral content exactly at the locations of the indents. From these data a correlation between local stiffness and mineral content could be extracted, for bone and calcified cartilage, respectively. The final result could be obtained already from the first sample, the two additional samples served as (successful) confirmation. For a clinical study, on the other hand, three samples would be ridiculous.

21.4.2
Storage

Another important point when studying biological materials is that in vivo loading always takes place at physiological temperatures with a certain water or fluid content. Since in-vivo SPM measurements are hardly possible, the question arises, how the material properties change after isolation from the organism. For animal bone there is usually already a time span between slaughter and the deposition into a suitable storage fluid and/or freezing the samples. This delay can conveniently be avoided when working with tooth material by putting the teeth into a suitable fluid immediately after extraction. From this point of view dentin has an advantage as sample material over bone for basic research on bone-like material. But what are the "ideal" storage conditions, or which storage is the best compromise? Despite many investigations on biological materials, not much has been published on this issue. A few observations have been reported on teeth and bone: Moscovich et al. [120] studied the in-vitro Vickers hardness of dentin following gamma-irradiation and freezing. Freezing is one of the most widely used methods for storage of bio-materials, ionizing radiation can be employed as well but is more common for the sterilization of heat-sensitive medical devices, pharmaceutical packing and raw materials. The latter method has the advantage of excellent tissue penetration, even under wet conditions or refrigeration. The experiments revealed that the frozen samples showed no property changes, whereas the irradiated ones evidenced a slight increase in hardness. Hengsberger et al. [122, 123] reported mechanical degradation of bone samples on the micron scale after having performed indentation tests for 12 h under fluid. Habelitz et al. [124] compared the influence of various storage fluids on human dentin and enamel. The fluids were deionized water, normal salt solution and Hank's balanced salt solution (HBSS), respectively, the latter containing carbonate and phosphate buffer. For the first two fluids, a pronounced degradation of hardness and elastic modulus was observed, after one week of storage in these media, degradations up to 50% occurred. The most probable explanation for this effect is a demineralization especially in the near surface regions. Storage in HBSS for up to two weeks, however, caused no noticeable change in hardness and modulus. Obviously, the phosphate and carbonate buffers help to avoid demineralization. In another study, Guidoni et al. [125] investigated the influence of freezing tooth and bone samples on the mechanical properties, using HBSS as the storage medium. Freezing of bone did not remarkably alter hardness and modulus, in contrast to human dentin, where freezing led to a decrease in indentation hardness and modulus by about 20% after thawing the samples. This effect could be reduced by carefully removing the water content by replacing the water step-by-step with alcohol before freezing. From these findings the authors concluded that the freezing damage of dentin with some water content is mainly caused by the expansion of water in the tubules during freezing, since such an effect could not be observed with bone, where no such tubules exist. A second observation in this study strongly confirmed the results by Habelitz et al. [124]: using HBSS for the storage of a second premolar tooth conserved its hardness and modulus for even two years! It should be noted at this point that the sample handling in this study was performed with extreme care: beginning with the choice of the samples (totally healthy, "young" second premolars), they were deposited into HBSS (with a temperature of 4°C) within a few seconds after extraction, and, moreover, during all preparation steps the samples were rinsed with HBSS.

21.4.3
Preparation

Apart from possible influences by the storage of samples, their preparation may have a detrimental influence on the results of small-scale mechanical tests. The first step usually is cutting with a saw, where mainly two points have to be borne in mind: material adjacent to the cutting line should not be deformed too extensively, nor should the sample material be exposed to high temperatures. A low cutting speed combined with water cooling is usually satisfactory. In this context it should be mentioned, that suitable tooth samples can only be obtained from carefully extracted teeth without any cracks, otherwise damage by overheating cannot be excluded with certainty. The next usual step is embedding the samples in polymer resin, where the samples are brought into contact with the resin either by heating at elevated pressure, or just by being moulded in a two-component resin. Even in the latter case, curing of the polymer causes a temperature increase due to the exothermic character of the process. The most common procedure to obtain flat and smooth surfaces is adopted from the classical metallographic preparation techniques, i.e. grinding in steps with decreasing grit and then polishing, also usually in steps with decreasing grit. The polishing agent may be either a diamond paste or, more commonly for biological samples, an alumina-based polishing agent. The quality of the surface finish can considerably depend on the material properties. This can be seen, for example, from a polished section of a tooth: the (hard) enamel is usually much smoother compared to the (softer) dentin. In general, it appears that softer materials are harder to prepare which is often observed also for a (final) preparation with an ultra-microtome. In many cases an ultra-microtomed surface is smoother compared to the results from grinding and polishing. However, exceptions may confirm the rules, and, moreover, sensitive methods like cutting with an ultra-microtome require a skilled and experienced operator. In seeming contradiction to the demand for smooth surfaces it may be desirable to still be able to obtain some information on the microstructure from the topography on the surface after preparation. This is extremely valuable, if by a scanning probe the topography is imaged before a local mechanical test, therefore allowing one to choose the test position very precisely. Ho and co-workers investigated the effect of various sample preparation techniques on the determination of structures and nanomechanical properties of human cementum hard tissues [126]. The preparation methods included cryofracturing, ultrasectioning and polishing. The ultrasectioned sample revealed the lowest surface roughness, moreover the fibre structures in this material could be imaged with an AFM with a similar quality as obtained from the cryofractured samples. The 10–50-μm wide cementum dentin junctions could be clearly observed in the ultrasectioned specimens but not in the polished ones. The differences in hardness and modulus obtained when indenting along lines over the structures were more distinct with the ultrasectioned samples compared to the polished specimens, giving clear evidence that material was smeared and distributed laterally over the surface by polishing. All these findings suggest that for this type of material ultrasectioning is superior to polishing. In addition to the above considerations, it may be important to regard alterations by chemical sample treatment as well. Such treatment may be necessary to study the influence of a chemical component on mechanical properties. Such an approach has been reported by Broomell et al. [127], where the role of zinc in the tip region of jaws of the marine worm *Nereis* was investigated. The removal of zinc as well as its reconstitution was proven successfully, but, despite the well documented and carefully performed experiments, additional effects on the chemistry or nanometre structure could not be excluded.

Regarding all these considerations, we conclude that great care has to be taken with the choice, storage and preparation of samples for small-scale mechanical tests on biological materials.

21.5
Experimental Conditions

21.5.1
Moisture

The in-vivo state of biomaterials is inherently linked to a certain fluid content, usually water or water-based fluids. Most commonly the exact fluid concentration in the material is not known. Thus, it is hard to decide, in which environment mechanical tests should be performed. In some cases at least reasonable assumptions can be made, e.g. for dentin, immersed in saliva from the environment in the oral cavity as well from the interior through the tubules, so that we may conclude that the material is saturated with fluid. The most straightforward experimental conditions are measurements either totally under fluid or doing the experiment in vacuum on completely dry samples. The second version has some convincing advantages: the sample condition is fairly well-defined; moreover certain types of in-situ experiments can be performed, observing the sample and its deformation in an SEM during a mechanical test. In-situ micro-bending and tensile tests on biological samples have been reported for example in [128]. Results from such experiments can be compared with parameters from indentation. In addition, as can be seen from the literature, the results of indentation tests on dry bone-like material and on enamel from different laboratories are very similar. For "wet" samples, however, the situation is a lot more complex: first, it is usually not well defined what is meant by "wet", the sample being either totally under water (including the upper surface), or only partly, the investigated surface being in air. Second, there is evidence that in a wet state the results of mechanical experiments depend remarkably on the indenter tip shape. Thus, it is not surprising that the results from different laboratories on similar materials in a "wet" state vary considerably. For example, the results from Hengsberger et al. [123] and Guidoni et al. [125] on bone and dentin agree quite well. All these experiments were performed with samples totally under fluid. However, if they are compared, e.g., with the results from Rho and Pharr and references therein

[129–132, 134, 135], large discrepancies are found between the first and the second group: Hengsberger et al. and Guidoni et al. observed a rather large difference between dry and wet state, modulus and hardness being lower in the wet state by a factor of about 1.5–1.7 and 2.5–3.3, respectively, whereas in [129–132, 134, 135] a factor of only 1.1–1.2 was reported. Schöberl and Jäger performed an experiment [136] where a tooth sample was stored under HBSS for several days. Then this sample was taken out of the fluid and immediately tested by a series of indentations. At the beginning the sample surface was still covered by a fluid film that evaporated within about 15 min. While at the beginning hardness and modulus agreed with the values reported in [123, 125], they increased continuously and approached the values reported in [129–132, 134, 135] after 10–15 min. Therefore, one may conclude that it makes a large difference if a wet material is investigated with the surface being covered with fluid or not. It should be noted, however, that such an experiment should be regarded with much caution. The reproducibility is rather poor mainly due to two effects: first, the evaporation of the fluid costs energy, usually leading to some decrease in temperature and, consequently, to thermal drift of the experimental setup. Second, the contact between probe tip and sample is surrounded by an amount of water causing adhesion forces which probably change during evaporation. In general it is found that at least most of the biological materials become softer and less stiff with increasing fluid uptake. In addition to hardness and stiffness, a high abrasion resistance is a very important property for many biotool materials, in particular those used for cutting and grinding. The results of abrasion tests on such materials strongly depend on the fluid content. Tests on dry samples often may lead to conclusions entirely different compared to those drawn from experiments under "wet" conditions (see the sections "Results" and "Examples from the Literature').

21.5.2
Temperature

Little or nothing has been reported on the influence of the temperature on the results of micron-scaled mechanical tests. Most of the investigations have been performed at the common laboratory temperatures (20°C or slightly more), whereas the in-vivo temperature of human bone is the well-known 37°C. Except for some influence on the viscous behaviour, no severe change of the mechanical properties is expected for a temperature difference of 15–17°C that can be considered negligible compared to the possible temperature rise of several tens of degrees during embedding and preparation.

21.5.3
Probe Tips

Now let us consider some basic aspects, concerning the determination of elastic modulus and hardness, the latter being no physical property on its own but related to the plastic flow stress in a complex way. It is well-known that even for standard materials like glass or simple metals, hardness and elastic modulus obtained from indentation experiments depend on the indenter tip shape. Similar observations have been reported for scratch, wear and friction tests. Since the mechanism of plastic deformation in biological tissues is not clear, the choice of a suitable indenter tip is more or less intuitive. The most common probe is a Berkovich-type with a large opening angle, causing mainly compressive stresses beneath the tip [137, 138]. This helps, to some extent, to avoid the formation of cracks, so that only "purely" plastic deformation occurs, whatever that may mean with the deformation of biological materials. Moreover, the results can easily be compared with data from the literature. The use of a Cube Corner tip makes sense, if either the lateral expansion of the plastic deformation shall be limited or if fracture toughness measurements are intended.

21.5.4
Test Velocity

Since almost all biological materials show viscous deformation behaviour it is obvious that the results of all types of mechanical tests depend on the loading rate. To gain information on the viscous behaviour, quasistatic indentation tests should be performed with a large variation of loading and unloading speed. However, the range of variations is limited: slow experiments are influenced mainly by long-term drift of the setup, fast experiments by the mass of the scanner-moved probe and the limited time resolution of data recording. If the quasistatic experiments are supplemented by dynamic measurements, a larger span of deformation velocities can be applied. Scanning wear tests are usually performed with scan speeds of some μm/s to $100\,\mu$m/s. The velocities of practical abrasion applied on biotools in vivo, however, are higher by some orders of magnitude; thus, direct conclusions from a test in a laboratory on the true abrasion situation in nature are at least doubtful. From the literature, e.g. [139], it is known for viscous materials, that with increasing abrasion velocity the wear resistance increases. For comparative abrasion studies a fairly good compromise can be reached by varying the scan speed as much as possible (usually two orders of magnitude) and by looking for trends in friction and wear rate. If no remarkable dependence of friction and wear rate on the scan velocity occurs, or the dependence shows a continuous behaviour, one may extend the conclusions drawn from the comparatively slow tests to the natural, "fast" friction and wear situation [136].

21.6
Results

21.6.1
Sources of Error

The interpretation of results from scanning probe or indentation measurements is crucially dependent on possible errors occurring with such experiments. Even though appearing trivial, first of all we mention two very basic sources of error: if the sample is not fixed properly or the indenter tip is contaminated, there is no way to do any reliable correction, in contrast to the errors discussed below (e.g. wrong compliance or area function), where usually a correction can be made in retrospect, if the original curves are recorded properly. When testing biomaterials, the errors caused by the limits of instrumental resolution are usually small compared to all others.

The primary data derived from indentation tests are load and displacement with resolutions of clearly below $1\,\mu N$ and $1\,nm$, respectively. In the literature, loads are typically at least $250\,\mu N$, and displacements at least $100\,nm$. Thus, the maximum error caused by the instrumental resolution is about or below 1%. The basic instrumental error in the displacement measurement is superimposed by additional influences, mainly by thermal drift of the whole setup. The drift can be minimized by enclosing the setup in a box protecting from air movements and acoustic (perhaps electrodynamic) influences and by waiting until potential temperature gradients have equilibrated. Moreover, stable long-term drifts can be easily taken into account by the software of most of the indentation devices. If, however, variations in the drift rate cannot be excluded for the whole measurement, an additional check may be performed, by introducing sections with constant load in the loading protocol. Especially for soft materials an additional problem occurs. Prior to an indentation experiment a preload of typically $1–10\,\mu N$ is applied, causing a displacement of the indenter tip into the sample surface. This may lead to some zero-point shift. Solutions for this problem are usually already included in the software of the indentation device. Fortunately, among the large variety of biomaterials, the materials forming tools are rather stiff and hard, so that the zero-point shift usually does not introduce severe troubles. In order to derive mechanical properties from the load-displacement curves, the knowledge of the projected contact area is most crucial. Even though several methods for calibrating the tip area function exist, all methods suffer from certain shortcomings. One of the most common methods is to perform a series of indentations on a material with well-known mechanical properties, then "imposing" one known mechanical property (commonly the elastic modulus) while making a fit to the results [79]. The calibration material should be isotropic, homogeneous and its deformation behaviour independent of the deformation velocity (no viscous behaviour). Fused silica is usually considered "the standard calibration sample". Because of its high hardness and modulus, however, its application is limited to measurements to rather shallow indentation depths, mainly for two reasons: the maximum load of some instrumentation setups is limited and, especially for small open angle pointed indenter tips (Cube Corner geometry, sharp conical indenters), cracks are formed at high load indentations, making the use of results at least doubtful. Most of the indentation experiments on biomaterials, however, are performed with rather large probing depths. As "soft" calibration standards aluminium or polycarbonate have been proposed. Al is elastically isotropic with a well-known modulus of elasticity. However, Al is rather stiff which results in a steep unloading curve, which is very receptive to influences on the displacement measurement, mainly by thermal drift, or to uncertainties in the machine compliance (see below). Second, indents in Al are surrounded by pronounced pile-ups, causing uncertainties in the true actual contact area. Several attempts have been proposed, e.g. in [140] to account for the pile-ups, but due to the unknown pressure distribution below the indenter (does the pile up material take up the same pressure as the "bulk" material?), such corrections may lead to even an "overcorrection". Polycarbonate is isotropic and quite homogeneous. The main problem here is caused by its viscosity which makes hardness as well as modulus results dependent on the loading protocol. The problem of tip shape calibration is, fortunately, moderated at least for pyramidal indenter geometries: an indenter tip shape consists, roughly spoken, of the basic theoretical geometry and the tip rounding. The latter can be determined by shallow indents in fused silica. If the basic geometry can be precisely maintained by the manufacturer, the tip area function is completely defined. In contrast to some other tip geometries, the opening angle of three-sided pyramidal tips is maintained within $0.05°$ by several manufacturers. Thus, the use of three-sided pyramidal indenter tips is strongly recommended, unless other, specific needs suggest or even impose the use of a different tip shape. One basic question shall be briefly considered: why not use the hardness, instead of the elastic modulus, as a measure for tip shape calibration, since it reacts more sensitively on errors of the contact area? The main disadvantage utilizing the hardness is the fact that it is not a true physical parameter and, especially in metals, depends on the size of the plastic zone around the imprint (well known as indentation size effect, see e.g. [141–148]) and, in addition, on the indenter shape [149]. Another common method to determine the true tip shape is to scan the tip over a spike with extremely high aspect ratio and minimum tip radius, whereby the observed topography reflects the shape of the probe tip. If such a spike is not available, one may perform indents in a soft material with minimum elastic deformation, e.g. in silver. The residual imprint now reflects the "turned over" tip shape which can be imaged by AFM with a sharp, high aspect ratio tip. In addition to the tip shape, the compliance of the experimental setup must be known, i.e. the elastic response of the device plus tip holder. Even though, as for tip shape calibration, several methods for compliance measurements can be found in the literature [79, 149], it is still not a trivial task. An additional check can be performed by determining the elastic stiffness of various materials spanning a wide range of elastic moduli known from the literature [149]. If the results for all materials, especially the stiffest materials, are consistent with the literature values, the compliance is confirmed quite safely. Taking into account the fact that biological materials (except for highly mineralized tissues) are rather low modulus materials compared to most metals or ceramics, we conclude that an error in the machine compliance will rarely have a severe impact on the results.

All considerations given above are valid more or less for all kinds of materials. In the following, sources of errors typical for biological samples and the corresponding problems occurring with sample preparation shall be discussed. First of all, biological materials show viscous deformation behaviour. For quasistatic tests the method by Oliver and Pharr is the most common one, using the initial slope of the unloading curve as a key parameter. If the material creeps, there is still creep at least during the first part of the unloading cycle. This becomes obvious, when in a test on a viscous material without hold time at maximum load the first part of the unloading curve shows a pronounced bend with an even negative slope, which is in contradiction to the basic assumptions for the Oliver–Pharr method [79]. This problem can be overcome (at least partly) by introducing a long hold period at maximum load before unloading which minimizes the creep rate and/or by refined analysis of the

Fig. 21.3. In-situ indentation in a SEM. During unloading (from *top* to *bottom*) the contact area decreases remarkably within a very short time

load-displacement curve [150–152]. In addition to creep, biological materials exhibit viscoelastic recovery that becomes evident upon introducing a holding period at constant (low) load after partial unloading: the displacement decreases at constant load, converging to a constant value with time. This time-dependent elastic recovery can be nicely observed when performing experiments while imaging the indentation process in a scanning electron microscope (SEM) [149]. Such observations show, in addition, that during unloading the contact area decreases pronouncedly for such materials, see Fig. 21.3.

Another, notable shortcoming of the O.P.-method or other methods analyzing the unloading curves should be mentioned: after having deformed the sample plastically by loading, the structures below the indenter may have been damaged, thus changes in the elastic interaction between indenter and sample during unloading cannot be excluded. As a solution, an analysis of the very first part of the loading curve before the onset of plastic deformation may be proposed (Hertzian fit). Unfortunately, this procedure cannot be applied to most biomaterials, since there the plastic deformation usually starts at very low loads. Considering the long list of shortcomings discussed in the literature [153, 154], it may seem surprising that the O.P.-method is still quite commonly applied on viscous materials. It is impressive how modulus and hardness determined for many materials in different laboratories and on various indentation devices agree reproducibly. Some authors [153–155] propose to determine the hardness by measuring the residual imprint with an AFM or a scanning indenter tip. However, since the most common and reasonable definition of hardness is the maximum load divided by the projected contact area at this maximum load, this method is only valid, if the contact area does not change during unloading (see above), a prerequisite usually not fulfilled in polymers or biomaterials.

The possible precision of indentation, friction, scratch and wear experiments is, often to a large extent, limited by surface roughness. All the geometric considerations on indentation experiments are done on the assumption of an ideally flat and smooth sample surface which usually does not hold in practice. Biological samples in particular are mostly hard to prepare towards a perfect surface finish. As a consequence, most indentation experiments on biological materials are performed with rather large penetration depths (from this point of view the denomination nanoindentation is often not justified). If the indentation device allows scanning the tip over the surface before the mechanical test, it is possible to choose suitable regions for the experiment and image the area around the imprint after the test, allowing at least an estimate of the error resulting from sample topography including eventual pile-ups.

The investigation of lamellar bone represents a good example for the possible influence of the topography on indentation results [123, 156]. After a standard sample preparation, usually classic metallographic grinding and polishing, the lamellar structure is clearly reflected by the topography with the lamellae protruding about 30 nm from the interlamellar regions. The interlamellar spacing is of the order of microns, thus the maximum indentation depth is severely limited. A comparison of mechanical data from the interlamellar region and the lamellae is superimposed by a systematic error due to the different contact situations: indentation in the interlamellar region represents an imprint "into a valley", whereas on the lamellae it is an imprint "on top of the hill". As a consequence modulus and hardness are overestimated for the former und underestimated for the latter. A method to minimize this systematic error has been proposed by Staedler et al. [156].

Measurements on biological materials are frequently conducted on areas with lateral dimensions of only microns, separated by pores or gaps etc. For handling and preparing (grinding, polishing) the samples they are usually embedded in polymer resin or similar material that partly fills any cavities. Another, similar situation occurs, if the experiments are performed on layered structures, where the thickness of individual layers is not drastically larger than the depth of the imprint. Several attempts have been made to include such a structural compliance into the analysis [157, 158, 169], a term that adds to the machine compliance, but unlike the latter the structural term can vary as a function of the position on the specimen [155]. For porous materials various attempts have been made to extract "true" material properties from indentation measurements [170–173], some of them applied for studies on bone.

21.6.2
Interpretation

Unfortunately, the modulus obtained from indentation measurements is not identical with the orientation dependent Young's modulus, measured anisotropies become "smeared" [41, 42, 174–177] due to the indentation geometry. Hardness, unfortunately, does not reflect a well-defined, physical property. Zhang and Sakai [178] investigated geometrical effects of pyramidal indenters with different face angles on the elasto-plastic contact behaviours of metals and ceramics. They found different hardness results for varying face angle. Moreover, they tried to derive a face angle dependent "true" hardness which was related to the flow stress by a constraint factor. Applying such considerations on biomaterials, however, is more or less hopeless, due to the lack of known constraint factors. Aside from hardness and stiffness, abrasion resistance is a very important property of biotools. Since hardness and modulus are the standard results from indentation measurements, it seems desirable to derive the abrasion behaviour from the former data. For metals, ceramics and coatings a high abrasion resistance is usually achieved by a combination of high hardness H at rather low elastic modulus E [179], the abrasion resistance frequently described as proportional to $H^{1.5} E^{-1}$. This proportionality factor has been taken into consideration in a number of studies on biotools, e.g. [12, 180–183]. Apart from copper, frequently Zn is found, for example in the tip region of the jaws of the marine worm *Nereis* [12] or the cutting edge of the desert locust mandible *Schistocerca gregaria* [136]. These mandibles represent an example for a high sensitivity of the mechanical behaviour on the water content as well as the fact that conclusions drawn from hardness and modulus results on the abrasion resistance can be erroneous. The mandibles have zinc-enriched (several at-%) regions at their cutting edge, i.e. in the material where high abrasion resistance is absolutely mandatory. Surprisingly, the abrasion rate depends strongly on the water content only in the cutting edge region. A similar observation was made for the abrasion behaviour of human teeth and the jaw of the marine worm *Nereis*: only for the material with the necessity of being highly abrasion resistant i.e. the tip of the jaw and the enamel, respectively, a high sensitivity on the water content was found [136]. The highest resistivity against abrasion of the locust mandibles at their zinc-enriched cutting edge was found for a certain water content. Hardness and modulus measurements showed that this is not at all the state of the largest value of $H E^{-1}$ or $H^{1.5} E^{-1}$. Scanning wear tests are basically not complicated experiments. However, in some cases they can be performed on mineralized materials only with some restrictions, e.g. on enamel which consists of mineral rods with protein in between. If the scanning area covers not only a single rod, but is larger, mineral particles may be pulled out which makes the definition of an abrasion rate almost impossible; results will often not be reproducible. A similar situation appears for the atacamite-enriched jaw regions of the marine worm *Glycera*, if a very sharp probe tip is used. In this case using a probe with larger tip radius and opening angle could help to achieve reproducible results [23].

21.7
Examples from the Literature

In the following some examples from the literature are given presented in order to elucidate the implications of the above considerations for scientific work on biotool materials. Although there is a vast variety of biotool tissues, here we shall focus on jaws and teeth and discuss the heavily mineralized version found in vertebrates as well as selected invertebrate tooth tissues containing little or no mineral.

A great number of studies have been performed on the mechanical properties of human teeth, e.g. [184–193]. In [193] area-scans of human coronal dentin were carried out in order to investigate correlations between local mechanical properties and the density, size and crystallinity of the mineral particles. The mineral content of dentin was found to decrease and the thickness of mineral crystals to increase towards the dentin–enamel junction (DEJ). Hardness and elastic modulus both decreased towards the DEJ. In a correlation analysis, the mineral content and, even more so, the thickness of mineral crystals were found to be crucial factors for the prediction of hardness. The dentin layer close to the DEJ exhibited a local minimum in hardness and elastic modulus, a configuration known to be an effective obstacle for crack propagation. It was concluded that the observed variations of mechanical and structural properties define crown dentin as a gradient material optimized for its mechanical function. Further studies on the dentin–enamel junction were performed for example in [194–197], where hardness, elastic modulus and fracture were studied as well as the functional width of the DEJ. Using AFM-based microscratching it could be shown that the width of the junction is only a few microns, much smaller than expected, being a nice example, how a complementary technique in addition to indentation allows deeper insight in such a system.

In stark contrast to vertebrate teeth based on large amounts of calcium mineral (enamel contains about 90% vol. hydroxyapatite), the jaws of the marine worm *Glycera* contain very low levels of a rare copper mineral (atacamite), the only one of its type known so far to be involved in biomineralization. The mineral is arranged in fibres running parallel to the outline of the jaw and was identified in 2002 by Lichtenegger et al. [180] who also performed instrumented indentation experiments in order to elucidate its function. Indentation was combined with local chemical analysis by microprobe measurements and SEM imaging on the same sample. A clear positive correlation between the local hardness and stiffness and the amount of atacamite was found, suggesting a structural function of the mineral. The performance index $H^{1.5} E^{-1}$ [198] was used as a rough estimate for abrasion resistance and found to correlate with the copper content. In a subsequent publication [182], H and E values of *Glycera* jaws and other materials were arranged into a double-logarithmic performance plot following the approach outlined by Ashby [198]. Data lying on a straight line with a slope of 2/3 should then represent materials with equivalent abrasion performance. From this plot it was, tentatively, suggested that compared to many materials, *Glycera* jaws have an excellent resistance to abrasion. Since H and E given in [180] were from dry samples embedded in polymer resin this estimate might not reflect the natural abrasion situation in sea water [153]. Indentation measurements performed by Moses et al. [182] on dry and wet *Glycera* jaw samples yielded similar spatial distributions of hardness and modulus in both cases, with an overall decrease of modulus by 25% and of hardness by 28% in wet samples. Furthermore, in this study samples were subjected to hydrolysis in boiling HCl for 48 h, which removed the proteinaceous portion of the jaw matrix as well as almost all Cu as evidenced by EDS (Energy-dispersive X-ray spectroscopy) measurements, but left the shape of the jaw intact. The residue, identified as melanin scaffold, exhibited a modulus reduced by 60% and a hardness reduced by 40% as compared to the untreated jaw.

Although it had been reported earlier that *Glycera* jaws also contain a certain amount of unmineralized Cu potentially coordinated with the histidin (His)-rich jaw matrix [199], it was only found in a recent study by Pontin et al. that the atacamite mineralized fibres are located underneath the jaw surface and are embedded in a layer rich in unmineralized copper [23]. In this work *Glycera* jaws were also subjected to wear tests. Selecting a tip with a comparatively large radius allowed the use of a relatively simple model for interpretation: a hard spherical indenter ploughing across a flat block of a perfectly plastic medium was assumed. The atacamite mineralized regions exhibited a wear depth about twice as large as the adjacent, copper-enriched but mineral-free matrix. It was also found that the wear rates increased by a factor of about three in all regions after hydration. Scratch tests revealed similar trends. Under both, wet and dry conditions, the regions with unmineralized copper emerged as most scratch resistant. The ratio between lateral and normal load (friction coefficient) was highest in the copper-free and the atacamite containing regions; hydration caused an increase by 30%, the increase of friction coefficient in copper-rich, unmineralized regions was about 22%.

These findings were explained with the coordination of Cu by the histidine-rich protein and melanin thus establishing inter-molecular crosslinks [23]. Such crosslinks are not only expected to enhance local hardness and stiffness, but may also reduce susceptibility to water, perhaps by occupation of intermolecular sites by Cu ions. The enhanced wear rate in atacamite-rich regions can be explained by chipping and cracking of the (brittle) mineral fibres, promoted by elastic/plastic mismatch between mineral and matrix. Even though the authors of this work presented some speculations, how the wear rate was increased by hydration, no real explanation for this behaviour could be given.

Similar studies were performed on the slightly different jaws of the marine worm *Nereis* [12] that contain Zn instead of Cu, albeit in entirely non-mineralized form and in lower amounts. Again, the jaw matrix is proteinaceous and rich in His, but in contrast to that of *Glycera* jaws it does not contain any melanin. The zinc is predominantly concentrated in the jaw tip, thus suggesting a mechanical function. The influence of Zn on stiffness and hardness was probed with a combination of indentation and microprobe analysis on dry jaws, yielding a positive correlation between local metal content and mechanical performance. The Zn content was shown to vary in parallel with the presence of Cl and local His levels. X-ray absorption experiments suggested a coordination of Zn by three His and one Cl in analogy to the arrangement found in Zn insulin [12]. In a recent study by Broomell et al. [127] *Nereis*

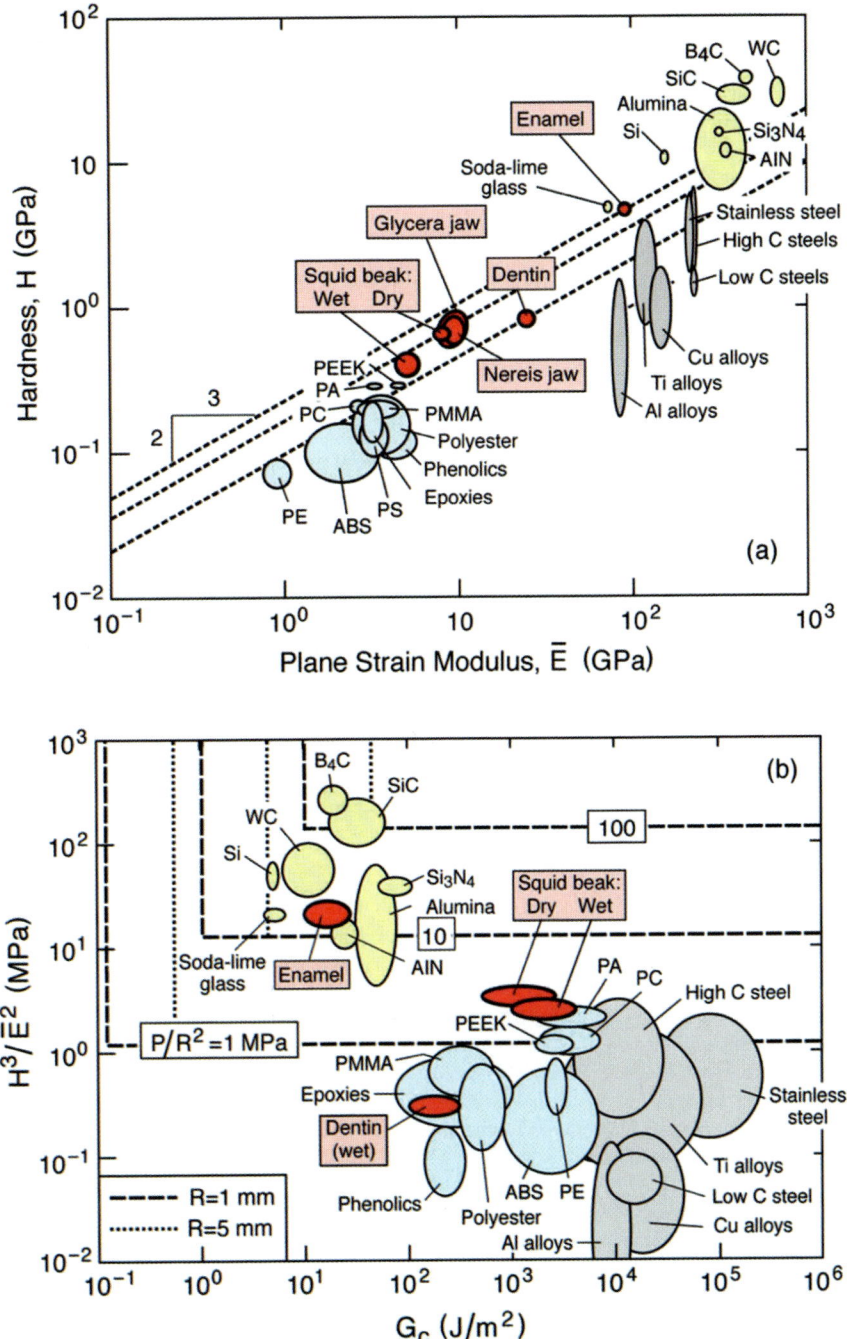

Fig. 21.4. Property maps showing (**a**) comparisons of hardness and stiffness of *Dosidicus* beaks with typical engineering materials and (**b**) critical loads for yielding and for cracking at contact with spherical rigid abrasives

jaws were investigated by indentation in dry and wet state and by element analysis using EDS. Samples were measured before and after treatment with EDTA for 96 h at 25°C that removed a large fraction of the Zn without causing changes in the organic constituents, as was confirmed by X-ray photoelectron spectroscopy (XPS) on pulverized jaws and secondary mass spectroscopy. In a further step, the samples were enriched in Zn again by incubation of pre-depleted jaw surfaces in $ZnCl_2$ for 72 h at 25°C. Not a complete, but partial reconstitution was evidenced by XPS and EDS. The authors could show that the measured hardness and modulus values correlated with the variations in Zn. The Zn removal resulted in over 65% reduction of hardness and stiffness in relevant areas and its reintroduction caused an appreciable recovery.

Finally, the beaks of the jumbo squid *Dosidicus gigas* represent a purely organic biotool material. Although the beaks contain neither metals nor biomineral, they exhibit a hardness and stiffness up to twice as high as that of the most competitive synthetic organic materials. In a study by Miserez et al. [181] indentation tests were carried out on squid beak samples in ambient air and under distilled water. The experiments were complemented by macroscopic fracture toughness measurements. Abrasion was assumed to be due to one of two failure mechanisms, either the onset of plasticity or the formation of cracks. The material was treated as flat plate and the abrading particle as a rigid solid of revolution, allowing treatment by classical contact mechanics. The normal load for yielding is then proportional to $H^3 E^{-2}$, whereas the critical load to develop a crack P_C was proportional to the shear modulus G. By plotting $H^3 E^{-2}$ vs. G, property maps were drawn (see Fig. 21.4), from which the deduced abrasion resistance was compared to the values of enamel, dentin, some polymers and metals, indicating a high abrasion resistance of the squid beaks. No direct measurements of the abrasion behaviour, however, were carried out.

References

1. Gorb SN, Sinha M, Peressadko A, Daltorio KA, Quinn RD (2007) Bioinsp Biomim 2:117
2. Otten A, Herminghaus S (2004) Langmuir 20:2405
3. Shan D, Wang S, He Y, Xue H (2008) Mat Sci Eng C28:218
4. Oulevey F, Burnham NA, Gremaud G, Kulik AJ, Pollock HM, Hammiche A, Reading M, Song M, Hourston DJ (2000) Polymer 41:3087
5. Pietrement O, Troyon M (2001) Surf Sci 490:L592
6. Jäger I, Fratzl P (2000) Biophys J 79:1737
7. Gao HJ, Jäger IL, Arzt E, Fratzl P (2003) PNAS 100:5597
8. Block RJ, Bolling D, Brand FC, Schein A (1939) J Biol Chem 128:181
9. Fraser RDB, Mc Rae TP, Rogers GE (1972) Keratins: their composition, structure and biosynthesis. Thomas CC, Springfield, IL
10. Vincent JFV, Wegst UGK (2004) Arthropod Struct Develop 33:187
11. Saito Y, Okano T, Chanzy H, Sugiyama J (1995) J Struct Biol 114:218
12. Lichtenegger HC, Schöberl T, Ruokalainen JT, Cross JO, Heald SM, Birkedal H, Waite JH, Stucky GD (2003) PNAS 100:9144
13. Sachs C, Fabritius H, Raabe D (2006) J Struct Biol 155:409
14. Evans LA, Alvarez R (1999) J Biol Inorgan Chem 4:166
15. Lee AP, Brooker LR, Macey DJ, Webb J, von Bronswijk W (2003) J Biol Inorgan Chem 8:256

16. Lichtenegger HC, Schöberl T, Bartl MH, Waite JH, Stucky GD (2002) Science 298:389
17. Lichtenegger HC, Schöberl T, Bartl MH, Waite JH, Stucky GD (2003) Science 301:5636
18. Rinnerthaler S, Roschger P, Jakob HF, Nader A, Klaushofer K, Fratzl P (1999) Calcif Tissue Int 64:422
19. Fantner GE, Rabinovych O, Schitter G, Thurner PH, Kindt JH, Finch MM, Weaver JC, Golde LS, Morse DE, Lipmann EA, Rangelow IW, Hansma PK (2006) Compos Sci Technol 66:1205
20. Gupta HS, Wagermaier W, Zickler GA, Aroush DR, Funari SS, Roschger P, Wagner HD, Fratzl P (2005) Nano Lett, 5: 2108
21. Gupta HS, Seto J, Wagermaier W, Zaslansky P, Boesecke P, Fratzl P (2006) PNAS 103:17741
22. Jackson AP, Vincent JFV, Turner RM (1988) Proc R Soc London Ser B 234:415
23. Moses DN, Pontin MG, Waite JH, Zok FW (2008) Biophys J doi:10.1529/biophysj.107.120790
24. Lichtenegger HC, Birkedal H, Waite JH (2008) in: Met Ions Life Sci 4 (Biomineralization. From nature to application)
25. Lee EH, Radok JRM (1960) J Appl Mech 27:438
26. Ting TCT (1966) J Appl Mech 88:845
27. Johnson KL (1985) Contact mechanics. Cambridge University Press, Cambridge
28. Oyen ML (2006) Phil Mag 86:5625
29. Oyen ML (2005) J Mater Res 20:2094
30. Bembey AK, Oyen ML, Bushby AJ, Boyde A (2006) Phil Mag 86:5691
31. Bembey AK, Bushby AJ, Boyde A, Ferguson VL, Oyen ML (2006) J Mater Res 21:1962
32. Vandamme M, Ulm F-J (2006) Int J Solids Struct 43:3142
33. Sasaki N, Enyo A (1995) J Biomech 28:809
34. Jäger IL (2005) J Biomech 38:1451
35. Jäger IL (2005) J Biomech 38:1459
36. Cowin SC (1999) J Biomech 32:217
37. Wang HF (2000) Theory of linear poroelasticity with applications to geomechanics and hydrogeology. Princeton University Press, Princeton, NJ
38. Agbezuge LK, Deresiewicz H (1974) Israel J of Technol 12:322
39. Selvadurai APS (2004) Int J Solids Structures 41:2043
40. Oyen ML, Bembey AK, Bushby AJ (2007) Mater Res Soc Symp Proc 975
41. Vlassak JJ, Nix WD (1996) Phil Mag A67:1045
42. Swadener JG, Rho JY, Pharr GM (2001) J Biomed Res 51:108
43. Butt HJ, Capella B, Kappl M (2005) Surf Sci Rep 59:1
44. Meyer G, Amer NM (1988) Appl Phys Lett 53:2400
45. Alexander S, Hellemans L, Marti O, Schneir J, Elings V, Hansma PK, Longmire M, Gurley J (1989) J Appl Phys 65:164
46. Albrecht TR Akamine S, Carver TE, Quate CF (1990) J Vac Sci Technol A8:3386
47. Sader JE (1995) Rev Sci Instrum 66:4583
48. Neumeister JM, Ducker WA (1994) Rev Sci Instrum 65:2527
49. Hazel JL, Tsukruk VV (1999) Thin Solid Films 339:249
50. Butt H-J, Siedle P, Seifert K, Fendler K, Seeger T, Bamberg E, Weisenhorn AL, Goldie K, Engel A (1993) J Microsc 169:75
51. Sader JE, Larson I, Mulvaney P, White LR (1995) Rev Sci Instrum 66:3789
52. Senden TA, Ducker WA, (1994) Langmiur 10:1003
53. Maeda N, Senden TJ (2000) Langmuir 16:9282
54. Notley SM, Biggs S, Craig VSJ (2003) Rev Sci Instrum 74:4026
55. Degertekin FL, Hadimioglu B, Sulchek T, Quate CF (2001) Appl Phsy Lett 78:1628
56. Holbery JD, Eden VL, Sarikaya M, Fisher RM (2000) Rev Sci Instrum 71:3769

57. Cleveland JP, Manne S, Bocek D, Hnsma PK (1993) Rev Sci Instrum 64:403
58. Bonaccurso E, Butt H-J (2005) J Phys Chem B109:253
59. Hutter JL, Bechhoefer J (1993) Rev Sci Instrum 64:1868
60. Gibson CT, Watson GS, Myhra S (1996) Nanotechnology 7:259
61. Van der Werf KO, Putman CAJ, De Groth BG, Greve J (1994) Appl Phys Lett 65:1195
62. Cappella B, Paschieri P, Frdiani C, Miccoli P, Ascoli C (1997) Nanotechnology 8:82
63. Koleske DD, Lee GU, Gans BI, Lee KP, Dilella DP, Wahl KJ, Berger WR, Whitman LJ, Colton RJ (1995) Rev Sci Instrum 66:4566
64. Radmacher M, Fritz M, Cleveland JP, Walters DA, Hansma PK (1994) Langmuir 10:3809
65. Baselt DR, Baldeschwieler JD (1994) J Appl Phys 76:33
66. Heuberger M, Dietler G, Schlepbach L (1995) Nanotechnology 6:12
67. Rosa-Zeiser A, Weilandt E, Hild S, Marti O (1997) Meas Sci Technol 8:1333
68. Krotil HU, Stifter T, Waschipky H, Weishaupt K, Hild S, Marti O (1999) Surf Interf Anal 27:336
69. Janshoff A, Neitzert M, Oberdörfer Y, Fuchs H (2000) Angew Chem Int Ed 39:3212
70. Zhang W, Zhang X (2003) Prog Polym Sci 28:1271
71. Rief M, Oesterheld F, Heymann B, Gaub HE (1997) Science 275:1295
72. Fritz J, Anselmetti D, Jarchow J, Fernadez-Busquets X (1997) J Struct Biol 119:165
73. Kikuchi H, Yokoyama N, Kajiyama T (1997) Chem Lett 11:1107
74. Hertz H (1882) J Reine Angew Math 92:156
75. Johnson KL, Kendall K, Roberts AD (1971) Proc R Soc London A 324:301
76. Derjaguin BV, Muller VM, Toporov YP (1975) J Colloid Interf Sci 53:314
77. Müller VM, Yushchenko VS, Derjaguin BV (1980) J Colloid Interf Sci 77:91
78. Müller VM, Derjaguin BV, Toporov YP (1983) Colloids Surf 7:251
79. Oliver WC, Pharr GM (1992) J Mater Res 7:1564
80. Doerner MF, Nix WD (1986) J Mater Res 1:601
81. Sneddon IN (1965) Int J Engng Sci 3:47
82. Bhushan B, Koinkar VN (1995) Surf Coat Technol 76–77:655
83. Bhushan B, Lowry JA (1995) Wear 190:1
84. Koinkar VN, Bhushan B (1996) J Appl Phys 79:8071
85. Bhushan B (1999) Diam Rel Mater 8:1985
86. Miyake S, Kaneto R, Miyamoto T (1992) Diamond Films Technol 1:205
87. Jiang Z, Lu CJ, Bogy DB, Bhatia CS, Miyamoto T (1995) Thin Solid Films 258:75
88. Kaneko R, Miyamoto T, Yandoh Y, Hamada E (1996) Thin Solid Films 273:105
89. Schiffmann KI (2008) Wear, in press
90. Degiampietro K, Colaco R (2007) Wear 263:1579
91. Zhao X, Bhushan B (1998) Wear 223:66
92. Pethica JB, Oliver WC (1989) in: Bravman JC, Nix WD, Barnett DM, Smith DA (eds) Mater Res Soc Symp Proc 130:13
93. Syed Asif SA, Pethica JB (1997) in: Gerberich WW, Gao H, Sundgren JE, Baker SP (eds) Mater Res Soc Symp 436:201
94. Li X, Bhushan B (2002) Mater Charact 48:11
95. Lawn BR, Fuller ER (1975) J Mater Sci 10:2016
96. Ponton CB, Rawlings RD (1989) Mater Sci Technol 5:865
97. Cook RF, Pharr GM (1990) J Am Ceram Soc 73:787
98. Lawn BR, Wilshaw TR (1975) J Mater Sci 10:1049
99. Evans AG, Charles EA (1976) J Am Ceram Soc 59:371
100. Frank FC, Lawn BR (1967) Proc R Soc London A299:291
101. Lawn BR, Swain MV (1975) J Mater Sci 10:113
102. Marshall DB, Lawn BR (1979) J Mater Sci 14:2001

103. Marshall DB, Swain MV (1988) J Am Ceram Soc 71:399
104. Lathabai S, Rodel J, Dabbs T, Lawn BR (1991) J Mater Sci 26:2157
105. Burns SJ, Chia KY (1995) J Am Ceram Soc 78:2321
106. Burns SJ, Chia KY (1995) J Am Ceram Soc 78:2328
107. Sglavo VM, Green DJ (1995) J Am Ceram Soc 78:650
108. Fischer-Cripps AC, Lawn BR (1996) Acta Mater 44:519
109. Pharr GM (1998) Mater Sci Eng A253:151
110. Lawn BR, Evans AG, Marshall DB (1980) J Am Ceram Soc 63:574
111. Anstis GR, Chantikul P, Lawn BR, Marshall DB (1981) J Am Ceram Soc 64:533
112. Casellas D, Caro J, Molas S, Prado JM, Valls I (2007) Acta Mater 55:4277
113. Lankford J, Davidson DL (1979) J Mater Sci 14:1662
114. Harding DS, Oliver WC, Pharr GM (1995) Mater Res Soc Symp Proc 356:663
115. Yan J, Karlson AM, Chen X (2007) Eng Fract Mech 74:2535
116. Lee JS, Jang J, Lee BW, Choi Y, Lee SG, Kwon D (2006) Acta Mater 54:1101
117. Bhushan B, Kulkarni AV, Bonin W, Wyrobek JT (1996) Phil Mag A74:1117
118. Fratzl P, Gupta HS, Paschalis EP, Roschger P (2004) J Mater Chem 14:2115
119. Gupta HS, Schratter S, Tesch W, Roschger P, Berzlanovich A, Schöberl T, Klaushofer K, Fratzl P (2005) J Struct Biol 149:138
120. Moscovich H, Dreugers NHJ, Jansen JA, Wolke JGC (1999) J Dent 27:503
121. Halls N (1992) Med Device Technol August/September:37
122. Hengsberger S, Kulik A, Zysset P (2001) Eur Cells and Mater 1:12
123. Hengsberger S, Kulik A, Zysset P (2002) Bone 30:1
124. Habelitz S, Marshall G, Balooch M, Marshall SJ (2002) J Biomech 35:995
125. Guidoni G, Denkmayr J, Schöberl T, Jäger IL (2006) Phil Mag 86:5705
126. Ho SP, Goodis H, Balloch M, Nonomura G, Marshall SJ, Marshall G (2004), Biomaterials 25:4847
127. Broomell CC, Mattoni MA, Zok FW, Waite JH (2006) J Exp Biol 209:3219
128. Orso S, Wegst UGK, Eberl C, Arzt E (2006) Adv Mater 18:874
129. Rho JY, Pharr GM (1999) J Mater Sci Mater Med 10:485
130. Evans FG, Lebow M (1951) J Appl Physiol 3:563
131. Evans FG (1973) Mechanical properties of bone. Charles C Thomas, Springfield IL, p 43
132. Yamada H (1970) Strength of biological materials. Williams & Wilkins, Baltimore MD, p 74
133. Townsend PR, Rose RM (1975) J Biomech 8:199
134. Currey JD (1988) J Biomech 21:439
135. Jameson MW, Hood JAA, Tidmarsh BG (1993) J Biomech 26:1055
136. Schöberl T, Jäger IL (2006) Advanced Eng Mater 8:1164
137. Chu JL, Lin HY, Wu, Lee S (2000) Mat Sci Eng A282:23
138. Lichinchi M, Lenardi C, Haupt J, Vitali R (1998) Thin Solid Films 312:240
139. Schallamach A (1957) Wear 1:384
140. Lee YH, Baek U, Kim YI, Nahm SH (2007) Mater Lett 61:4039
141. Elmustafa AA, Stone DS (2002) Acta Mater 50:3641
142. Abu Al-Rub RK (2007) Mechan Mater 39:787
143. Abu Al-Rub RK, Voyiadis GZ (2004) Int J Multiscale Comput Eng 3:50
144. Chiu YL, Ngan AHW (2002) Acta Mater 50:2677
145. Durst K, Backes B, Franke O, Göken M (2006) Acta Mater 54:2547
146. Fleck NA, Muller GM, Ashby MF, Hutchinson JW (1994) Acta Metal Mater 42:475
147. Qu S, Huang Y, Nix WD, Jiang H, Zhang F, Hwang KC (2004) J Mater Res 19:3423
148. Chudoba T, Schwarzer N, Richer F, Beck U (2000) Thin Solid Films 377:366
149. Chudoba T, Schwaller P, Rabe R, Breguet JM, Michler J (2006) Phil Mag 86:5265
150. Feng G, Ngan AHW (2001) Mater Res Soc Symp Proc 649:7.1.1

151. Kermouche G, Loubet JL, Bergheau JM (2007) Mech Mater 39:24
152. Ngan AHW, Wang HT, Tang B, Sze KY (2005) Int J Solids Struct 42:1831
153. Schofield RMS, Nesson MH, Richardson KA (2002) Naturwissenschaften 89:579
154. Schofield RMS, Nesson MH (2003) Science 301:1049a
155. Jakes JE, Frihart CR, Beecher JF, Moon RJ, Stone DS (2008) J Mater Res 23:1113
156. Staedler T, Donnelly E, Van der Meulen MCH, Baker SP (2003) Mater Res Symp Proc 778:79
157. Stone DS (1998) J Mater Res 13:3207
158. Stone DS, Yoder KB, Sproul WD (1991) J Vac Sci Technol A9:2543
159. Yoder KB, Stone DS, Hoffman RA, Lin JC (1998) J Mater Res 13:3214
160. Soifer YM, Verdyan A, Kazakevich M, Rabkin E (2005) Mater Lett 59:1434
161. Hodzic A, Stachurski ZH, Kim JK (2000) Polymer 41:6895
162. Downing TD, Kumar R, Cross WM, Kjerengtroen L, Kellar JJ (2000) J Adhes Sci Technol 14:1801
163. Lee SH, Wang S, Pharr GM, Xu H (2007) Compos Part A-Appl Sci 38:1517
164. Gerber CE (1968) Contact problems for the elastic quarter-plane and for the quarter space. Stanford University, Palo Alto, CA, p 100.
165. Hetenyi M (1960) Trans ASME Series E, J Appl Mech 27:289
166. Hetenyi M (1970) Trans ASME Series E, J Appl Mec. 37:70
167. Keer LM, Lee JC, Mura T (1984) Intl J Solids Struct 20:513
168. Popov GY (2003) Mech Solids 38:23
169. Schwarzer N, Hermann I, Chudoba T, Richter F (2001) Contact modelling in the vicinity of an edge. Elsevier, San Diego, CA, p 371
170. Hengsberger S, Enstroem J, Peyrin F, Zysset P (2003) J Biomech 36:1503
171. Rice JC, Cowin SC, Bowman JA (1988) J Biomech 21:155
172. Sevostianov I, Kachanov M (2000) J Biomech 33:881
173. Hoffler CE, Moore KE, Kozloff K, Zysset PK, Brown MB (2000) Bone 26:603
174. Fratzl P, Gupta HS, Paschalis EP, Roschger P (2004) J Mater Chem 14:2115
175. Gindl W, Schöberl T (2004) Composites Part A35:1345
176. Gindl W, Reifferscheid M, Adusumalli RB, Weber H, Röder T, Sixta H, Schöberl T (2008) Polymer 49:792
177. Gindl W, Konnerth J, Schöberl T (2006) Cellulose 13:1
178. Zhang J, Sakai M (2004) Mater Sci Eng A 381:62
179. Bhushan B (1999) Principles and application of tribology. Wiley, New York
180. Lichtenegger HC, Schöberl T, Bartl MH, Waite H, Stucky GD (2002) Science 298:389
181. Miserez A, Li Y, Waite JH, Zok FW (2007) Acta Biomater 3:139
182. Moses DN, Harreld JH, Stucky GD, Waite JH (2006) J Biol Chem 281:3482
183. Moses DN, Mattoni MA, Slack NL, Waite JH, Zok FW (2006) Acta Biomater 2:521
184. Ho SP, Marshall SJ, Ryder MI, Marshall GW (2007) Biomaterials 28:5238
185. Zioupos P, Rogers KD (2006) J Bionic Eng 3:19
186. Nizam BRH, Lim CT, Cheng HK, Yap AUJ (2005) J Biomech 38:2204
187. Cheng HK, Ramli HN, Yap AUJ, Lim CT (2005) J Dentistry 33:363
188. Lippert F, David M, Parker DM, Jandt KD (2004) J Colloid and Interface Sci 280:442
189. Balooch G, Marshall GW, Marshall SJ, Warren OL, Asif SAS, Balooch M (2004) J Biomech 37:1223
190. Cuy JL, Mann AB, Livi KJ, Teaford MF, Weihs TP (2002) Archs Oral Biol 47:281
191. Finke M, Hughes JA, Parker DM, Jandt KD (2001) Surf Sci 491:456
192. Habelitz S, Marshall SJ, Marshall GW, Balooch M (2001) J Struct Biol 135:294
193. Tesch W, Eidelman N, Goldenberg F, Roschger P, Klaushofer K, Fratzl P (2001) Calcif Tissue Int 69:147

194. Nalla RK, Balooch M, Ager III JW, Kruzic JJ, Kinney JH, Ritchie RO (2005) Acta Biomater 1:31
195. Mahoney EK, Rohanizadeh R, Ismail FSM, Kilpatrick NM, Swain MV (2004) Biomaterials 25:5091
196. Kinney JH, Balloch M, Marshall SJ, Marshall GW, Weihs TP (1996) Archs Oral Biol 41:9
197. Angker L, Swain MV, Kilpatrick N (2003) J Dentistry 31:261
198. Ashby MF (1999) Materials selection in mechanical design, 2nd edn, Butterworth-Heinemann, Oxford
199. Waite JH, Lichtenegger HC, Stucky GD, Hansma PK (2004), Biochemistry 43:7653

22 Nanomechanics and Microfluidics as a Tool for Unraveling Blood Clotting Disease

D.M. Steppich · S. Thalhammer · A. Wixforth · M.F. Schneider

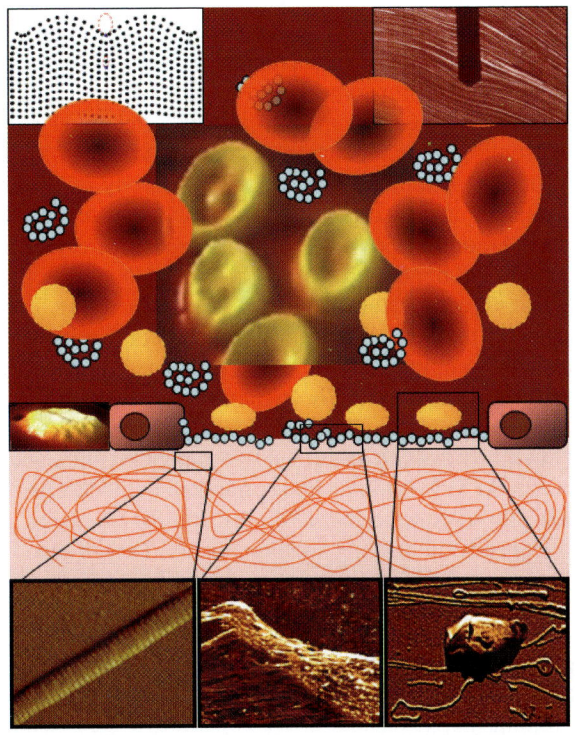

Key words: Blood Clotting, Von Willebrand Factor, Collagen Assembly, Nanomechanics, Erythrocyte Elasticity, AFM on Biomolecules, Lab on a Chip, Microfluidics

22.1
Introduction

One of the key advantages of Atomic Force Microscopy (AFM) over a variety of other techniques in life science is the freedom to operate in almost any medium.

Especially for the investigation of cells, proteins, and membranes and their complex interplay, a prerequisite for reliable and biologically significant results is the choice of the right medium. Small changes from natural conditions in pH, temperature, salt concentration, etc., can make a huge difference for the interpretation of obtained data and for assigning them to a peculiar biological process in vivo. The use of an AFM is not limited to vacuum or a gaseous medium and further provides the possibility to operate in fluid media with variable buffer conditions.

In recent years, with a continuous increase in expertise for probing soft matter, AFM has developed into a decisive tool for approaching questions in life sciences. Scientists benefit from AFM in cell imaging, questions in regard to cell elasticity, the measurement of both specific lock-and-key and non-specific adhesion in a wide range of biological, physical, and also medical questions. On the following pages, we present a selection of results providing a more detailed understanding of blood clotting and hemostasis, which is obtained by a very inventive and non-conventional use of the AFM technique. A special challenge in this task was the consideration and integration of fluid flow, mimicking blood flow conditions, into the experimental setup. As mentioned above the medium is an extremely important factor in approaching biological questions. To further complicate things, the dynamic conditions inside a blood vessel like flow pattern, flow velocity, and velocity gradients influence and sometimes generate circumstances, which barely can be probed or reproduced in a static environment. Atomic Force Microscopy, as a rather direct technique of observation, further does have limited access into a sealed blood vessel or any common flow chamber trying to copy natural flow conditions. Therefore, unraveling secrets in blood clotting with the help of an AFM requires combination with a pumping system, which is able to mimic variable blood flow conditions and is accessible to the cantilever tip at the same time. These framework conditions led to the unique combination of a planar Surface Acoustic Wave (SAW)-driven microfluidic reactor, mimicking the blood flow on an optical transparent biochip, and an AFM. This microfluidic reactor–AFM hybrid system is presented in the last part of this chapter. To illuminate the role of Atomic Force Microscopy for the investigation of blood clotting we present results in imaging and probing elasticity of blood cells, platelets, and compounds of the subendothelial matrix exposed to different flow conditions. Subsequently, we introduce the SAW-driven microfluidic pumping system leading to a physicist's equivalent of a blood vessel. This approach is the direct precursor to "lab-on-a-chip" systems for probing blood clotting diseases. In the last section we present the microfluidic reactor–AFM hybrid combining the before mentioned techniques in a unique and valuable way. This hybrid system provided the means to unravel direct evidence for the formation and internal dynamics of protein networks responsible for blood clotting under elevated shear stress.

22.2
Topography

The invention of Atomic Force Microscopy opened the field for imaging surfaces with a lateral and vertical resolution in the range of Angstroms [1]. The resolution in imaging soft biological matter like membranes or cell surfaces often is far from that value and strongly depends on the kind of sample under investigation. For example, it may vary strongly according to wall and adhesion properties of different cells and the tip. The main contribution to the lack of resolution in cells is maybe the tip-induced deformation or even penetration of soft parts of proteins, membranes, and cells [2]. Nevertheless, the AFM technique proved to be a crucial tool in exploring biological questions. In the following sections, we will present a short scheme of the course of blood clotting and how AFM is applied to elucidate exciting features of the most important compounds involved in this process.

22.2.1
Little Story of Blood Clotting

Blood coagulation is a very complex process involving more than a dozen blood coagulation factors. We will sketch the course of wound healing in small arterioles as much as necessary for a proper understanding of the underlying basic processes while keeping it as simple as possible to motivate the basic questions under investigation in the AFM experiments. The focus will be set on vWF (von Willebrand Factor) network formation and platelet binding to the damaged vessel walls, which oppose high shear stress arising from the rapid flowing blood.

Wound healing is an extremely important adjustment of nature preventing continuous bleeding into the tissue after damage to the blood vessels. A normal human being is bleeding continuously in small arterioles and veins without even being aware of it. The first and an often neglected mechanism to prevent excessive blood loss is the contraction of the injured blood vessel, which can change the streaming properties of the blood and the forces on every single compound involved in blood clotting. Such bleeding, especially in these high shear regimes, is continuously sealed by an instantaneous repair mechanism (Fig. 22.1).

22.2.1.1
von Willebrand Factor Binding

An essential role for wound healing especially in small arterioles comprising high shear forces plays the biopolymer von Willebrand Factor. Its participation in wound healing has been known for a long time. The specific lock-and-key binding sites of vWF molecules to both extracellular matrix and damaged endothelial cells, exposed at sites of vascular injury, were characterized, but the mechanism of the activation of vWF's binding potential remained unclear. We could recently show directly that a conformational change in vWF molecules from a coiled to an activated unrolled state triggers its biological function [3]. In this activated conformation vWF's ability to bind to both exposed extracellular matrix at a site of vascular damage and to platelets out of the blood stream is rapidly increased. The extracellular matrix is mainly composed of different kinds of collagens (see also Sect. 22.2.2.1), where binding sites to vWF have already been identified [4]. This coiled–unrolled transition can be induced solely by mechanical stress on the molecule experienced due to shear stress of the blood stream. Comparable to the stream lines in a pipe, the velocity of blood in the blood vessel decreases from the center towards the endothelial cell surface. This velocity difference between contiguous layers of fluid results in

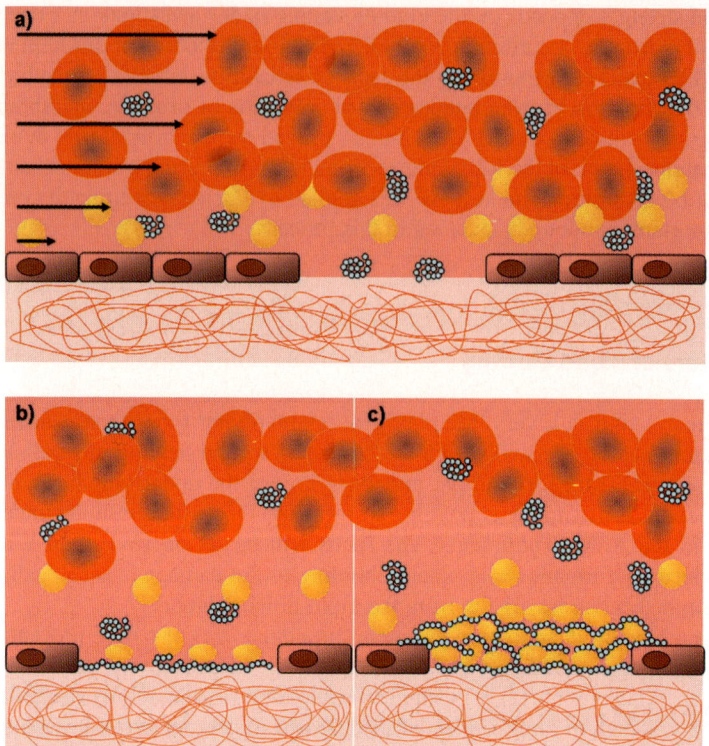

Fig. 22.1. (**a**) Erythrocytes (*red*), platelets (*yellow*), and von Willebrand Factor (vWF) (*blue*) facing a site of damage to the endothelial cell layer which seals the blood vessel against the extravascular tissue. vWF binding to collagens in the extravascular connective tissue is the initial step for wound healing in vessels with high fluid flow velocities. The *arrows* indicate the different flow velocities inside a blood vessel with a maximum speed in the center of the vessel and a minimum at the vessel wall. This velocity gradient induces shear stress on any compound in the blood stream. (**b**) Because of the increased shear stress at the vessel wall, immobilized vWF molecules are mechanically unrolled from their coiled into an activated stretched conformation building a sticky protein network. In this unrolled and activated conformation platelets can bind to vWF at the site of vascular damage, whereas there is no binding of vWF and platelets in the normal blood stream. (**c**) In the course of wound healing more and more platelets aggregate forming multiple layers stabilized by the vWF glue till a plaque has formed which is able to prevent further bleeding into the tissue

a shearing effect. The shear rate $\dot{\gamma} = \frac{\partial v}{\partial r}$ defined as the flow velocity gradient with respect to the vessel radius describes the acting shear field. The highest wall shear rates of approximately 470–4,700 1/s occur at the vessel walls of small arterioles (10–30-μm diameter) [5] and can induce vWF's activation.

From a physical perspective, in these high shear rate areas it seems extremely counterintuitive for any protein or cell to adhere to sites of vessel damage opposing the drag forces of the rapid surrounding flow. To solve this dilemma, evolution developed vWF, initializing its enhanced adhesion abilities exactly in this high shear environment, which makes commonly known adhesion mechanisms extremely unlikely or even impossible. The key in understanding vWF's binding potential for both compounds of the extracellular matrix and platelets is hidden at least in part in its enormous multimeric size. Large vWF molecules can be made up of up to a hundred identical subunits and its total mass can exceed 20,000 kDa [6]. In its coiled conformation a huge amount of vWF's binding sites are buried inside the protein whereas an unrolling of this biopolymer directly results in a drastic increase of accessible binding sites and therefore in the activation of the protein. The adhesive strength of vWF to exposed collagen at a vessel damage site or a damaged endothelial cell increases proportionally to its free binding sites.

22.2.1.2
Platelet Binding

Activated and bound to the extracellular matrix, vWF comprises binding sites for platelets too. Platelets are blood cells without a nucleus whose function is to prevent bleeding. While the vWF network represents the basis for wound healing, platelet aggregation to this network is the starting point for the growing plug, which finally seals the vessel damage. In this regard it is worth mentioning that platelets are enriched near the vessel wall while erythrocytes for example are found mainly in the center of the vessel. This condition enables a higher probability for platelets to adhere to sites of vascular damage and therefore seal lesions more effectively [7].

Platelets adhere to a vWF network at a site of injury in a two-step mechanism. First a weak bond with a high dissociation constant is formed. Because of the torque induced by the fluid streaming and the weak adhesion to the surface, platelets are decelerated and roll on the vWF network, giving a stronger and irreversible bond with a low dissociation rate enough time to form [8]. The process of rolling is conditioned by repeated forming and breaking of these weak bonds. Sometimes these weak adhesion contacts do not break while the platelet still translates over the surface. Because of the soft character of the platelets a membrane tube from the adhesion point on the vWF surface to the cell body along the direction of the fluid flow can be formed [9]. The morphology, formation, and detachment kinetics from the anchor point of these membrane tethers depend strongly on the acting shear field. At high shear rates the probability for tether formation and its growth rate are drastically increased while the lifetime is significantly lower. Dopheide et al. [10] speculated that the extension of membrane tethers may be a very important step in mediating strong adhesion by slowing platelets down and reducing the level of shear stress experienced by individual receptor–ligand interactions.

After a tight bond is formed, platelets tend to spread on the surface and therefore comprise even more binding sites for vWF while reducing the opposing drag forces at the same time. The stabilizing effect of platelet spreading is often accompanied with the formation of several filopodia. Although similar in their structure, filopodia are distinctly different from membrane tethers and protrude onto the surrounding surface without respect to the flow pattern.

As more and more platelets aggregate on the site of vascular damage, vWF serves also as a glue between multiple layers of platelets. This platelet plaque grows as long as it is sufficient to prevent further loss of blood. If the rent in the vessel wall is small,

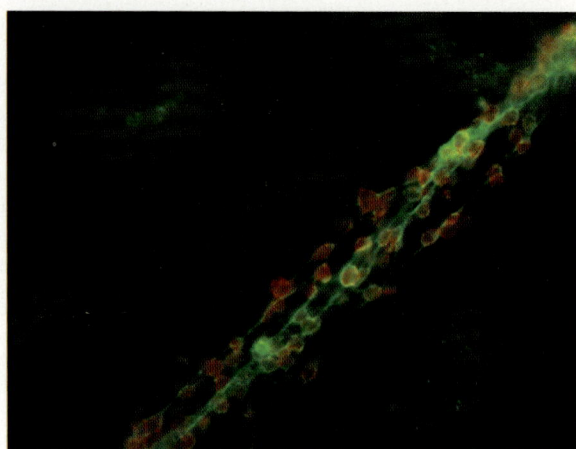

Fig. 22.2. The fluorescence image points out the functional activity of unrolled vWF (*green*) in platelet (*red*) binding. Platelets exhibit sizes in the order of 10 μm, which also highlights also the enormous size of the vWF network (Figure reprinted from George T, Kleinerüschkamp F, Barg A et al., Microfluidics reveals generation of platelet strings on tumor-activated endothelium, Thromb Haemost 2007; 98:285 with permission from Schattauer)

this platelet plug by itself can stop blood loss. A larger hole requires further stabilization of the platelet aggregate. In addition a meshwork of fibrin can interconnect platelets and blood cells forming a blood clot [11]. Figure 22.2 illustrates the platelet binding potential of a huge vWF network.

22.2.2
High-Resolution Imaging

22.2.2.1
Collagen Fibrils

Collagen molecules are a major compound of the extracellular matrix and are exposed to the blood stream at a site of vascular damage, building the first contact point for the initiation of blood clotting by network forming proteins. They consist of three polypeptide chains (α-chains), which form a unique triple-helical structure. More than 20 genetically distinct collagens exist in mammalian tissue, where collagen types I, II, III, V, and XI self-assemble into D-periodic cross-striated fibrils. Collagen molecules, forming the fibril, consist of an uninterrupted right-handed triple helix called tropocollagen [12], approximately 300 nm in length and 1.5 nm in diameter. The collagen self-assembles in cross-striated fibrils that normally occur in the extracellular matrix of connective tissues. These fibrils are stabilized by covalent cross linking of specific lysines and hydroxolysines of the collagen molecules, which are ordered parallel in a D-periodic pattern [13]. The stagger of molecules gives rise to a characteristic band pattern of light and dark regions when negatively stained and viewed using an electron microscope [14, 15]. Fundamental experiments for the understanding of the morphogenesis of collagen units and fibrils and influencing factors such as temperature, ionic strength, pH value and so on, were performed by Gross, Wood, Keech et al. [15–21] and further work was done by Bard and Chapman [22, 23].

Studies on collagen molecules by cryo-AFM [24] and normal AFM studies of segment-long spacing (SLS) crystals of collagen [25] have revealed variations in the structure. AFM investigations carried out by Paige et al. [26] show native fibrils and fibrous long spacing fibrils (FLS-fibrils) [27]. Cocoon-like fibrils, which are hundreds of nanometers in diameter and 10–20 mm in length, were found to coexist with mature FLS fibrils. On the basis of detailed AFM studies a stepwise process in the formation of FLS collagen was proposed. Different pH ranges were investigated to clarify the various stages in the self-assembly process of the fibrils. In early stages below pH 4 thin non-banded fibrils were formed. Filamentous structures showed protrusions at pH~4 and at pH>4.6 mature fibers emerged [28]. Gutsmann et al. observed that collagen fibrils from tendons behave mechanically like tubes [29]. They concluded that the collagen fibril is an inhomogeneous structure composed of a relatively hard shell and a softer, less dense core. For high-resolution AFM microscopy the collagen samples can be investigated immediately after self-assembly and drying on freshly cleaved mica. A novel collagen preparation technique [30], based on a dialysis system, allows the reproducible production of single collagen fibrils with different banding patterns (Fig. 22.3).

Automated electron tomography studies, performed on corneal collagen fibrils showed that collagen molecules are organized into microfibrils ($\approx$4 nm diameter) in a 36-nm diameter collagen fibril, which are tilted at $\approx$15° to the fibril long axis in a right-handed helix. Analysis of the lateral structure demonstrated that the microfibrils exhibit regions of order and disorder within the 67-nm axial repeat of the collagen fibrils [31].

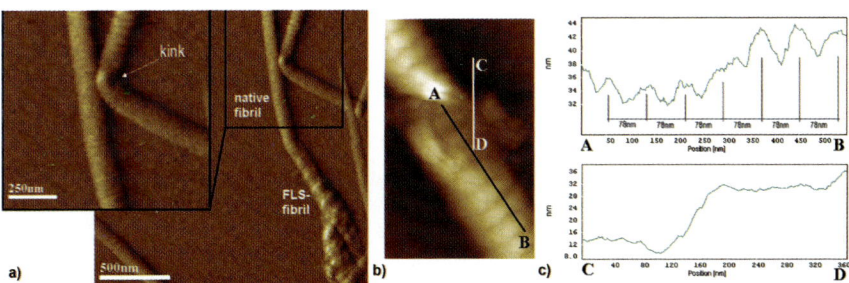

Fig. 22.3. (**a**) High-resolution AFM image of a single collagen fibril with a kink: due to kinks collagen are sometimes compared to tubes. The other fibril displays the polymorphism of collagen with native and FLS-parts in one fibril. (error signal, *scale bar* 500 nm). (**b**) AFM image of a microdissected collagen fibril: The core of the collagen fibril is revealed and the banding pattern can be recognized inside the fibril. The *arrows* indicate the overlap zones of the collagen molecules that arise during the self-assembly process of the collagen fibril (topography signal, *scale bar* 250 nm). (**c**) Line measurements on the microdissected collagen fibril **b**: Between point A and B the banding can be recognized on the shell as well as in the core. The banding pattern can be determined to be approximately 78 nm. Between point C and D the height difference of around 16 nm between the substrate and the level of the cut area can be seen

In contrast to other high-resolution imaging methods, like scanning electron microscopy (SEM) or transmission electron microscopy (TEM), which are operating in a vacuum with dried samples and conducting surfaces, the preparation for AFM investigations is very simple and does not alter the sample properties in most cases. For structural and mechanical investigations in the submicron range the AFM proved to be a suitable tool for manipulation and probing the mechanical properties of biological macromolecules in the nanometer range. Besides the manipulation of proteins like collagen it is also possible to probe and manipulate genetic samples. Thalhammer et al. demonstrated that it is possible to use the AFM tip as a new tool like a mechanical nanoscalpel and a nanoshuffle where minute amounts of material can be extracted and further processed [32]. The AFM-based microdissection on collagen fibers can be applied to reveal the inner structure of the specimen. During manipulation the shell and upper parts of the fibril are scratched away, to image the remaining core in a high-resolution mode (Fig. 22.3b, c). The dissection is carried out in a defined angle to the fibril axis in order to exclude artifacts originated by an orthogonal or parallel (to the fibril axis) scratching procedure. It could be shown that there are no major geometrical differences of the banding on the shell and in the core of native single collagen fibrils. The banding pattern inside the fibril fits to that on the shell in width and distance. The fibril has a banding pattern of 78 nm, a height of 30 nm, and a width of 270 nm. The cut area, shown in Fig. 22.3b, is located in the core of the fibril, which was confirmed by line measurements. Between the measured distance A and B the banding can be recognized on the shell as well as in the core. Among points C and D the height difference between the substrate and the level of the cut area can be seen. The height difference between the substrate and cut area was determined to be 16 nm [33]. The AFM could also be used to compose nanoscopic collagen matrices by orientating individual collagen fibers of self-assembled collagen layers [34]. AFM nanodissection of big FLS-fibrils with a width of about 1.7 μm and a banding pattern of 270 nm showed the FLS banding also in the core of the fibril [35]. For FLS-fibrils a different assembly pathway and structure are postulated. The characteristic banding mainly arises from the attachment of α1-acid glycoprotein in FLS-fibrils [28]. The characteristic banding of native fibrils is determined by the repetition of overlap and gap zones [36].

22.2.2.2
Elasticity

Several investigations using the AFM as a tool for measuring the tensile modulus of collagen fibrils and subunits revealed details of the protein assembly. Graham et al. calculated force elongation/relaxation profiles of single collagen fibrils using the AFM. The elongation profiles showed, that in vitro assembled human type I collagen fibrils are characterized by a large extensibility. It was shown that the fibrils are robust structures with a significant conservation of elastic properties [37]. Gutsmann et al. probed the crosslinks on a lower level of organization using an AFM cantilever to pull substructures out of the assembly. Two different rupture events were determined; the first with a strong bond and a periodicity of 78 nm (bonds between subunits) and a second weaker one with a periodicity of 22 nm (between molecules) [38]. Bozec and Horton [39] studied trimeric type I tropocollagen molecules by AFM, both topologically and by force spectroscopy, showing multiple stretching peaks on the molecular level similarly as shown by Gutsmann et al. [38]. Fratzl et al. and Puxkandl et al. investigated the fibrillar structure, viscoelastic and mechanical properties of collagen by recording stress/strain curves [40, 41]. The stress/strain curves can be divided into several regions [37]. At first crimps [42] and kinks [43] are removed, before a linear region is seen where the collagen triple helices are stretched, along with increase of the gap zones compared with the overlap zones. Slippage is first seen within fibrils at crosslink deficient collagen, and then higher strains lead to a disruption of the fibril. The mechanical behavior of native single fibrils was tested by recording force-distance curves on the shell and in the core of the fibrils to gain insights into the collagen assembly and mechanical properties [33]. The evaluation of the adhesion forces of the spectroscopy data indicates a higher adhesion in the core of the fibril (Fig. 22.4a). In Fig. 22.4b and c the height of the snap-out effect calculated from the retrace curves is shown. The average value was calculated for the shell to 5 nN and for the core to 6 nN, which points to a higher adhesion in the core of the fibril. Gutsmann et al. suggested the presence of more highly crosslinked collagen molecules near the fibril surface compared to the central region [29]. This could lead to a higher amount of binding capacity for the tip and cause higher adhesion forces during the measurement. However, the results could have been also influenced by the scratching process, which could have led to a rupture of molecules and destruction of crosslinks. The destroyed crosslinks and ruptured molecules could stick to the tip and increase the measured adhesion force. Moreover, the collagen fibrils were investigated in a dried state of preservation, which could have also an influence on the mechanical properties. The mechanical properties of collagen-rich tissues, e.g. tendons, are largely determined by the collagen structure [40]. An inhomogeneous assembly of collagen fibrils was published by several authors; Sarkar et al. [44] proposed that fluid domains in the collagen allow molecules to slip relative to one another, which was shown by Mosler et al. [45], in order to relieve applied stress. It has been proposed that, in case of high forces, the stiff outer shell of the collagen fibrils could break while the fluid core remains intact and might be used in the repair of the shell [29]. These morphological results as well as the statistical evaluation of the Young's modulus using the indentation measurements could not confirm a "fluid" core or different structures of core and shell. Solely the adhesion measurements show differences between core and shell.

22.2.2.3
von Willebrand Factor Network Formation and Platelet Binding

The multimeric von Willebrand Factor molecule is the key element in mediating adhesion of platelets to sites of injury at rapid flow conditions. Because of a conformational change from a globular state into an unrolled state, vWF is able to bind to various types of collagen from the extracellular matrix. Single vWF molecules and their adhesion and binding properties have been investigated by AFM technology. In early studies, Marchant et al. examined the adhesion and structural properties of vWF both on hydrated and dehydrated mica surfaces [46]. They experienced problems similar to those researchers are still facing in imaging soft matter under buffer conditions today. On hydrophilic mica, vWF is only interacting weakly with the

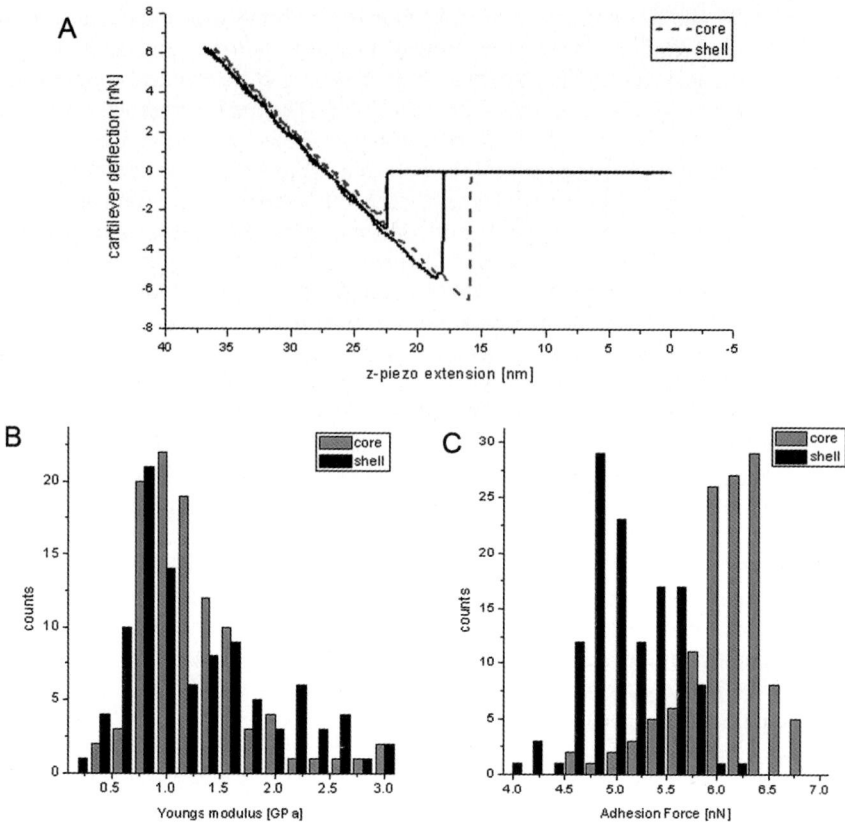

Fig. 22.4. (a) Typical sample of force-distance curves recorded on the shell and the core of a single collagen fibril. The *dark curve* shows a spectroscopy curve on the shell, whereas the *gray dashed curve* was recorded in the core. The slope of both curves is nearly the same and indicates identical elasticity. Measurements on the shell and in the core have to be performed in the same sample height to exclude "thin layer effects." The Hertzian model was fitted to the positive cantilever deflection range. (b) Elasticity measurement of a single collagen fibril. Force-distance curves were recorded on the shell and the exposed core. On both measuring points more than 100 curves were recorded. The diagram displays the Young's moduli of the force-distance curves versus the frequency. The nanoindentation experiment indicates no measureable difference between core and shell. (c) Adhesion measurement on a single collagen fibril. The data display the evaluation of the adhesion forces calculated from the height of the snap-out of the retrace curve. The values show a higher adhesion in the core of the microdissected fibril

surface and thus easily moved by the tip into a perpendicular orientation to the fast scanning direction. In contrast imaging of the protein in a dehydrated (but not very physiological) case is performed easily due to a tighter molecule immobilization to the surface (Fig. 22.5).

Marchant et al. speculated that the absence of water molecules in between the vWF protein and the mica substrate in the dehydrated sample accounts for the stronger protein–surface interaction. Raghavachari et al. extended this

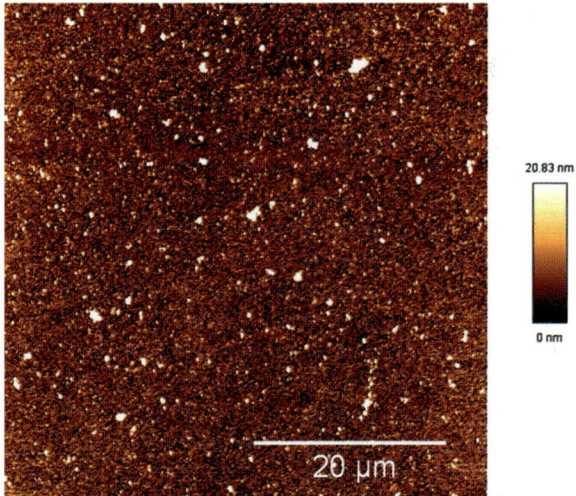

Fig. 22.5. Globular vWF molecules deposited on a thin collagen layer imaged in tapping mode in air. A broad distribution of vWF sizes is visible

surface-dependent adhesion study of vWF to hydrophobic substrates under physiological conditions [47]. Their data may provide a closer approach to the surface-dependent functional behavior of vWF molecules. On a hydrophobic OTS (octadecyltrichlorosilane) surface, vWF undergoes adsorption and molecular spreading. In contrast, vWF is loosely adsorbed and presents more extended structures on hydrophilic mica, which the authors explain as an optimization of hydration, electrostatic and van der Waals interactions. The first AFM study of unrolled vWF molecules was performed by Siedlecki et al. also on hydrophobic OTS [48]. The authors used a rotating disc system to unroll vWF on the substrate and imaged elongated vWF molecules using modified scanning probe tips and mathematical morphology modeling techniques. But also in this study, tip-induced irreversible deformation and elongation of globular vWF molecules occurred, which could be distinguished from unrolled vWF due to shear stress.

The AFM technique was proven to be a significant tool for investigating surface and conformation-dependent adhesion mechanisms and delivered valuable hints for both the driving forces for adhesion to different surfaces and simultaneously for the forces for the vWF monomer–monomer interaction, which keep the protein in its globular state under low shear conditions. In order to understand the blood clotting process it is important to extend the experiments presented above. Whereas the unrolling and activation of a vWF molecule on a site of vascular injury is the initial step for wound healing at high shear conditions, a single extended molecule would never have been able to diminish or even stop bleeding. Only the whole entity of a multitude of activated vWF molecules, self-assembling and forming huge networks, gains the adequate potential to bind enough platelets out of the rapid blood stream to seal a lesion. Imaging such networks is also a demanding task considering the elastic properties of protein assemblies and fibers within the network.

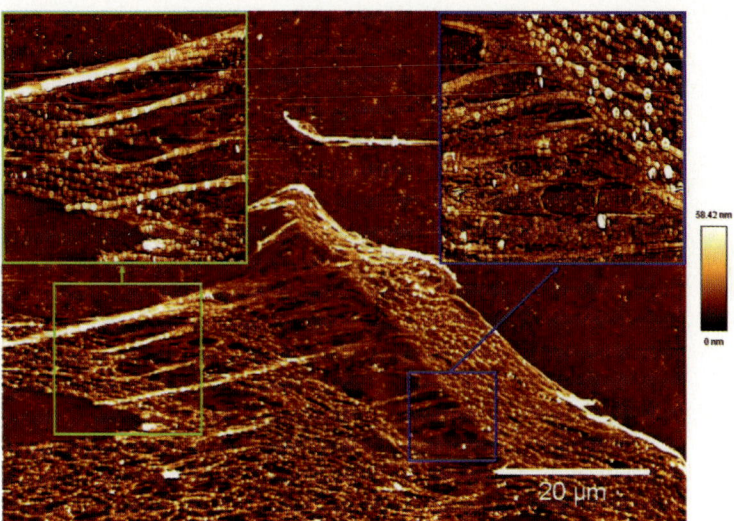

Fig. 22.6. A huge vWF network formed under shear conditions on a thin collagen layer. After dehydration the network was imaged in tapping mode in air. Both *insets* show different regions of the network exhibiting similar and also very different features. Both elongated thick vWF fibers (*left inset*) and globular vWF molecules (*right inset*) can be distinguished within the network

These networks can exhibit a broad variety of sizes dependent on the injury. In vitro networks with diameters in the range from $1\,\mu m$ (and below) up to the order of mm were observed (Fig. 22.6). Both the very small and the almost macroscopic vWF networks proved to be hemostatically active and could mediate platelet binding.

Under high shear flow conditions platelet binding to a site of injury, or to the pre-formed vWF network to be precise, is accomplished in a two-step process. In a first initial tethering, platelets bind via their GP *Ibα* receptor to the A1 domain of vWF. This bond forms rapidly but has also a high dissociation rate and therefore a short half-life. This weak adhesion decelerates the platelet and keeps it near the surface, where a rolling motion is induced by the surrounding flow. The reduced velocity accompanied with an increased residence time on the surface enables a slower but tighter bond to be formed via the $\alpha_{IIb}\beta_3$ platelet binding site and the C1 domain of vWF in the second step [8]. The rolling and terminal stopping of the platelet is often accompanied by the formation of an extended cell membrane tether. These tethers are pulled from small localized contact points formed between the cell body and the vWF matrix via $\alpha_{IIb}\beta_3$ and the A1 domain. Dopheide et al. found also kinks in membrane tethers, indicating that multiple bond formation within a tether is possible [10]. Tether formation further increases the residence time of platelets on the vWF matrix compared to rolling platelets. Shao et al. addressed the physiologically important fact that compared to platelets without tethers, tether formation significantly reduces the force, experienced from the surrounding flow, which is working against the bond between platelet and vWF matrix [49]. Thus, a cell membrane tether further increases the residence time of platelets on the surface and increases the probability of tight bonding between platelets and the vWF surface. The observed tether formation is

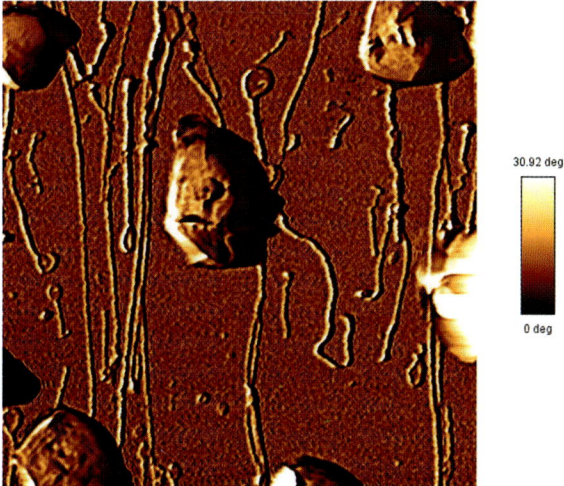

Fig. 22.7. Platelets were imaged in buffer conditions after a shear stress of 20,000 Hz was applied and subsequent formaldehyde fixation (image size $17 \times 17\,\mu m$). Membrane tethers were predominantly found in the direction of flow. But also kinked and broken tethers can be observed indicating multiple binding sites within the tether

not a biochemically driven rearrangement of the cytoskeleton but is a function of the shear rate of the surrounding liquid. Figure 22.7 illustrates immobilized platelets on a vWF-coated surface exhibiting various membrane tethers.

After a tight bond has formed platelets are activated and spread on the surface, exhibiting a smaller contact area for the surrounding flow and more binding sites for the next layer of platelets stabilized via vWF between adjacent platelets. Fritz et al. followed the activation process of platelets and were able to image the drastic reorganization of the cytoskeleton by tuning the scanning force of the tip to visualize internal structures of the cytoskeleton through the softer outer membrane [50].

22.3
Lab-on-a-Chip

In the previous section, the course of blood clotting was followed and investigated by AFM step-by-step from the exposure of subendothelial matrix at a site of vessel damage till platelet adhesion and thrombus formation occurred. On the following pages we will sketch how the AFM technique can be utilized in determining both distinct blood diseases and the age of bloodstains as an especially accurate measurement in forensic science. In the second part of this section we will highlight Surface Acoustic Wave technology as a relatively new and powerful tool for pumping small amounts of liquid. This technique is perfectly suited for mimicking blood flow conditions in vitro and can be combined to a variety of well-established methods for answering a broad variety of questions.

22.3.1
Nanomechanical Diagnostics

Atomic force microscopy has been widely applied in imaging biological samples under physiological conditions (for a review see [51]). The shape of red blood cells is characteristic of particular diseases and there is a growing interest in the identification of ultrastructural features associated with diseases. In early publications, images of fixed erythrocytes were presented [52, 53]. Easily recognizable erythrocytes were observed but little subcellular detail was seen. Häberle and coworkers immobilized the separated cells on a micropipette, which was mounted on the scanner. The imaging was performed on the protruding part of the cell and they achieved a resolution in the 10-nm range [54]. Several groups demonstrated the imaging of red blood cells in physiological media. At high resolution, the underlying cytoskeleton of the blood cell has been resolved and flaws in the cytoskeleton could be observed [2]. The membrane skeleton structures, without major distortions or deformations of the cell surface, were observed by Kamruzzahan and coworkers using tapping mode under near-physiological conditions [55]. With this approach significant differences in the morphology of red blood cells from healthy humans and patients with systemic lupus erythematosus were detected on topographical images. The surface of red blood cells from systemic lupus erythematosus patients showed characteristic circular-shaped holes approximately 200 nm in diameter under physiological conditions. Blood cells are subject to various kinds of stresses in flow fields. Hemolysis is the phenomenon in which a higher stress than normal damages the erythrocyte membrane and results in the leakage of its contents. Even if the stress is not strong enough to cause cell lysis, however, the cell membrane may sustain some damage. The fine surface structure of sheared erythrocytes was observed by Ohta and coworkers [56].

Blood analysis including blood cell measurement is one of the most basic items of hematological testing which is indispensable in health examination, disease diagnosis, and treatment. The common lab test is used to diagnose and monitor the body's response to diseases. Some tests measure the components of blood itself and others examine substances found in the blood to identify abnormal function of various organs. In forensic sciences the examination of bloodstains represents a major application during crime scene investigation. There exist a lot of reliable methods for the detection and identification of blood spots. For the evaluation of suspected bloodstains solutions such as phenolphthalein, tetramethylbenzidine can be used, as they change color when they come into contact with peroxidase or hemoglobin in the blood [57]. For the detection of even minute amounts of blood traces the presumptive luminol chemiluminescence test is widely used in forensic practical work [58]. It is further possible to unambiguously attribute the blood to a certain individual by using molecular biological techniques, such as genetic fingerprinting [59]. However, the determination of the age of a blood spot remains an unsolved problem in forensic routine work. For more than a hundred years forensic scientists have been engaged in finding a methodology, which allows determining the exact age of a dried bloodstain. Since then several approaches were proposed to have potentially solved this important problem during crime scene investigation, but none of these could ever be established in forensic practice. An attempt was undertaken by evaluating the time differences of the solubility of bloodstains of different ages [60]. In control experiments it turned out, that both methods considerably diverge from the real age of an investigated bloodspot. More recently, new methods were tested such as remission spectrophotometry [61] or electron-spin-resonance measurements [62], which can detect an age-dependent increase of signal intensity of meth-hemoglobin, non-hem-iron molecules, and organic radicals. Concordantly, these methods proved to have a high error rate, which only allows a rough estimation of the age of a bloodstain and does not excuse the high technical expense. In more recent times chromatographic methods were tested for use in dating blood spots. Inoue et al. used high-pressure liquid chromatography (HPLC) to measure the quantitative compound of the globin chains of the red blood dye hemoglobin [63]. They found a decrease of the α-chains related to the heme, the color defining prosthetic group of the red blood dye hemoglobin, with increasing age of a blood spot. The measurements of the standardized bloodstains revealed high deviations and revealed not to be practicable for routine forensic work. Moreover, a mobile application of this method for the use directly at a crime scene investigation seems not to be realizable. Taken together, all described methods proved to be not applicable in routine forensic work, as the results of the age measurement of the bloodstains provided too high deviations compared to the real age. Most of these attempts rely on advanced technical equipment and, therefore, do not allow application directly at a crime scene. Additionally, a part of the sometimes rare trace material has to be consumed for the use of these approaches, so that it will be lost for further important analysis, like molecular genetic investigations. A new AFM-based methodology, which could be used for the age determination of dried bloodspots during crime scene investigation, was introduced by Strasser and coworkers [64]. An AFM was used as a nanoindentor to monitor age-related changes of the elasticity of bloodstains under standardized conditions. Since the invention of the AFM, indentation experiments are generally applied in measuring elastic properties. Nanoscaled materials are probed with the AFM due to its high lateral and vertical resolution down to 0.01 nm. In contrast to other hardness testers the force resolution of the AFM can reach ranges of 10^{-4} nN [65]. The application of AFM-force spectroscopy is widespread and used for measuring polymer systems [66,67], bone elasticity [68], collagen [69,70], or cells [71–73]. For the analytic procedure a fresh blood spot was applied on a glass slide and the AFM detection started after drying of the blood drop. In a first step, an overview image was generated showing the presence of several red blood cells, which could easily be detected due to their typical "doughnut-like" appearance. The consecutive morphological investigations in a timeframe of 4 weeks could not show any alterations. Secondly, AFM was used to test the elasticity by recording force-distance curves. The measurements were performed immediately after drying, and at 1.5 h, 30 h, and 31 days. The conditions were kept constant at room temperature (20°C) and a humidity of 30%. The obtained elasticity parameters were plotted against a timeline and repeated several times. The elasticity pattern showed a decrease over time, most probably influenced by the alteration of the blood spot during the drying and coagulation process. The preliminary data demonstrates the capacity of this method for use in the development of calibration curves, which can be used for estimation of bloodstain ages during forensic investigations [64] (Fig. 22.8).

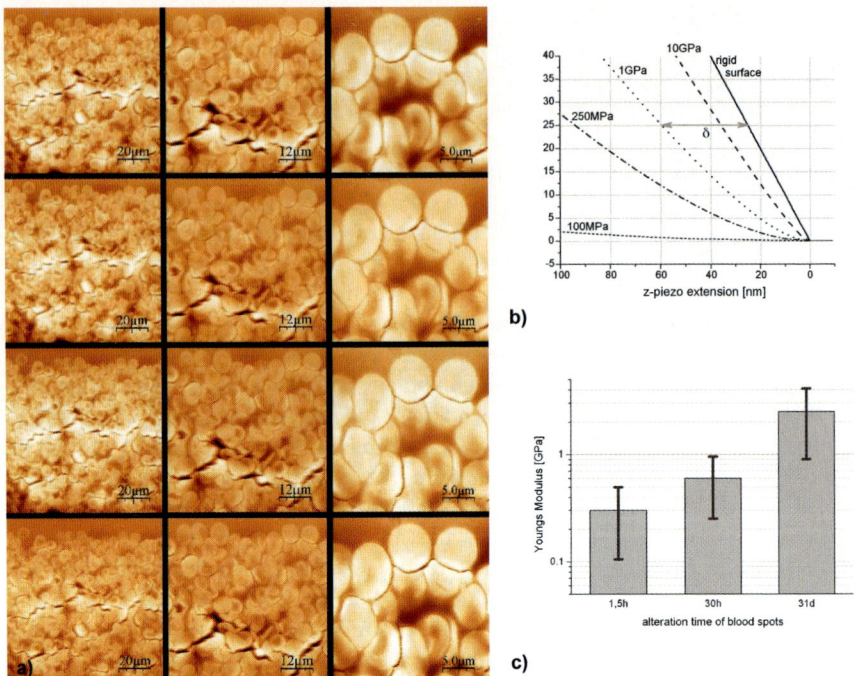

Fig. 22.8. (**a**) Results of the morphological investigations of a dried blood spot by AFM. The illustration is divided into rows and columns. The AFM images in the rows correspond to the individual measuring days. The columns display identical scan areas depending on the alteration time of the blood spot. The first series (first row) was scanned after 1.5 h, the second series after 1 week, the third series after 2 weeks, and the fourth series after 4 weeks. The morphology independent of the scan range (100, 60, and 25 μm) does not change with the alteration time; (**b**) Force-distance curves of various elastic samples calculated with the Hertzian theory. At rigid surfaces the AFM tip cannot penetrate into the sample, therefore the cantilever deflection corresponds to the extension of the z-piezo. The softer the sample the more penetrates the tip into the sample, therefore the slope is very smooth. To get the elasticity modulus (Young's modulus, E) the Hertzian or similar models have to be fitted to the measurement curves; (**c**) Elasticity change versus alteration time. The evaluation of the Young's moduli indicates an elasticity change depending on the alteration time. The bloodstain was measured after 1.5 h, after 30 h, and after 31 days. Every column was calculated from more than 100 individual force-distance curves. The *whisker* displays the standard deviation of the force curves

22.3.2
Mimicking Blood Flow Conditions on a Surface Acoustic Wave-Driven Biochip

So far, only static experiments on blood compounds after adhesion or fixation have been presented. The blood stream itself is a very dynamic process and caution has to be exercised in interpreting data of the functional behavior of its cells, proteins, etc. when achieved under static conditions. Apart from AFM there are a variety of

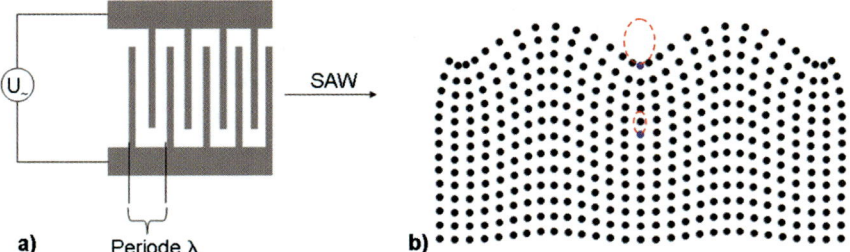

Fig. 22.9. (a) A voltage signal on a piezoelectric material results in a propagating mechanical deformation of the crystal. By adding a higher number of electrodes after one another and applying an alternating signal, matching the resonance frequency $f = \frac{v}{\lambda}$ (v: sound velocity of the piezoelectric material; λ: periodicity of the electrodes), the mechanical impulse is enhanced and a Surface Acoustic Wave is emitted. (**b**) Side view of a piezoelectric material excited via an IDT (not shown). The *red lines* display the elliptic course of the blue lattice points within one period of movement. The reason for the terminus Surface Acoustic Wave accounts for the exponential decrease into the substrate of the amplitude of an oscillating lattice point in the crystal

other techniques for investigation of blood clotting. SAWs offer an extremely potent method of driving small amounts of fluids and mimicking the blood vessel system on a planar substrate. The concept of Surface Acoustic Waves has been known since the late 19th century, when Lord Rayleigh first gave a theoretical description of earthquake waves [74]. Since that time SAWs have gone a long way. About 80-years later physicists started to apply this theoretical work to a scale Lord Rayleigh might not have thought about [75]. By lithographical methods Interdigital Transducers (IDTs) (Fig. 22.9) were deposited on piezoelectric substrates like $LiNbO_3$, where an alternating voltage signal excites a propagating mechanical high-frequency wave on the order of 100 MHz, an amplitude of 1 nm, and a wavelength of a few μm. The energy of the SAW is confined to the surface of the piezo and decreases almost totally within one wavelength into the material.

Dependent on the boundary conditions of the piezoelectric substrate, shear- or Rayleigh-waves, or a mixture of both wave modes, are excited. Only Rayleigh-waves exhibit an "out of plane" movement of the oscillating parts of the substrate, which can be applied for driving small amounts of fluids [76]. The surface oscillation dissipates energy into the liquid, which results in a pressure gradient within the fluid and hence induces fluid streaming. All further applications and experiments presented here are performed on $LiNbO_3$ chips and exploit the dissipated energy (leaky wave) of a Rayleigh-wave to induce acoustic streaming in a fluid [77].

There are a variety of advantages to using a planar SAW-driven microfluidic device compared to commonly used flow chambers. Because of the extremely small amount of sample volume (down to a few μl) even expensive or rare materials can be processed in contrast to conventional flow chambers with a typical working volume of some ml. By adjusting the voltage on the IDTs the flow velocity and therefore the shear rate of the fluid can be set very precisely. The 2D and optical transparent chip further offers the possibility to combine the setup with imaging techniques like fluorescence microscopy or AFM (see also Sect. 22.4). Furthermore, by

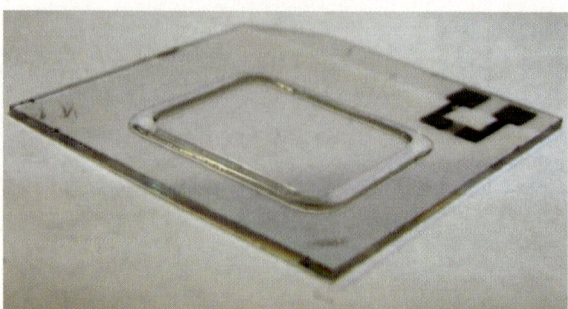

Fig. 22.10. Mimicking a blood vessel. The IDTs (*upper right*) on the 2D biochip (1 × 1 cm) drive the fluid clockwise within a hydrophobic-hydrophilic track. This track is deposited onto the chip by means of photolithography. A variety of different hydrophobic-hydrophilic structures can be produced to mimic even the smallest ramifications of blood vessels (Figure reprinted from Schneider MF, Guttenberg Z, Schneider SW et al., An Acoustically Driven Microliter Flow Chamber on a Chip for Cell-Cell and Cell-Surface Interaction Studies, ChemPhysChem 2008; 9: 641 with permission from Wiley-VCH)

means of standard optical lithography almost any flow geometry can be produced as a hydrophobic-hydrophilic track on a SAW-driven biochip. Using such a SAW-driven biochip, flow geometry and therefore the flow pattern can be fitted exactly to the physical, chemical, or medical problem under investigation. Figure 22.10 displays a hydrophobic-hydrophilic functionalized SAW-chip with a rectangular water channel representing a physicist's equivalent of a blood vessel. Full experimental freedom is achieved by the possibility to grow cells or coat the surface with a variety of different materials within the fluid track.

22.4
The Lab on a Chip – AFM – Hybrid

The SAW-driven artificial blood vessel provides the basis for testing variable blood flow conditions. In order to provide a first insight into the structural and dynamic properties of vWF-networks and -bundles we combined a microfluidic reactor, mimicking the hydrodynamic conditions in the microcirculatory system of our body [78] with an atomic force microscope for minute protein manipulation. After presenting the experimental setup and pointing out some basic advantages compared to commonly used AFM setups, we will present the hybrid setup in action, investigating the dynamic network properties of the blood clotting protein von Willebrand Factor in an amazing new way.

22.4.1
Experimental Setup

The completely planar reactor layout and fluid actuation system gives us the technological freedom to create a continuous flow in any desired geometry. No drain and hence no supply is necessary, allowing us to monitor the identical solution of

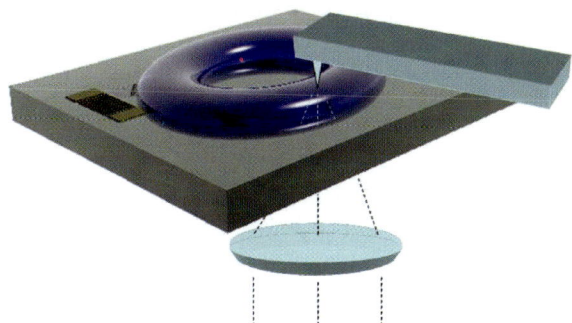

Fig. 22.11. The microfluidic set up. The unique combination of a microscope, an AFM tip, and a SAW-driven pumping system enables both the study of the formation of macromolecular VWF networks and conglomerates in real time and the mechanical manipulation of VWF aggregates and bundle pulling in particular (Figure reprinted from reference [79] with permission from Elsevier)

proteins, cells, vesicles, etc. without any loss of material or a change in any crucial parameter for long periods of time (~1 h). Although the fluid streaming induces a lot of vibration on the cantilever and hence too much noise for imaging, force spectroscopy under certain circumstances is still possible (Fig. 22.11).

Nevertheless, this setup provides distinct advantages in imaging too. Especially the measurement of the time course of sample alterations, induced by external stimuli, has always been a problem for AFM users. There are two significant deficits of normally operated devices in investigating samples before and after the application of either chemical or mechanical reagents or forces. One is the time scale till a reagent makes an impact on the sample and the second is the recovery of the region of interest (ROI) after this impact has taken place. Whereas chemical interactions such as changes in pH or salt concentration can be induced without moving either sample or tip, and thus do not affect the recovery of the ROI essentially, the equilibration process and therefore the reaction of the system is diffusion limited. Depending on the particle size of the added reagent the equilibration of the system may be extremely long. Without additional external mixing, equilibration of pH or salt concentration can be achieved within several minutes as a function of the sample volume. This may be sufficient for a variety of special applications, but is far from an ideal situation, where an immediate response is desirable to distinguish between a reagent effect and simple sample aging. An investigation of the influence of an enzyme of protein, for example a blood clotting protein, to a target sample would only be measurable as recently as several hours after addition depending on the protein size. Most biological samples would degrade within that time. The SAW pumping system provides mixing ability for a great variety of fluids with different viscosities, emulsions with particles of different sizes, and establishes equilibrium almost immediately. A tremendous advantage in doing so is that the sample does not have to be taken out of the AFM setup or moved at any time. Such actions increase the probability for sample contamination by other chemicals or dirt particles and open the door for other unpredictable interferences to the system in every experiment.

Furthermore, in the hybrid setup the cantilever can rest on the ROI and does not even have to be retracted during the few seconds of mixing. Thus, the fretful problem of recovering the actual scanning site is eliminated completely. Apart from mixing and driving the system into equilibrium, measuring the effect of shear stress on single cells, platelets of proteins, mimicking blood flow conditions or just adding mechanical forces on the sample is almost impossible in a common AFM-setup. Removal of the sample, putting it into an external flow chamber or any other device and mounting it back into the AFM, makes it extremely difficult to recover the ROI, and furthermore would cause a high degree of uncertainty in the system, if qualitative and quantitative conclusions can still be drawn at all.

The planar setup, providing full optical accessibility of the light path between the AFM tip and the fluidic track, is completed by an inverted microscope for recording the microfluidic channel and the AFM tip simultaneously [79]. This optical control system enables the online monitoring of any optically accessible event and thus provides a reliable means for reacting immediately with the AFM to every important change of the system. Furthermore, additional optical images of stained samples to a height profile for instance can often facilitate data interpretation a lot. In this way optical microscopy provides a direct reference system for AFM results. This hybrid system, combining the advantages of three different experimental approaches, does not simply serve as an inbuilt reference system. The assembled techniques supply information from different perspectives for the same question creating a unique and more complemented picture of the scientific problem. Thus, applying the AFM technique and optical microscopy simultaneously whilst maintaining fluid streaming to a particular sample, the acquired information is maybe more than just the sum of results of single experiments performed one after the other. The challenges of measuring a single sample in different devices after one another have already been discussed earlier.

This newly developed hybrid setup features high potential in every field where there is benefit or simply the necessity of the combination of the high-resolution imaging or the accuracy of force measurements of an Atomic Force Microscope, fluid mechanics, and optical accessibility.

In the next two sections we will proceed one more step and present the microfluidic reactor–AFM hybrid system illuminating new correlations in blood clotting by using the AFM cantilever not for measuring minute interaction forces or height profiles but as a hook for manipulating vWF networks.

22.4.2
Bundle Relaxation

vWF networks are a prerequisite for platelet aggregation at a site of vascular injury. The formation of vWF networks and conglomerates was shown to be shear-dependent and is thought to follow the same mechanism as the adhesion of vWF molecules to the extracellular matrix. Above a critical shear flow vWF molecules become activated, unroll and spontaneously self-aggregate with one another [3, 80]. On the other hand, there is only little knowledge about the actual binding properties and kinetics of vWF molecules inside such a network. Its reaction to an external induced force or the relaxation of single bundles, pulled out of such a network, can deliver valuable information about how such a network is made up on a molecular and energetic level.

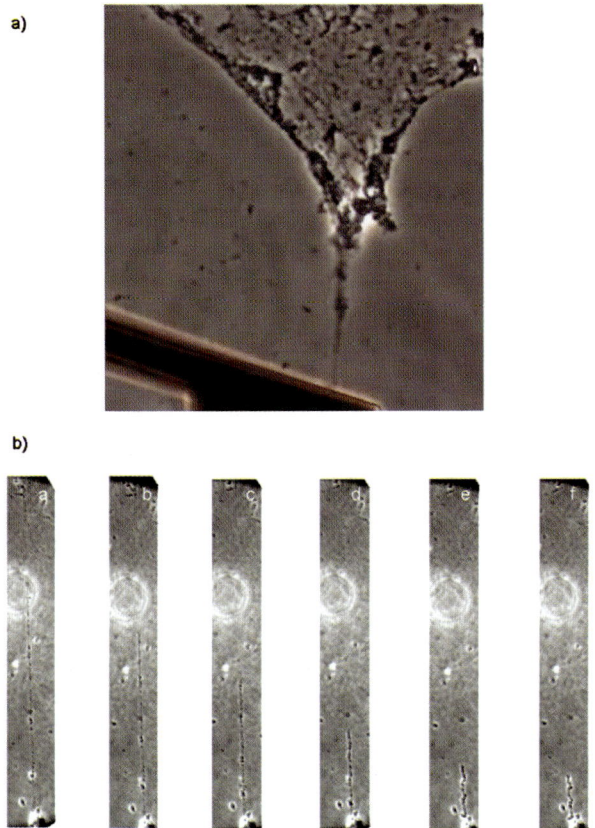

Fig. 22.12. (a) ULVWF immobilized on the AFM tip. (b) Relaxation after rupture (each image of the sequence is approximately 840 μm in height and 120 μm in width) (with courtesy of Elsevier) (Figure reprinted from reference [79] with permission from Elsevier)

Therefore, in the course of an experiment the AFM cantilever tip is approached to a protein network after the SAW-induced vWF conglomeration process is finished. vWF bundles are gently pulled by an untreated AFM tip until they finally rupture (Fig. 22.12). Acquired data of the time course of the relaxation process were analyzed using standard software. The hybrid setup enables us to study the relaxation of ultra-large vWF (ULVWF) bundles of different lengths pulled from vWF networks freshly formed by hydrodynamic stress in the microfluidic reactor while keeping all external parameters constant. In this set of experiments the AFM was solely used as a hook to pull vWF bundles until they finally rupture and not to quantify the required rupture-forces.

On a more quantitative basis, the dynamics of the system were monitored by measuring the end-to-end length of the bundle as a function of time. The maximum length L_0 was determined by the distance between the AFM tip and the point where rupture occurred. A typical trace of the end-to-end distance (i.e. the distance from

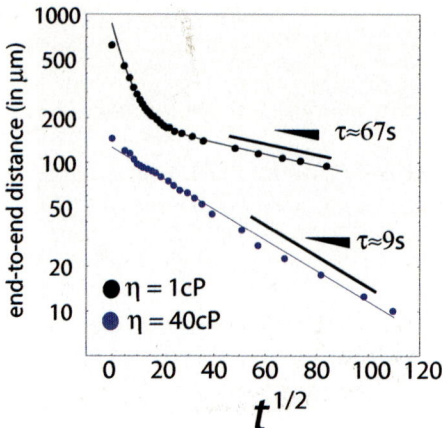

Fig. 22.13. End-to-end distance as a function of $\sqrt{t}$ plotted in a log-linear scale for different viscosities. The upper (*black*) curve corresponds to a viscosity of 1 cP, and the lower (*blue*) curve corresponds to a viscosity of 40 cP. In the low viscosity regime two distinct relaxation time scales could be observed (with courtesy of Elsevier) (Figure reprinted from reference [79] with permission from Elsevier)

the AFM tip to the free end of the ULVWF) as a function of time is presented in Fig. 22.13. We note that the relaxation has two characteristic time scales, and both of these periods are well described by stretched exponentials of the form: $L \propto e^{(t/\tau)^{\beta}}$ with $\beta \sim 0.5$. The fact that the relaxation process follows a stretched exponential implies that the underlying dynamics is governed by hopping events between random minima in a rough energy landscape. Equivalently, the relaxation spectrum is composed of a broad (random) distribution of exponential processes [81]. Such relaxation behavior is very common in nature, and has been found in different contexts. In particular, it has been observed in the electric birefringence relaxation of ensembles of synthetic polyelectrolytes [82], in the relaxation of DNA [83], and in proteins [84], to name a few. In these studies, the origin of the relaxation spectrum arises from the averaging over a multitude of parallel non-interacting (presumably exponential) relaxation processes. Theoretically, stretched exponential and power-law behavior has been extensively studied in the area of critical dynamics and glasses [85]. In the context of single polymer relaxation, Cherayil et al. have shown [86] that after inclusion of memory effects to the dynamics of the system, the relaxation is a stretched exponential with an exponent of $\beta \sim 0.5$. These memory effects included in this work accounted for the fact that a given conformation at earlier times affects the distribution of conformations at a later time along the relaxation process.

vWF bundles present new features compared to those described previously: (1) it has two clear relaxation time scales, and (2) the relaxation of the individual molecules is coupled to the one of all others in a non-trivial fashion. In this sense, we can say that vWF bundle relaxation is a more complex problem than those studied before. One of the striking features of our findings is that the longest relaxation rate k, corresponding to the inverse of the longest relaxation time τ, seems not to be dependent on the initial length of the bundle (data not shown). This implies that the dynamics of relaxation is dominated by conformational constraints, as well as effective internal friction limiting the rate of deformation of the object itself.

To further unravel the physical origin of the observed phenomenon the effect of viscosity of the solvent was studied on the long and short timescales of the bundle relaxation by adding a thickening agent, such as glycerol. In Fig. 22.13 we show two different traces of the end-to-end distance at different viscosities: 1 and 40 cP. As can be seen, the initial temporal decay is slowed down such that the full relaxation can be well described by a single stretched exponential. Interestingly, the slowest relaxation rate is now faster than the one for pure buffer solution (with lower viscosity). Further discussion on the origin of the double-time relaxation spectrum, and on the apparent enhancement of the long-time relaxation rate is given below.

In a high viscous environment, surprisingly, only a single relaxation time is observed. However, the relaxation spectrum found here is still different from the one of a single collapsing polymer. There, the relaxation of the end-to-end distance is supra-exponential, or in other words, the rate of compactification grows as the polymer relaxes [87]. If at all, we only find evidence of such a relaxation type during the first few 100 ms after rupture. This observation is not too surprising, as the ULVWF consists of many single fibers of different lengths relaxing at the same time in a complex fashion. Comparing it to the dynamics of a single polymer seems to at the least be doubtful or not valid at all. The complexity of our system arises because the relaxation of one fiber affects the relaxation of the other fibers, and vice versa.

For a better feeling of why the observed relaxation is a stretched exponential, it is probably better to describe the bundle as a single entity relaxing in a rough potential landscape determined by the configurations of the vWF single fiber constituents. The manifold of configurations is presumably strongly dependent on the number of self-association contacts, and hopping from one state to the other can only be done through thermally activated processes. The average barrier height will be reduced in the highly stretched state because of entropic forces, and thus the rate of relaxation should be faster at the beginning. Nonetheless, this argument would imply that one would see a continuous change in the relaxation of the bundle, and not two clear relaxation times as is found here. An alternative explanation is that there are two attractive interactions of different origins, which in this case would presumably correspond to self-association and hydrophobic attraction among monomers, driving the relaxation in each of these two regimes. To this date it is still unclear which of these interactions is the dominant one in the collapse of vWF, but we might argue that at long times, the hydrophobic interactions must become important since in principle the majority of the possible self-association contacts within the bundle should be present and only unfavorable surface interactions with the solvent and entropy can drive the further decrease in size.

With this unique combination of a Surface Acoustic Wave-driven planar microfluidic reactor, an optical microscope, and an AFM cantilever used in an unorthodox way, it was possible to unravel new aspects and concepts in protein network and bundle assembly. Beyond the scope of Imaging and Force Spectroscopy the AFM technique was successfully implanted into a fluidic system and functioned as a minute instrument in manipulating protein assemblies. Whereas bundle pulling was performed manually and no rupture forces were quantified, the sophisticated AFM control was a valuable tool for approaching and intruding protein aggregates and networks with distinct forces. In this case, the AFM technique, applied in a non-conventional manner, provided an ideal means to achieve more details on a very complex non-trivial question.

22.4.3
Stream Line Manipulation and Flow Sensoring

22.4.3.1
Stream Line Manipulation

The logical extension of the system presented above is to operate the AFM while exposed to fluidic streaming. Limitations in imaging due to an increased noise level can be compensated by benefits in other areas. This approach promises auspicious potential for real-time investigation of von Willebrand Factor, whose activity is triggered by the surrounding shear flow. With the help of the AFM cantilever, streamlines can be manipulated in a decisive way (Fig. 22.14), vWF network formation can be induced and moreover fluid streaming velocities can be measured with tip bound polymers.

Any stationary body inside a streaming fluid modifies the stream lines in a certain way. As far as the cantilever is thin and long enough, the overall flow pattern in the vicinity of the cantilever tip can be regarded as almost undisturbed. Around the tip the almost parallel flow lines are bent. Considering no slip boundary conditions at the tip, an extremely high shear rate can be induced dependent on the flow velocity of

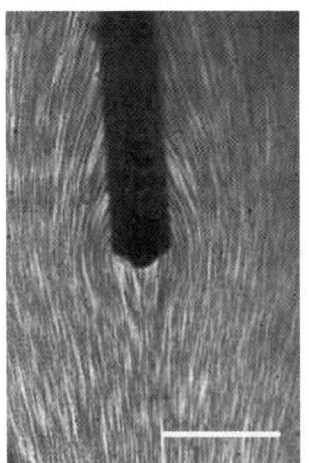

Fig. 22.14. An AFM cantilever tip (*scale bar* ~70 μm) is exposed to parallel and perpendicular fluid streaming. Because of the variable geometry of the planar SAW-driven microfluidic reactor and its adjustable pumping power, user-defined flow directions and velocities can be realized. Around the tip at the very end of the cantilever, the streamlines are softly deformed. The streamlines in the figures **a** and **b** are an overlay of a set of pictures from a high-speed camera, showing beads moving along the cantilever in a distinct way

the surrounding fluid. The shape, direction, and magnitude of the shear flow pattern including the shear rate acting on the tip can be varied both by an adjustment of the external voltage on the SAW-pumping system and by a change of the AFM cantilever position by the stepper control system in x-, y-, and z-direction relative to the variable stream lines.

Damage to a blood vessel also induces a mechanical obstacle and thus a disturbance of the flow pattern to the stream lines. Although a daring simplification from the point-of-view of a medical doctor, the effects of an injury to the vessel wall in the blood stream and of an AFM tip in the microfluidic reactor resemble each other in the basic aspects. Both create an increased shear rate in their close vicinity and hence are able to induce a conformational change of vWF molecules, if this shear rate is high enough. As discussed before, this conformational change is accompanied by an activation of vWF's binding potential resulting in a network formation both at the vascular injury and the AFM tip (Fig. 22.15).

The simulation of the actual flow conditions around the AFM tip derived from the experimentally detected stream lines for different flow conditions will soon make it possible to estimate the forces around the tip. Knowing the precise behavior of vWF under shear conditions will allow using vWF as a probe for sensing forces in a fluid with and without the AFM tip. This will not only enable us to compare field and point-like forces, but also to determine the forces on soft objects under flow, a field which is only very little explored due to the lack of experimental data.

(a)   (b)

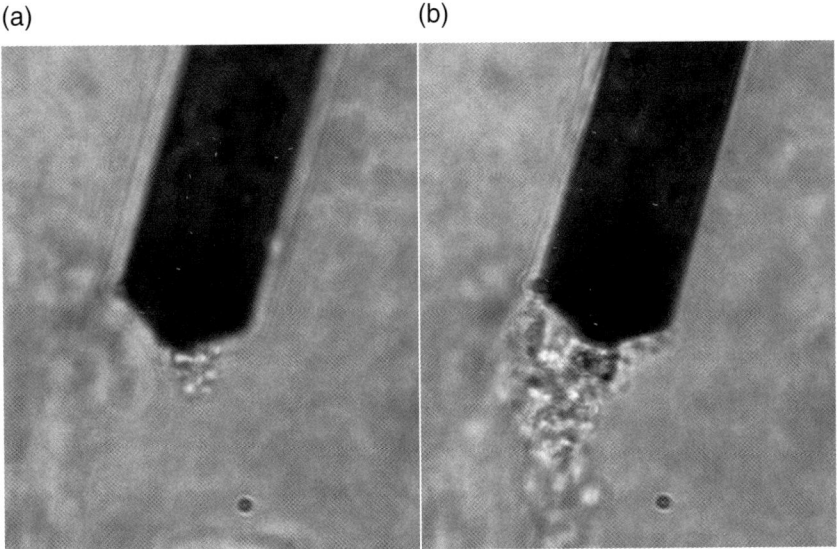

Fig. 22.15. The cantilevers were exposed to a vWF solution in a (**a**) low shear and (**b**) high shear regime for several minutes. Whereas in the low shear regime almost no or only minor protein adhesion could be observed, vWF self-associates continuously in the high shear regime forming a sticky three-dimensional conglomerate (cantilever width ~$30\,\mu$m)

22.4.3.2
Flow Sensoring

Despite producing whole vWF conglomerates on the tip, it is much more difficult to bind single proteins or polymer fibers with one end onto the AFM tip, while keeping the rest of the protein unconstrained. This cannot be achieved simply by creating a high shear rate around the tip and waiting for spontaneous adhesion. Defined protein fibers are a prerequisite for investigating quantitative protein reactions to an external fluidic stream on one hand, and could also be a valuable tool for probing streaming properties by a protein "wind gauge" waving in the stream lines of a microfluidic reactor on the other hand.

Similar to AFM Force Spectroscopy measurements, where the tip scans for a protein on the surface first, protein fibers are "collected" from a vWF covered surface. The tip penetrates a protein conglomerate and a fiber or bundle of variable thickness is bound. Normally, a lot of such binding attempts have to be undertaken to pick up a suitable protein fiber. In contrast to the bundle relaxation experiments another prerequisite for a successful experiment is that the fiber does not rupture during the pulling sequence. Relaxation of a thin fiber to the cantilever tip will result most probably in an undefined and irreversible adhesion of the protein around the tip. In order to bind a loose and mobile vWF fiber with one end onto the AFM tip, it must not lose contact to another small and spatial unconfined body, acting as a contact point for external pulling forces. This approach is similar to the method of optical tweezers [88,89], where a protein is pinned to a solid surface (the AFM tip in this case) and experiences forces mediated by a bead tightly bound to the protein's second end. The role of the bead is taken over by a stiff conglomerate of vWF in the experiment shown in Fig. 22.16. In the course of the experiment, an extremely thin vWF fiber (below the optical resolution of the microscope) was linked to the AFM tip and remained attached to a freely moving vWF conglomerate ($\sim$ vWF-"bead"). The force on the vWF assembly due to the fluid streaming elongated the vWF fiber. Throughout the experiment, the SAW-driven microfluidic system provided constant

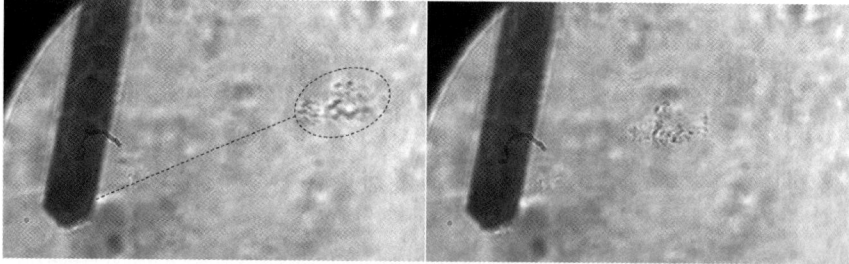

Fig. 22.16. A tip bound vWF fiber experiencing alternating forces by the fluid stream mediated by vWF assembly. The vWF assembly is highlighted by the spherical *inset* to guide the eye. The *dashed line* illustrates the optically not resolved extremely thin vWF fiber and furthermore points out the direction of the flow. Whereas on the picture on the left side the vWF fiber is stretched in a fast flowing regime, the fiber length drastically decreases under a critical flow velocity (cantilever width $\sim$30 μm)

power to pump the liquid. However, in particular the streaming velocity is not constant over the whole channel but is a function of the actual position inside the fluid. To check for the elongation of the vWF fiber as a function of the fluid streaming, the cantilever is retracted in 50-µm steps from its starting position far away from any wall. With increasing relative cantilever distance, the flow velocity and therefore the acting force on the fiber steadily decreased, resulting in a slight reduction of the fiber length. Above a critical height, or in other words falling below a critical force on the vWF-"bead," the protein fiber contracted in a non-linear fashion (Fig. 22.17). Note the continuous decrease of the flow velocity at the same time!

Similar experiments may help in understanding phase transitions in systems of fractal geometry. From a mathematical point-of-view, the protein fiber can only be considered as one-dimensional in a first approximation, but neither is it a two-dimensional object. The divergence from the one-dimensionality of the vWF fiber arises from multiple contact points of the linear molecule forming a protein ball of yarn. Therefore, probing the reversible elongation and contracting properties of such

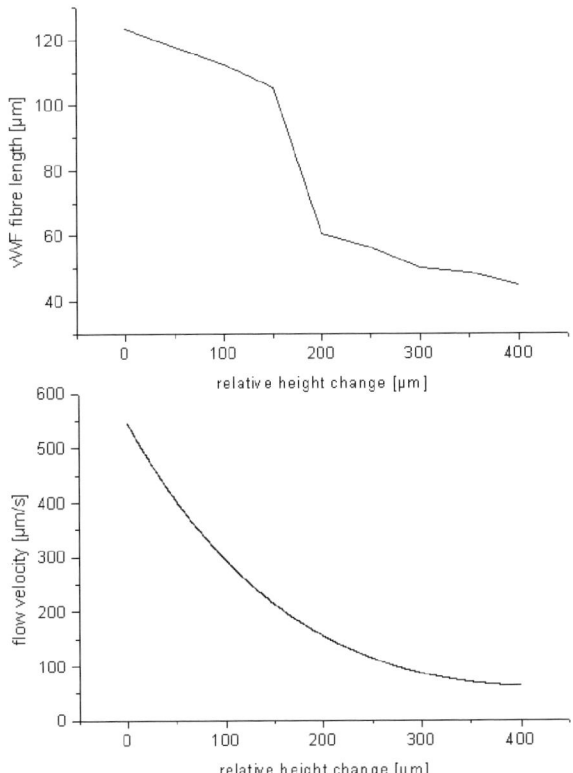

Fig. 22.17. While increasing the relative height the flow velocity steadily decreases whereas there is a significant discontinuous reduction of the vWF fiber length between 150 and 200 µm. Note that in first approximation the flow velocity is direct proportional to the acting force on the fiber

biopolymers may lead to a deeper insight into advanced mathematical concepts for a description of protein assemblies.

Taking advantage of almost unlimited possibilities in protein engineering for the replication of proteins [90–92], exhibiting similar dynamic conformational features as vWF, industry can apply these findings for the production of a new kind of flow sensor. Artificial proteins bound to an AFM tip on one end and to a bead on the other end can sense both the flow direction and the flow velocity, indicated by the polymer length. Such a sensor would be able to scan directly a three-dimensional vector field. Furthermore, a modified AFM tip comprising a distinct binding position for such a protein fiber would be of additional value. As this was not the case in the experiment presented above, the minute force detection of the AFM could not be taken into account. With an exactly known fiber contact point it would be possible to calculate acting forces from the bending of the cantilever, which could serve as an inbuilt reference measurement for the fiber elongation.

22.5
Summary and Outlook

In this chapter we presented AFM as a multi-functional tool for answering questions concerned with blood clotting. With the help of the AFM technique, phenomena like vWF network formation and its intrinsic dynamic structure, platelet adhesion to vessel lesions and pathological altered blood cells could be addressed and their underlying concepts elucidated. The combination of a Surface Acoustic Wave-driven two-dimensional microfluidic reactor with an AFM setup enabled the investigation of the dynamic properties of the mechanically activated blood clotting protein von Willebrand Factor not only under physiological buffer conditions but also under physiological mechanical flow conditions. From its start in 1986 AFM has gone a long way and its journey is far from over. Especially in combination with microfluidics the AFM technique is still at the very beginning and there are a lot of fields of activity left. Researchers will be able to probe dynamic correlations between components nobody has thought of yet. A demanding, but with current techniques possible task would be the investigation of blood clotting under atherosclerotic conditions. The properties of blood coagulation factors like vWF and also platelets depend strongly on the streaming pattern of the blood flow. Mimicking atherosclerotic conditions on a SAW-driven biochip would require only the production of a hydrophobic-hydrophilic track with a set of different "vessel" diameters which is easily achievable considering the recent state of the technology. Imaging of the resulting protein networks under these different physiological flow conditions can lead to a better understanding of thrombus formation, which may be a first step in preventing it. But this is just one example within the framework of blood flow. The essential concepts that form the basics of the results presented here will be extended to a far broader set of experiments and future applications.

Acknowledgments. The authors would like to thank Stefan Bössinger and Jennifer Angerer for laboratory work. Financial support by the Elite Netzwerk Bayern (ComplInt) and the German Excellence Initiative via the "Nanosystems Initiative Munich" (NIM) is gratefully acknowledged.

References

1. Binnig G, Quate CF (1986) Phys Rev Lett 56:930
2. Nowakowski R, Luckham P, Winlove P (2001) BBA 1514:170
3. Schneider SW, Nuschele S, Wixforth A, Gorzelanny C, Alexander-Katz A, Netz RR, Schneider MF (2007) Proc Natl Acad Sci 104:7899
4. Sadler JE (1998) Annu Rev Biochem 67:395
5. Tangelder GJ, Slaaf DW, Arts T, Reneman RS (1988) Am J Physiol 254:1059
6. Sadler JE (1991) J. Biol. Chem. 266:22777
7. Perkkio J, Wurzinger LJ, Schmid-Schonbein H (1987) Thromb Res 45: 517
8. Ruggeri ZM (1997) J Clin Invest 99:559
9. Reininger AJ, Heijnen HFG, Schumann H, Specht HM, Schramm W, Ruggeri ZM (2006) Blood 107:3537
10. Dopheide SM, Maxwell MJ, Jackson SP (2002) Blood 99:159
11. Guyton AC, Hall JE (2000) Textbook of medical physiology, 10th edn. W.B. Saunders, Philadelphia
12. Ramachandra GN, Karthan G (1955) Nature 176: 593
13. Kadler KE, Holmes DF, Trotter JA, Chapman JA (1996) Biochem J 316:1
14. Williams BR, Gelman RA, Poppke DC Piez KA (1978) J Biol Chem 253:6578
15. Gross J, Highberger JH, Schmitt FO (1951) J Appl Phys 22:112
16. Gross J, Highberger JH, Schmitt FO (1952) P Soc Exp Biol Med 80:462
17. Gross J, Highberger JH, Schmitt FO (1954) Proc Natl Acad Sci 40:679
18. Gross J (1956) J Cell Biol 2:261
19. Gross J, Kirk D (1958) J Biol Chem 233:355
20. Wood GC, Keech MK (1960) Biochem J 75:588
21. Wood GC (1960) Biochem J 75:598
22. Bard JBL, Chapman JA (1968) Nature 219:1279
23. Bard JBL, Chapman JA (1973) Nature-New Biol 246:83
24. Shattuck MB, Gustafsson MGL, Fisher KA, Yanagimoto KC, Veis A, Bhatnagar RS, Clarke J (1994) J Microsc 174:Rp1
25. Fujita Y, Kobayashi K, Hoshino T (1997) J Electron Microsc 46:321
26. Paige MF, Rainey JK, Goh MC (1998) Biophys J 74:3211
27. Morris VJ, Kirby AR, Gunning AP (1999) Atomic force microscopy for biologists. Imperial College Press, London
28. Rainey JK, Wen CK, Goh MC (2002) Matrix Biol 21:647
29. Gutsmann T, Fantner GE, Venturoni M, Ekani-Nkodo A, Thompson JB, Kindt JH, Morse DE, Fygenson DK, Hansma P.K. (2003) Biophys J 84:2593
30. Strasser S, Zink A, Heckl WM, Thalhammer S (2006) J Biomech Eng 128:792
31. Holmes DF, Gilpin CJ, Baldock C, Ziese U, Koster AJ, Kadler KE (2001) Proc Natl Acad Sci 98:7307
32. Thalhammer S, Stark RW, Muller S, Wienberg J, Heckl WM (1997)J Struct Biol 119:232
33. Strasser S, Zink A, Janko M, Heckl WM, Thalhammer S (2007) Biochem Biophys Res Commun 354:27
34. Jiang FZ, Khairy K, Poole K, Howard J, Muller DJ (2004) Microsc Res Tech 64:435
35. Wen CK, Goh MC (2004) Nano Lett 4:129

36. Petruska JA, Hodge AJ (1964) Proc Natl Acad Sci 51:871
37. Graham JS, Vomund AN, Phillips CL, Grandbois M (2004) Exp Cell Res 299:335
38. Gutsmann T, Fantner GE, Kindt JH, Venturoni M, Danielsen S, Hansma PK (2004) Biophys J 86:3186
39. Bozec L, Horton M (2005) Biophys J 88:4223
40. Fratzl P, Misof K, Zizak., Rapp G, Amenitsch H, Bernstorff S (1998) J Struct Biol 122:119
41. Puxkandl R, Zizak I, Paris O, Keckes J, Tesch W, Bernstorff S, Purslow P, Fratzl P (2002) Philos T Roy Soc B 357:191
42. Diamant J, Arridge RGC, Baer E, Litt M, Keller A (1972) P Roy Soc Lond B Bio 180:293
43. Fratzl P, Fratzl-Zelman N, Klaushofer K (1993) Biophys J 64:260
44. Sarkar SK, Hiyama Y, Niu CH, Young PE, Gerig JT, Torchia DA (1987) Biochem 26: 6793
45. Mosler E, Folkhard W, Knorzer E, Nemetschekgansler H, Nemetschek T, Koch MHJ (1985) J Mol Biol 182:589
46. Marchant RE, Lea AS, Andrade JD, Bockenstedt P (1991) J Colloid Interface Sci 148:261
47. Raghavachari M, Tsai HM, Kottke-Marchant K, Marchant RE (2000) Colloids Surf, B 19:315
48. Siediecki CA, Lestini BJ, Kottke-Marchant KK, Eppell SJ, Wilson DL, Marchant RE (1996) Blood 88:2939
49. Shao JY, Ting-Beall HP, Hochmuth RM (1998) Proc Natl Acad Sci 95: 6797
50. Fritz M, Radmacher, Gaub HE (1994) Biophys J 66:1328
51. Hoerber JK, Miles MJ (2003) Science 302:1002
52. Butt HJ, Wolff EK, Gould SAC, Dixon Northern B, Peterson CM, Hansma PK (1990) J. Struct Biol 105:54
53. Gould SAC, Drake B, Prater CB, Weisenhorn AL, Manne S, Hansma HG, Hansma PK, Missie J, Longmire M, Elings V, Dixon Northern B, Mukergee B, Peterson CM, Stoeckenius W, Albrecht TR, Quate CF (1990) J. Vac. Sci. Technol. A8:369
54. Häberle W, Hoerber JK, Binnig G (1991) J Vac Sci Technol B9:1210
55. Kamruzzahan ASM, Kienberger F, Stroh CM, Berg J, Huss R, Ebner A, Zhu R, Rankl C, Gruber HJ, Hinterdorfer P (2004) Biol. Chem. 385:955
56. Ohta Y, Otsuka C, Okamoto H (2003) J Artif Organs 6:101
57. Cox M (1991) J Forensic Sci 36:1503
58. Quickenden TI, Cooper PD (2001) Luminescence 16:251
59. Gill P (2002) Biotechniques 32:366
60. Schwarzacher W (1930) Zeitschrift der Gesellschaft für Gerichtliche Medizin 15:119
61. Lins G, Blazek V (1982) Zeitschrift Fur Rechtsmedizin-J Legal Med 88:13
62. Miki T, Kai A, Ikeya M (1987) Forensic Sci Int 35:149
63. Inoue H, Takabe F, Iwasa M, Maeno Y and Seko Y (1992) Forensic Sci Int 57:17
64. Strasser S, Zink A, Kada G, Hinterdorfer P, Peschel O, Heckl WM, Nerlich AG, Thalhammer S (2007) Forensic Sci Int 170:8
65. Burnham NA, Colton RJ, Pollock HM (1991) J Vac Sci Technol A 9:2548
66. Tomasetti E, Legras R, Nysten B (1998) Nanotechnology 9:305
67. Vanlandingham MR, McKnight SH, Palmese GR, Elings JR, Huang X, Bogetti TA, Eduljee RF, Gillespie JW (1997) J Adhes 64:31
68. Thompson JB, Kindt JH, Drake B, Hansma HG, Morse DE, Hansma PK (2001) Nature 414:773
69. Bozec L, Horton M (2005) Biophys J 88:4223
70. Strasser S, Zink A, Janko M, Heckl WM, Thalhammer S (2007) Biochem Biophys Res Commun 354:27
71. A-Hassan E, Heinz WF, Antonik MD, D'Costa NP, Nageswaran S, Schoenenberger CA, Hoh JH (1998) Biophys J 74:1564
72. Goldmann WH (2000) Biotechnol Lett 2:431

73. Touhami A, Nysten B, Dufrene YF (2003) Langmuir 19:4539
74. Rayleigh LJWS (1885) Proc London Math Soc 17:4
75. White RM, Voltmer FW (1965) Appl Phys Lett 7:12
76. Wixforth A, Strobl C, Gauer C, Toegl A, Scriba J, von Guttenberg Z (2004) Anal Bioanal Chem 379:982
77. Sritharan K, Strobl CJ, Schneider MF, Wixforth A, Guttenberg Z (2006) Appl Phys Lett 88: 054102
78. Frisch T, Verga (2002) *Phys Rev E Stat Nonlin Soft Matter Phys* 66:041807
79. Steppich DM, Angerer JI, Sritharan K, Schneider SW, Thalhammer S, Wixforth A, Alexander-Katz A, Schneider MF (2008) Biochem Biophys Res Commun 369:507
80. Alexander-Katz A, Schneider MF, Schneider SW, Wixforth A, Netz RR (2006) PRL 2006.
81. Phillips JC (1996) Rep Prog Phys 59:1133
82. Degiorgio V, Bellini T, Piazza R, Mantegazza F, Goldstein RE (1990) Phys Rev Lett 64:1043
83. Hong MK, Narayan O, Goldstein RE, Shyamsunder E, Austin RH, Fisher DS, Hogan M (1992) Phys Rev Lett 68:1430.
84. Hagen SJ, Eaton WA (1996) J Chem Phys 104: 3395
85. Metzler R, Klafter J (2000) Phys Rep 339:1
86. Cherayil BJ (1992) J Chem Phys 97:2090
87. Frisch T, Verga (2002) PhylRev E 65:041801
88. Wang MD, Yin H, Landick R, Gelles J, Block SM (1997) *Biophys J* 72:1335
89. Arya M, Anvari B, Romo GM, Cruz MA, Dong JF,.McIntire LV, Moake JL, López JA (2002) Blood 99:3971
90. Huston JS, Levinson D, Mudgett-Hunter M, Tai MS, Novotný J, Margolies MN, Ridge RJ, Bruccoleri RE, Haber E, Crea R (1988) Proc Natl Acad Sci 85:5879
91. Itzhaki LS, Otzen DE, Fersht AR (1995) J Mol Biol 254:260
92. Matouschek A, Kellis JT, Serrano L, Fersht AR (1989) Nature 340:122

23 Atomic Force Microscopic Study of Piezoelectric Polymers

Hyungoo Lee · Ke Wang · Taekwon Jee · Hong Liang

Key words: Polymer, Polyvinylidene Fluoride (PVDF), Atomic Force Microscope (AFM), Adhesion, Friction

23.1
Piezoelectric Materials

Materials play an important role in atomic force microscopic (AFM) analysis. Active materials, in particular, present interesting challenges due to their nature. Active materials possess properties or behaviors that change due to the vibration of energy through stress, temperature, magnetic, or electrical fields. In this chapter, we discuss the uniqueness of piezoelectric materials characterized using the AFM.

Piezoelectricity was firstly discovered in 1756 [1]. Materials made for piezoelectricity were reported in 1880 [2, 3]. To date, piezoelectric effects have been widely used in industrial and civilian applications, such as, transducers, sensors, actuators, power generators, piezo motors, and fuel cells, among others [4–7]. Many materials have piezoelectric properties; common ones are listed in Table 23.1. As seen here, these materials are in single crystal polycrystal, polymeric form.

Among the materials listed in Table 23.1, poly(vinylidene fluoride) (PVDF) is a unique polymer that has high pyro- and piezoelectric properties. PVDF has wide engineering applications [5–8]. Besides the high piezoelectric coefficient, advantages such as flexibility, bio-compatibility, lightness, and low acoustic and mechanical impedance make PVDF a favorable material for bio- and MEMS (microelectromechanical systems) applications.

Compared to piezoelectric ceramics, PVDF has higher voltage sensitivity and lower acoustic impedance [12]. As a semicrystalline polymer, PVDF has five crystallographic forms, α, β, γ, δ, and ε. Of those, the latter four crystalline structures possess permanent dipole moment. The dipoles associated with individual molecules are parallel to each other in the unit cell; as a result, overall PVDF exhibits non-zero dipole moment. The β phase exhibits the strongest piezo-, pyro-, and ferroelectric properties. In the α phase, the molecular dipoles are antiparallel in each unit cell resulting in no-net dipole present. The polar phases can be obtained from the nonpolar α phase by different processes such as applying tensile stress (α phase $\rightarrow$ β phase) [13, 14], poling under external electric fields (α phase $\rightarrow$ β and δ phase)

Table 23.1. Piezoelectric materials

	Materials	Piezoproperties	References
Crystals	Quartz (SiO_2)	Low loss dielectric	[8]
	Berlinite ($AlPO_4$)	Insensitive permittivity	[9]
Ceramics	Lead zirconate titanate (PZT)	Large dielectric constant	[10]
Polymer	Polyvinylidene fluoride (PVDF)	Strong ferroelectric	[11]

[15–17], and annealing or crystallization at high temperatures (α phase $\to \gamma$ and ε phase) [18,19]. Among the four well-defined crystal structures, phase transformation between crystalline and amorphous exists [20]. The β phase has been proven to have the best piezoelectric response among the four major types of polymorph: $TG^{+}TG^{-}$ in α and δ phases, all trans(TTT) planar zigzag in β phase, and $T_3G^{+}T_3G^{-}$ in γ and ε phases. Such molecular and phase structures make this material of interest for study using AFM.

23.2
Atomic Force Microscopy

Since its invention in 1984, the atomic force microscope has been widely used in characterization and manipulation of almost all types of materials [21–23]. The working principles of an AFM have been discussed in detail throughout this book. Here a brief review is provided with a schematic figure, as shown in Fig. 23.1. In this figure, the signal variation from laser movement due to the cantilever movement or vibration can be detected by a photo diode sensor when the probe travels over the sample surface. Through a feedback loop, this signal variation is collected and imaged by a software. There are two major types of scanning modes, contact and close contact (or tapping/vibrating mode). Contact-mode AFM is more useful for making clear topographical images of hard materials with low average roughness rather than soft materials.

A close-contact mode or tapping mode is useful for phase detection and non-destructive imaging. The tapping mode eliminates the problems associated with friction, adhesion, and electrostatic forces caused scanning contact. The tapping-mode imaging is obtained through oscillation of the cantilever assembly at the cantilever's resonant frequency in an ambient environment. Once the approaching oscillating tip is in contact with the surface, the change of cantilever oscillation is reflected by the surface properties of the sample disk. Typically, the tip oscillates at a high frequency of 200 kHz. Vibrating (tapping) mode makes it possible to measure the topography and phase of soft materials without damage to the material under study.

Beside surface and phase imaging, the AFM has been used for lateral and adhesive force measurement. With the contact mode, the friction can be measured sensing the left and right deflection of the probe during scanning. The adhesion force can be

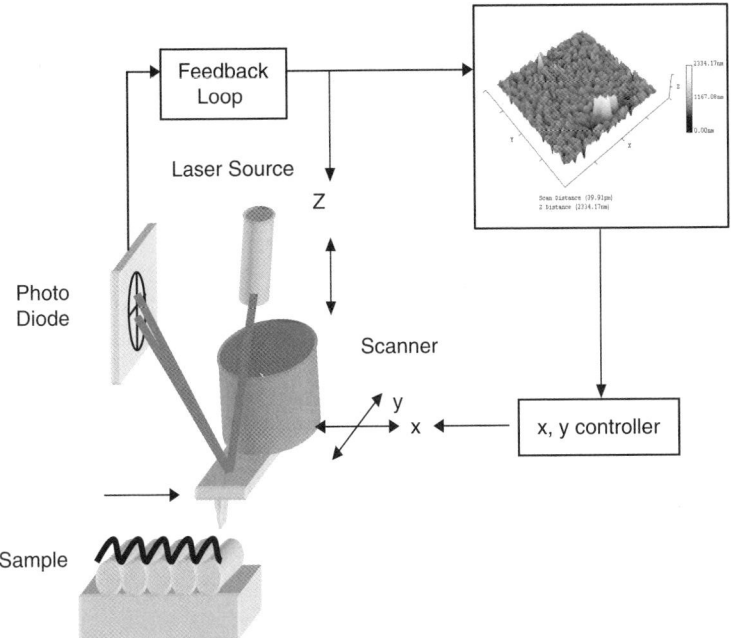

Fig. 23.1. A schematic diagram of AFM operation (courtesy: Taekwon Jee)

measured by the force displacement curve under the contact mode. Figure 23.2 shows the principles of such measurement. In the figure, the probe is approaching from A to C with increasing attractive force and detaching from C to D due to the repulsive force that caused a sudden pull-off from D to E. The distance between B and D is caused by adhesion force. Adhesion force can be calculated using Newton's Law:

$$F_{Adhesion} = k \times \Delta x \qquad (23.1)$$

where k is the spring constant (nN/nm) of the AFM probe.

23.3
Atomic Force Microscope Measurement

The scanning probe microscope (SPM) has been used for the study of ferroelectric materials [24–27]. SPM could be used to map polarization in a local domain. The spontaneous polarization, ferroelectric phase transition, surface phase transition, and pyroelectric change were observed through SPM in thin-film ferroelectric polymers [28–30]. One typical example among the prior studies was AFM probe-induced change. It was reported by Kimura et al. that the orientation of PVDF crystals and molecules could be controlled through AFM, as shown in Fig. 23.3 [31].

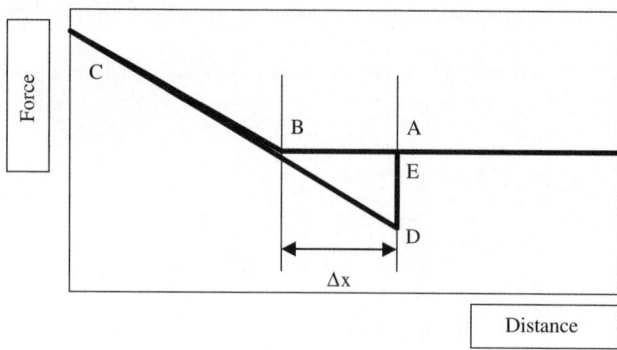

(a) Force-displacement curve

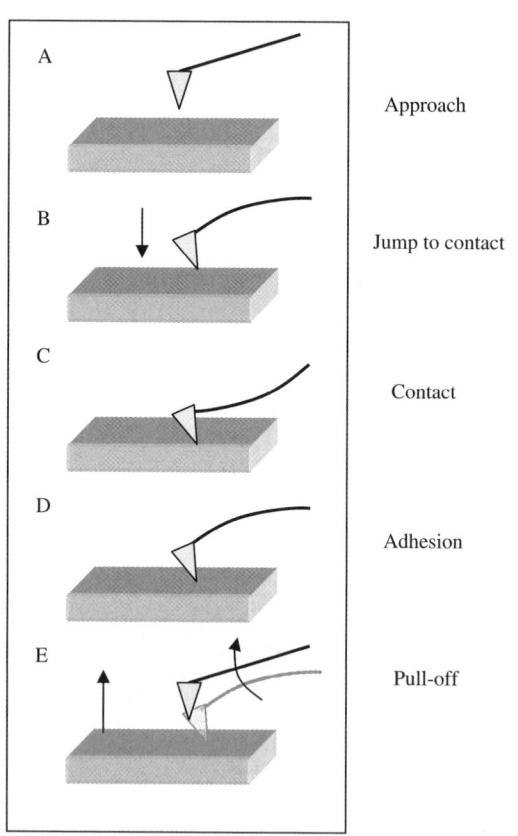

(b) Movement of cantilever

Fig. 23.2. A schematic diagram showing measurement of adhesion force; (**a**) force-displacement curve; (**b**) movement of the cantilever (courtesy: Taekwon Jee)

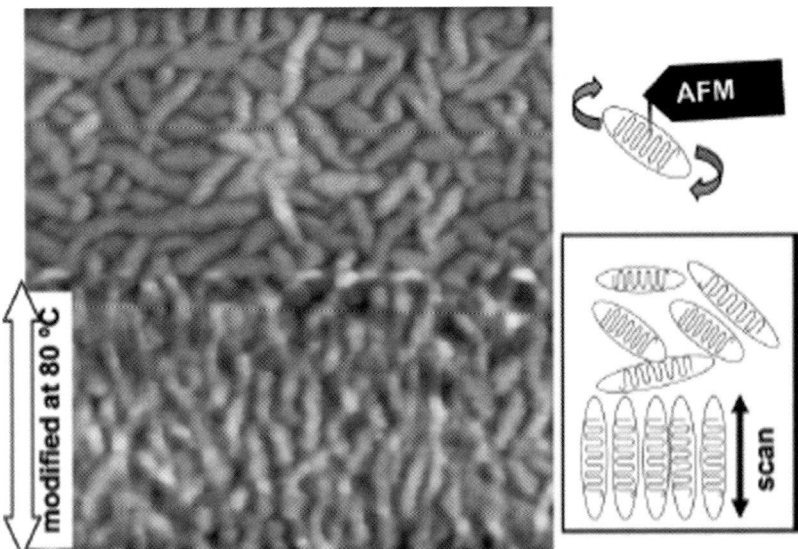

Fig. 23.3. The AFM was used to align the orientation of PVDF's microstructures through scanning at 80°C. The *left* shows morphological imaging and *right* is the schematic illustration. With permission from the Applied Physics Letters, American Institute of Physics

23.3.1
Sample Materials

The PVDF thin films were prepared for AFM analysis. Granular PVDF (Goodfellow Inc.), its solvent dimethylsulfoxide (DMSO), and other chemicals (EMD) were used to synthesize films. Granular PVDF was mixed with acetone (80 ml) and DMSO (20 ml) solutions. The concentration of PVDF was set at 40, 60, 80, and 100 g/L each for various viscosities. The solution was heated and stirred at 40°C for about 30 min till all solid particles were dissolved completely. The solution was then spin-coated (using the SCS P6204 spin coater) at the speed of 3,000 rpm for 20 s. Samples were then annealed at room temperature (23°C), 40, 60, and 80°C with corona poling of 30 kV for 2 min. These samples were labeled as A, B, C, and D, respectively. Different viscosities of PVDF solutions were spin-coated on a gold-coated silicon substrate (SCS P6204) at different speeds. Films were then heated at various temperatures before, during, and after 30 min with in situ corona poling at 30 kV for 2 min (DB-20AC, Electro Technic Products Inc.). The microstructural analysis was then conducted using wide-angle X-ray diffraction (D8 Advance, Bruker Axs Inc., USA) and a FTIR. Details will not be discussed here.

23.3.2
Surface Image Analysis

The morphology and phase images were obtained using an Atomic Force Microscope (DI Nanoscope IIIa), as shown in Fig 23.4. There are two images in this figure.

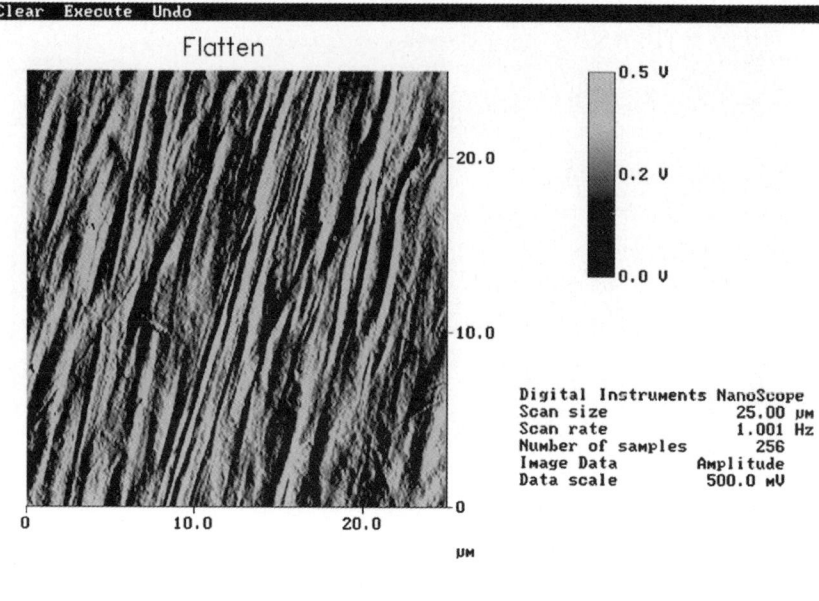

072905_p.000

(a) Phase image of commercial PVDF thin film

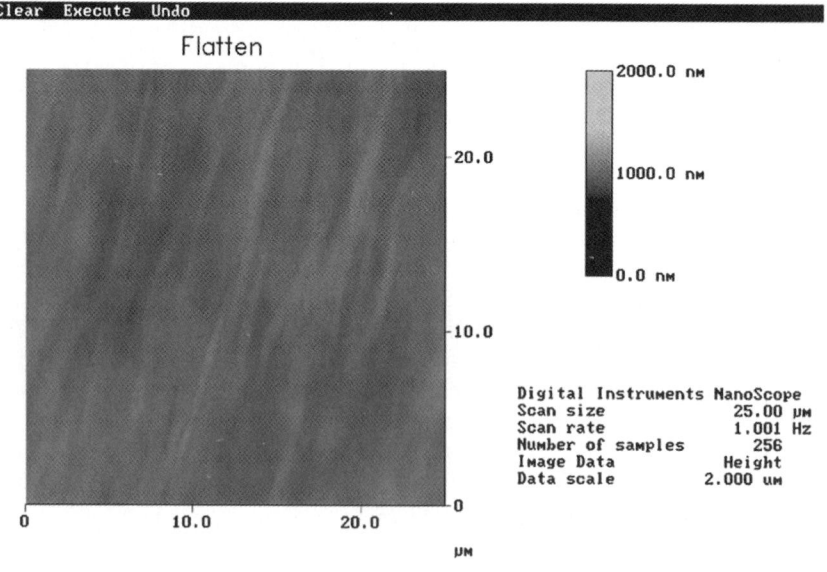

072905_p.001

(b) EFM Image of commercial PVDF thin film

Fig. 23.4. (a) Phase image and (b) EFM image of commercial PVDF thin film [32]

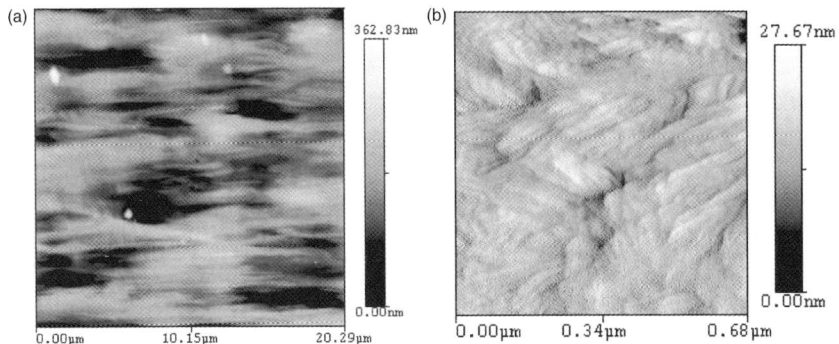

Fig. 23.5. (a) Stretching of PVDF film triggered by high spinning motion. (b) Biaxial stretching of PVDF film in nano-scale

Figure 23.4a represents the phase image while Fig. 23.4b is the Electrostatic force microscope (EFM) image. In Fig. 23.4a the stretched structure is clearly seen. It was caused by the corona process that stretches the PVDF in the direction indicated. The electrical output due to electrical force gradients is shown in Fig. 23.4b. The intensities are uniform across the sample indicating a stable electrical charge.

Morphological images of PVDF samples were acquired and results are shown in Fig. 23.5. Figure 23.5a shows the stretching direction horizontally. The spherulites and fibers seem to be connected from one spherulite to another. Figure 23.5b is a sample with bi-axial stretching. There are two stretching directions seen, radial and tangential to the spinning motion direction.

One sample that contains a mix of γ and β phase was analyzed using an AFM (Pacific Nanotechnology, Inc.). There are at least two different morphological patterns shown in Fig. 23.6. In Fig. 23.6a, where morphological images are shown, two distinguishable colors appeared. At a smaller scale (Fig. 23.6b), it is seen there are two types of structures; there are crushed smaller "grains" and stretched longer ones. Figures 23.6a and b are a mix of α, β, or γ phase. For comparison, the sample containing mostly α phase is shown in Fig. 23.6c. In this figure, structures are uniform and only one phase is shown.

23.3.3
Surface Force Measurements

Surface forces, such as lateral (or friction), adhesion (sticktion), attraction, and repulsion, are important for MEMS and NEMS (nanoelectromechanical systems) devices that have high aspect ratio. Adhesion (sticktion) and friction can readily destroy the electrical connection during operation [32–34]. The adhesion force can be measured using the lateral force microscope (LFM), based on the equation $F_{Adhesion} = k \times \Delta x$. Four samples previously discussed were analyzed using an AFM (PNI).

Results are shown in Fig. 23.7. For sample A (see previous materials session), the Δx is 310.96 nm (Fig. 23.7a). The adhesion force is then 1.8 (nN/nm) × 310.96 nm

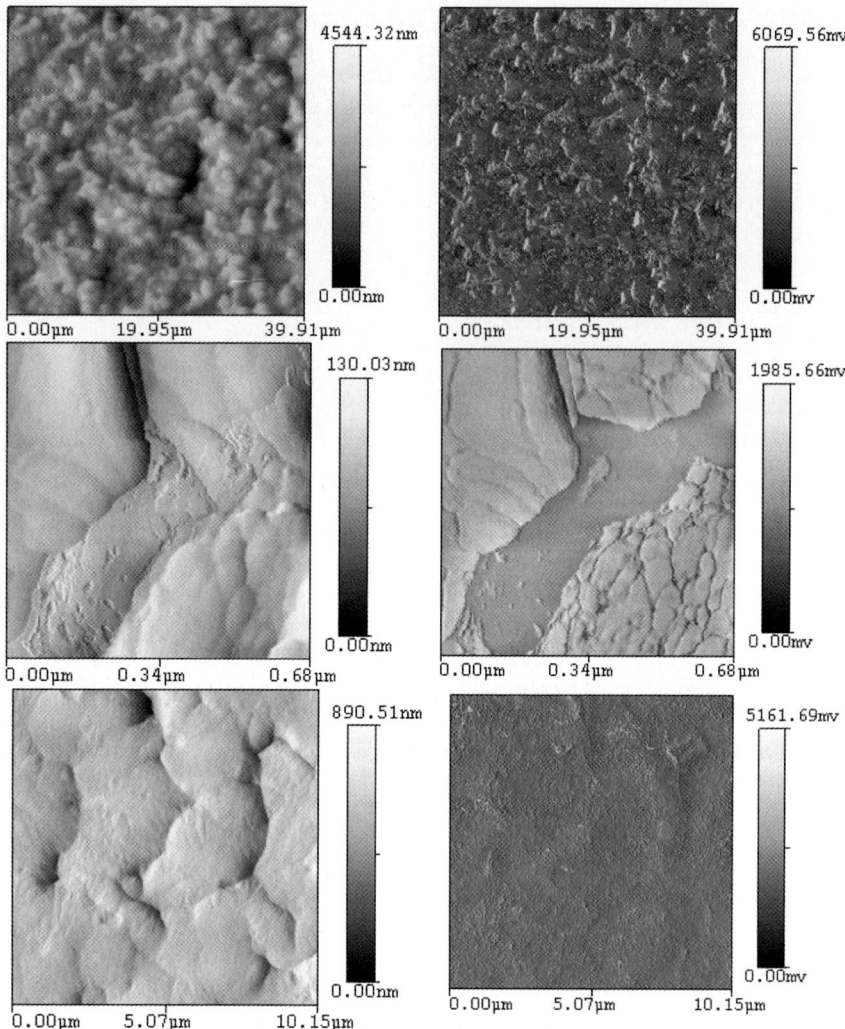

Fig. 23.6. Topography (*left*) and phase (*right*) image of sample A in (**a**) 40 μm; (**b**) 680 nm; (**c**) α phase sample

equal to 732.6 nN. Here k is the force constant of the cantilever, 1.8 nN/nm. According to this figure, the adhesion force of α and mixture of β and γ are almost the same, however sample D, where most phases are γ, has the highest value of adhesion (Fig. 23.7b).

It is believed that the amount of β phase reduces adhesion force due to electrostatic force. The β phase is less electrostatic than that of γ or α. The non-polar α phase sample, on the other hand, has a high adhesion force due to the effect of electrostatic

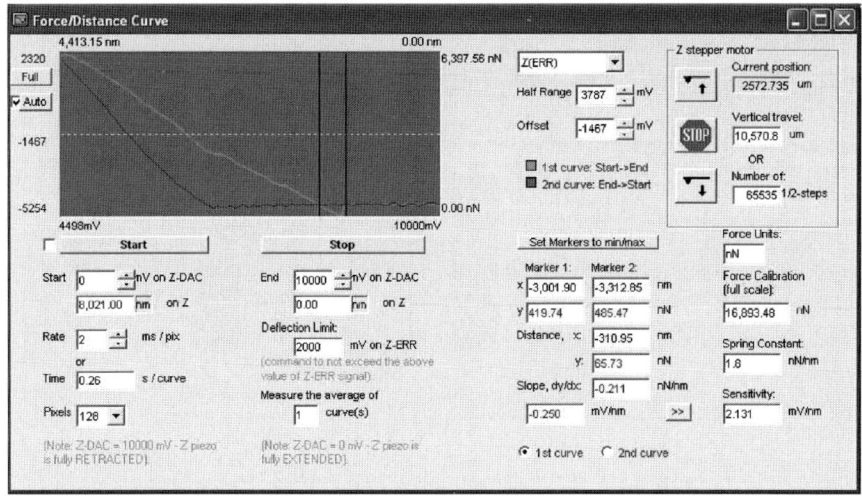

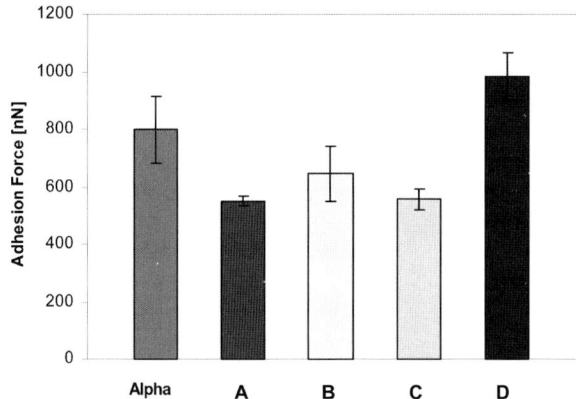

Fig. 23.7. (**a**) Force-displacement curve of sample A. (**b**) Plot of adhesion force of PVDF samples

force. The existence of a high electrostatic force from a non-polar alpha phase has been observed in the past.

The friction values of PVDF samples were measured using the L-R deflection mode of LFM. Figure 23.8 is the comparison of friction values. It shows that the samples A and B, which contain a high amount of β phase, provide a high value of friction. The mixed phases of β and γ lead to a high roughness value compared with samples where a single phase dominates such as α alone. The friction of samples with mixed phases is high compared to samples with only a single phase. Overall, a PVDF sample with a pure beta phase has either a low adhesion or a low friction value.

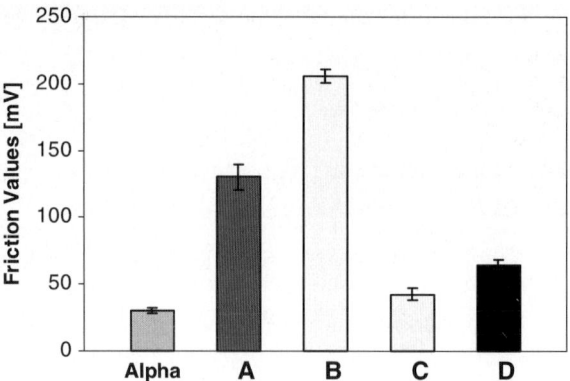

Fig. 23.8. Comparison of friction values through AFM

23.3.4
Nano-Piezoelectricity

As mentioned earlier active materials present interesting and challenging behavior during AFM analysis. This uniqueness can be detected using the AFM with in situ monitoring. A PVDF sample was metalized by sputtering a Ni/Cu film on both sides. The sample was mounted on an AFM (PNI) holder with one loose end. The charge signal of the PVDF due to mechanical bending can be measured directly through the AFM reading, as shown in Fig. 23.9. In this figure, the y-axis is the output voltage recorded by the AFM and the x-axis is time in seconds. Each peak shown in this figure represents one bending motion. Under mechanical stress (bending), a charge

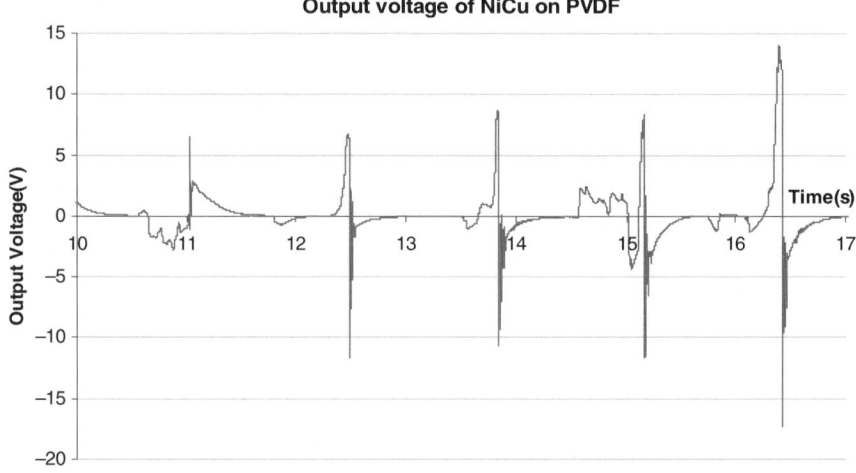

Fig. 23.9. In-situ observation using an AFM of piezoelectricity under bending motion

output was generated immediately with the motion and was released right after. The highest output charge reached 13 mV while the average is around 7.

The microstuctures of the PVDF and surface morphology were analyzed using an AFM (PNI). Images are shown in Fig. 23.10. Of the four images shown, the height images are the left two and phase images are on the right. The height images represent the surface topography. The phase images indicate varying phases of the material. From the figures shown, material grains are around 5 μm in size. In the same figure, the two subfigures on the top were made with no electrical potential applied. The bottom two subfigures were obtained when 5 V was applied across the thickness of the material. It is interesting to see that when an electrical voltage is applied, the space between grains decreases. As highlighted by the circles in the figure, areas of uneven surface height seem to be "squeezed" due to the applied potential. This "squeeze" function is due to the dipole alignment inside the piezo-material. Under an applied potential, the dipole alignment affects interactions of individual grains leading to the expansion of the material. This is thus seen on the surface.

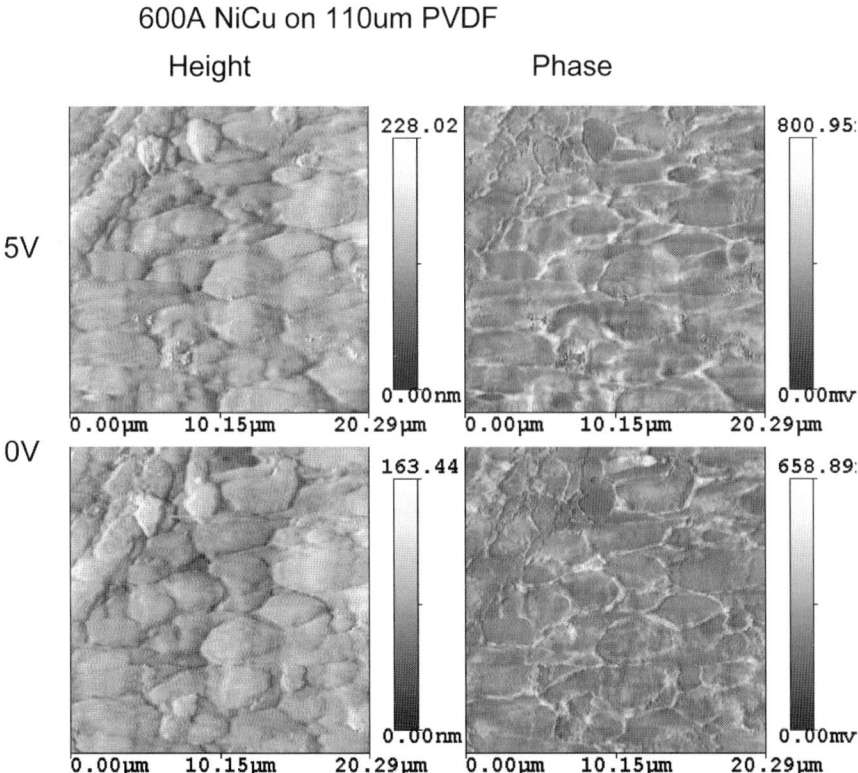

Fig. 23.10. AFM surface analysis of PVDF under the influence of applied voltage [35]

(a)

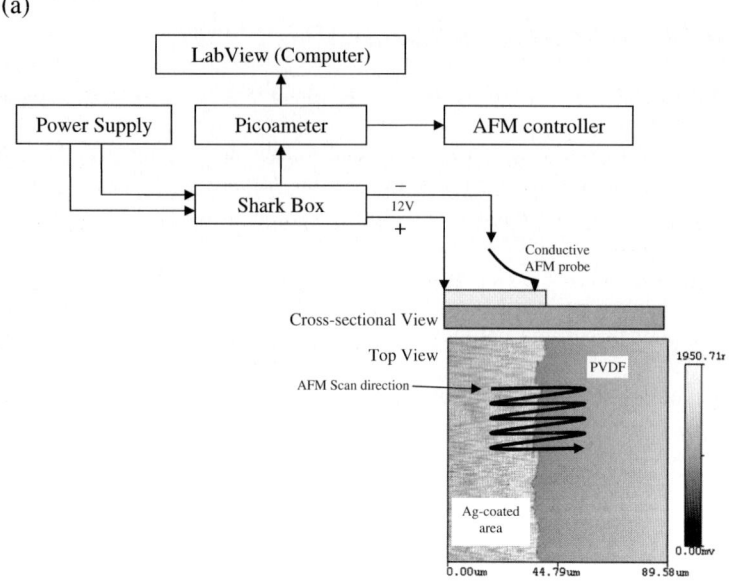

(b)

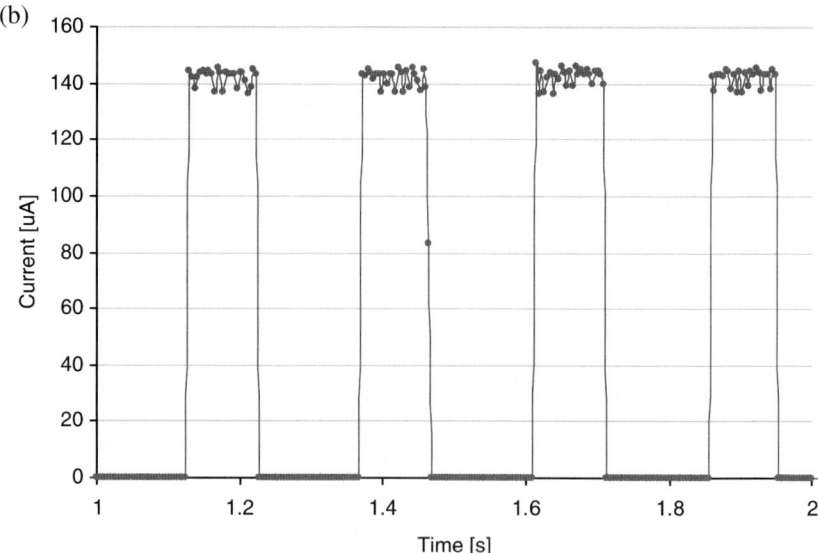

Fig. 23.11. Electrical conductivity measurement for a PVDF sample. (**a**) The simple diagram of the experimental setup to measure the conductivity of PVDF. (**b**) Current output through the Ag-coated area on the PVDF sample

23.3.5
Conductivity Measurement of PVDF Samples

AFM can also measure the conductivity of a material. We have recently designed and built a special configuration, as shown in Fig. 23.11. In this figure, Fig. 23.11a illustrates the AFM setup to measure the conductivity. The setup contains an external power supply, a picoameter, a labview PC system, and a shark box. The shark box functions as a splitter. It distributes electrical potentials and passes a current from samples to a picoameter.

A silver film was coated on a half area of a PVDF sample. Electrical potentials (12 V) were applied. The positive end was connected to the Ag layer on the PVDF sample, and the negative one to an AFM conductive probe. The AFM probe was scanned by reciprocating between the Ag-coated area and PVDF regions. A current was generated during scanning when the probe was above the Ag-layer. Current flows to the picoameter displaying and recording its amplitude. The amplitude of the generated current depends on the conductivity of the surface as shown in Fig. 23.11b. Here the y-axis is the amplitude of the current and x-axis is the scanning time. The average amplitude of generated current is $140\,\mu A$.

The conductivity measurement can also be presented in a map form. The AFM image in Fig. 23.11a is actually a conductivity image. The left area shows a higher output than the right. The current curve in Fig. 23.11b corresponds to the scanning step in Fig. 23.10a.

23.3.6
Time-Dependent Study

The change of piezo-properties can also be analyzed using an AFM. In one of our recent studies using the AFM, it was found that the PVDF has a phase transformation due to stress and this transformation is time dependent [36]. Using an AFM (PNI), the surface morphology and phase images of each deformed PVDF film were observed. The close-contact mode was used with SiN_3 tips. For a quality image, the scanning rate was set at 0.5 Hz with 512 resolutions.

Figure 23.12 shows the AFM results. In the figure, to the left are 3D views, the middle ones are surface morphology images, and the right images show the phases. The initial AFM scan was performed on the samples as soon as bending was applied, i.e. $T = 0$ min. Maintaining the same bending deformation, the AFM scan was repeated at time (T) equal to 6, 13, 25, 35, 43, 61, and 68 min. Figure 23.11a is the reference sample prior to bending. The 3D (left), surface morphology (middle), and phase (right) images are shown with a scanning area of $20 \times 20\,\mu m$. The highest peak in the scanned region is 123.29 nm. Immediately after the same sample was bent, the AFM scan was performed on a smaller area of $10 \times 10\,\mu m$ at the center region of the Fig. 23.12a. Results are shown in Fig. 23.12b displaying a peak with a height of 750.51 nm. The phase image shows island-like structures. When compared with Fig. 23.12a, both the topographic and phase images confirm there is a drastic surface microstructure change due to bending deformation. In order to track the change, a series of scans were conducted at time points (T) of 6, 13, 25, 35, 43, 61, and 68 min.

Figure 23.12c ($10 \times 10\,\mu$m) shows that after 6 min, the upper island-like structures were dwindling, but the lower ones were enlarging. The fringes between these island-like structures are less distinctive than that in Fig. 23.12b. In the phase image, two piles of "islands" came apart and the upper one eventually disappeared (Figs. 23.12d–h).

At the 13th minute, shown in Fig. 23.12d ($10 \times 10\,\mu$m), the upper island-like structures continued to shrink and the lower counterparts showed a decreasing trend; the fringes became clearer. The transformation in phase is clearly revealed by comparing the phase images of Figs. 23.12c and d. The rest of the AFM scans, given in Figs. 23.12e through i, showed continued dwindling with time. As time went on, the protruding area on the surface was gradually relaxed. A signal shift was expected after each measurement. However, the change observed in relative peak height was not related to the shift. Our results indicated that under deformation, the

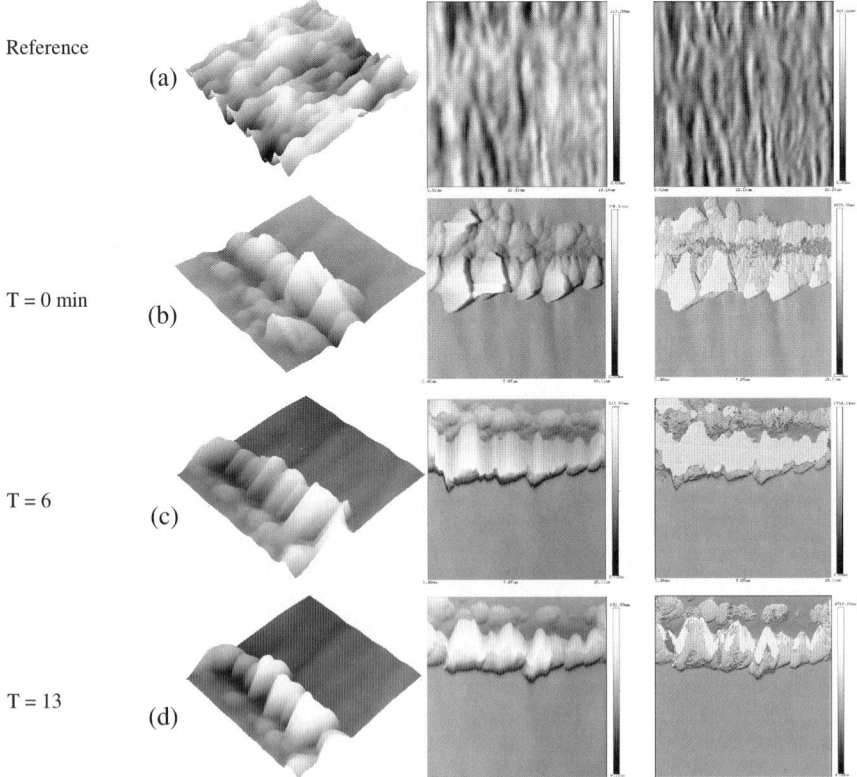

Fig. 23.12. AFM scans of PVDF film under deformation at (**a**) Reference sample without deformation; (**b**) $T = 0$ min; (**c**) $T = 6$ min; (**d**) $T = 13$ min; (**e**) $T = 25$ min; (**f**) $T = 35$ min; (**g**) $T = 43$ min; (**h**) $T = 61$ min; and (**i**) $T = 68$ min. *Left*: 3D images; *Middle*: surface morphology images; *Right*: phase images

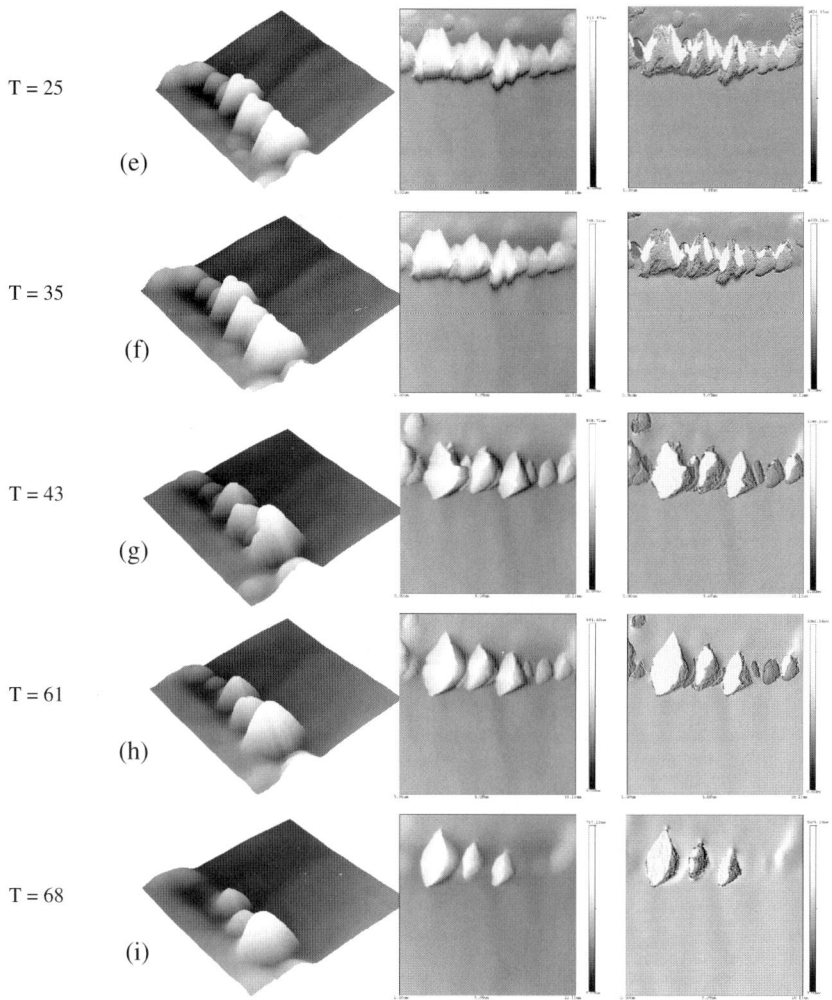

Fig. 23.12. (continued)

PVDF sample underwent surface morphology and phase change. The change was time-dependent and possibly related to a phase transformation.

This chapter discussed new AFM analysis of an active piezoelectric polymer. Some recent results were presented herein. The PVDF piezoelectric polymer is typical for a group of active materials. Their special requirement for AFM analysis apparently furthers the application of AFM. The conductivity measurement and time-dependent study were achieved based on this aspect. Future development of the AFM-based technique depends on the advancement of materials.

Acknowledgments. Funding was provided by NSF 0515930. This chapter was reviewed by Dr. Paul West from the Pacific Nanotechnology Inc. and Dr. Sergei Magonov from Agilent Inc.

References

1. Gautschi G (2002) Piezoelectric sensorics. Springer, Berlin.
2. Taylor GW, Gagnepain JJ, Meeker TR, Nakamura T, Shuvalov LA (1985) Piezoelectricity, Gordon and Breach Science Publishers, New York.
3. Ikeda T (1990) Fundamentals of piezoelectricity. Oxford University Press, Oxford.
4. Wang TT, Herbert JM, Glass AM (1988) The applications of ferroelectric polymer. Blackie, New York.
5. Arnau A (2004) Piezoelectric transducers and applications, Springer, Berlin.
6. Singh J, (2005) Smart electronic materials, Cambridge University Press, Cambridge.
7. Yoo M, Frank CW, Mori S, and Yamaguchi S (2004) Chem Mater 16:1945–1953.
8. Kawashima SYU, Nishida M, Ueda I, Ouchi H (1983) J Am Cer Soc 66(6):421–423.
9. Record MC, Goiffon A, Giuntini JC, Philippot E (1990) J Matls Sci Lett 9:895–897.
10. Rouquette J, Haines J, Bornand V, Pintard M, Papet P, Bousquet C, Konczewicz L, Gorelli FA, Hull S (2004) Phys Rev B 70:014108.
11. Bauer F (1991) Ferroelectrics 115:247–266.
12. Li JC, Wang CL, Zhong WL, Zhang PL (2002) Appl Phys Lett 81:2223.
13. Sobhani H, Razavi-Nouri M, Yousefi AA (2007) J Appl Polym Sci 104:89.
14. McGrath JC, Ward IM (1980) Polymer 21:855.
15. Davis GT, McKinney JE, Broadhurst MG (1978) J Appl Phys 49:4998.
16. Naegele D, Yoon DY, Broadhurst MG (1978) Macromolecules 1:1297.
17. Das DB, Doughty GK (1977) Appl Phys Lett 31:585.
18. Weinhold S, Litt MH, Lando JB (1980) Macromolecules 13:1178.
19. Lovinger AJ (1981) Macromolecules 14:322.
20. Vinson HB, Jungnickel BJ (1998) Ferroelectrics 216:63.
21. Binnig G, Quate CF, Gerbre Ch (1986) Phys Rev Lett 56:930.
22. Binnig G, Rohrer H (1985) Sci Am 253:50.
23. Albrecht TR, Quate CF (1987) J Appl Phys 62:2599.
24. Güthner P, Dransfeld K (1992) Appl Phys Lett 61:1137.
25. Franke K, Besold J, Haessler W, Seegebarth C (1996) J Vac Sci & Technol B 14:602.
26. Hidaka T, Maruyama T, Saitoh M, Mikoshiba N, Shimizu M, Shiosaki T, Wills LA, Hiskes R, Dicarolis SA, Amano J (1996) Appl Phys Lett 68:2358.
27. Chen XQ, Yamada H, Horiuchi T, Matsushige K (1998) Jpn J Appl Phys 37:3834.
28. Palto S, Blinov L, Dubovik E, Fridkin V, Petukhova N, Sorokin A, Verkhovskaya K, Yudin S, Zlatkin A (1996) Europhys Lett 34:465.
29. Bune A, Fridkin VM, Ducharme S, Blinov L, Palto SP, Sorokin AV, Yudin SG, Zlatkin A (1998) Nature 391:874.
30. Choi J, Dowben P, Pebley S, Bune AV, Ducharme S, Fridkin VM, Palto SP, Petukhova N (1998) Phys Rev Lett 80:1328.
31. Kimura K, Kobayashi K, Yamada H, Horiuchi T, Ishida K, Matsushige K (2003) Appl Phys Lett 82(23):4050–4052.
32. Tai YC, Fan LS, Muller RS (1998) Proc. IEEE MEMS, 1–6.
33. Deng K, Collins RJ, Mehregany M, Sukenik CN (1995) Proc. MEMS 95, Amsterdam, Netherlands, January–February.
34. Trimmer WS (1997) Micromechanics and MEMS: Classic and Seminal Papers to 1990, Wiley-IEEE Press.
35. Lee H, Cooper R, Mika B, Clayton D, Garg R, González GM, Vinson SB, Khatri S, Liang H (2007) IEEE Sensors J 7(12):1698–1702.
36. Wang K, Lee H, Cooper R, Liang H (2008) Time Resolved, Stress Induced, and Anisotropic Phase Transformation of a Piezoelectric Polymer, J App Phys, A, accepted.

24 Quantitative Analysis of Surface Morphology and Applications

Maria Cecília Salvadori

Abstract. The quantitative analysis of surfaces using atomic force microscopy and scanning tunneling microscopy is described. Roughness, correlation length, fractal dimension, and power spectral density are discussed. A number of applications are presented, showing quantitative morphological analyses. The applications include investigations of the dynamic growth of thin films, periodicity of lines nanolithographed by AFM in PMMA, electrical resistivity of nanostructured thin films, and grain size analysis.

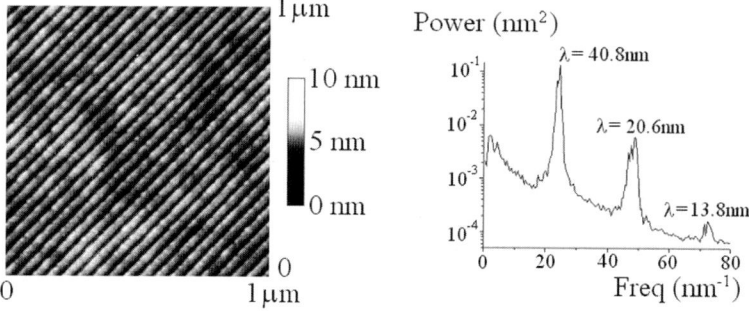

Key words: Roughness, Correlation length, Fractal dimension, Power spectral density, Dynamic growth, Nanolithography, Electrical resistivity, Grain size

24.1 Introduction

Atomic force microscopy (AFM) and Scanning Tunneling Microcopy (STM) are powerful tools for imaging surface morphology at high spatial resolution. The morphological analysis of surfaces can be extremely complex, however, it is not uncommon for these techniques to be only partially employed.

In this chapter we discuss quantitative parameters used to describe or analyze surfaces using AFM and STM. Parameters for quantification of surface morphology are introduced. Roughness is the most widely used parameter for quantitative characterization of surface morphology, but it adds limited information about the surface and is scale dependent, meaning that in terms of scanning probe microscope (SPM) images, the derived roughness depends on the scan size (ℓ). The parameter ξ, defined as *correlation length*, is another important parameter used to quantitatively describe the surface. The area $\xi \times \xi$ is the minimum area of the surface that is representative of the entire surface. The concept of fractal dimension, d_f, is introduced and discussed as another quantitative parameter for surface morphology analysis. But the most complete method to quantify a surface morphologically is the Power Spectral Density (PSD). This technique involves basically a Fourier transform of the surface topography, displaying a graphic of the squared modulus (amplitude) of the Fourier transform (scaled by an appropriate constant) as a function of the morphological wavelength (or frequency). A number of applications are then presented, demonstrating some quantitative morphological analyses. The first application is the dynamic growth of thin films. This approach allows us to investigate various growth processes and associate them with universality classes. In addition, dynamic growth studies provide us with some important parameters for quantitatively describing the film surface, such as the correlation length ξ and fractal dimension d_f. The second application is a periodicity analysis of lines nanolithographed by AFM in PMMA. Nanolithography performed by AFM has been used to generate reproducible and predictable periodic morphology that is mainly dependent on the scan size and the number of scan lines. A PSD characterization provides in this case a powerful method to identify the morphological wavelengths present in the nanolithographed pattern. The third application is a study of the electrical resistivity of nanostructured thin films. The electrical resistivity of thin films is best described by a quantum model, when appropriate conditions are satisfied. In this model, conduction electrons are scattered by the surface morphology and quantitative morphological analysis of the film surface is fundamental to the model. Related to this application, we also explore the resistivity anisotropy that can be induced by film surface morphological anisotropy. Finally, the last application that we consider is the quantitative grain size analysis of surface morphology and its importance in some specific areas. In this context we define and discuss "morphological grain size" and "crystallographic grain size."

24.2
Quantifying Morphology

Roughness is the most commonly used parameter for characterizing surface morphology quantitatively, but surprisingly it provides incomplete information about the surface. To exemplify the limitations of this parameter, let us calculate the roughness of a simple surface (see Fig. 24.1) defined by a sinusoidal profile in the x direction and flat in the y direction. In this case the roughness can be taken as

$$\omega_{RMS} = \sqrt{\frac{\int_{x_1}^{x_2} [f(x)- <f(x)>]^2 dx}{x_2 - x_1}}$$

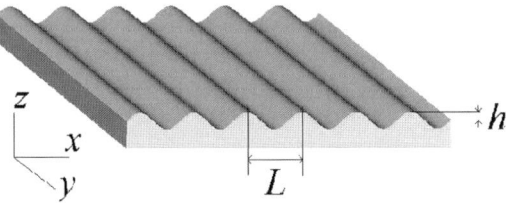

Fig. 24.1. Surface defined by a sinusoidal profile in the x direction and flat in the y direction. The amplitude of the sinusoidal profile is h and its morphological wavelength is L

where $f(x) = h\sin(2\pi x/L)$, h being the amplitude of the sinusoidal profile and L the morphological wavelength. Integrating over an integral number of cycles, we obtain $\omega_{RMS} = {}^{h}/_{\sqrt{2}}$. This result indicates clearly that the roughness of the surface shown in Fig. 24.1 depends only on the amplitude h and is independent of the morphological wavelength L. Embarrassingly, we conclude that two surfaces like that shown in Fig. 24.1, having the same amplitude h but with different wavelengths L, for example with cross sections as shown in Fig. 24.2, will have exactly the same roughness.

Although roughness is thus a parameter carrying limited information, it nevertheless still plays an important role in morphology characterization. Let us consider roughness further.

For a self-affine surface [1–12], as is the case for most "real surfaces," the roughness is scale-dependent. In terms of SPM image acquisition, the roughness depends on the scan size (ℓ). Figure 24.3 illustrates the behavior of self-affine surface roughness as a function of ℓ. We verify that the roughness increases as ℓ increases and that the roughness saturates for $\ell > \xi$. This distance ξ is a parameter named *surface correlation length* [1–13]. The correlation length ξ is a meaningful parameter for the quantitative description of a surface.

As will be seen in what follows, for $\ell << \xi$, the roughness $\omega(\ell, t) \sim \ell^{\alpha}$, where α is a parameter called *growth exponent* [1–12] and is related to the fractal dimension d_f of the surface by $d_f = 3 - \alpha$. The fractal dimension is itself an important parameter for quantitative surface morphology analysis and will be discussed below.

The most complete method to quantify surface morphology is through its Power Spectral Density (PSD) [14]. This technique utilizes a Fourier transform of the surface topography, displaying a graphic of the squared modulus (amplitude h) of the

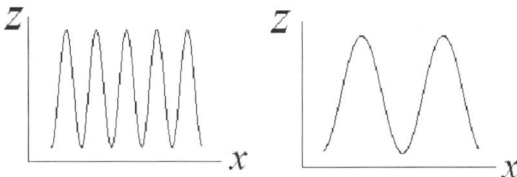

Fig. 24.2. Cross sections of two different surfaces with the same amplitude h and with different wavelengths L. These surfaces have the same roughness $\omega_{RMS} = {}^{h}/_{\sqrt{2}}$

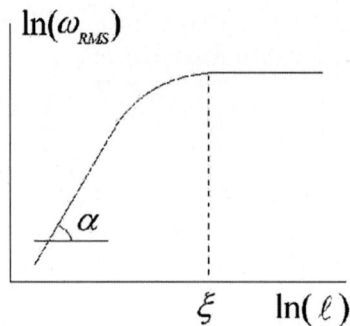

Fig. 24.3. Illustration of the behavior of self-affine surface roughness as a function of the image scan size (ℓ), where ξ is the surface *correlation length* and α is a growth exponent

Fourier transform (scaled by an appropriate constant factor) as a function of the morphological wavelength (or frequency). The PSD is expressed in squared length units (for example nm^2) and can be related to the roughness ω_{RMS} (associated with a given wavelength) and to the amplitude h of the surface Fourier transform by PSD $\propto \omega_{RMS}^2 \propto h^2$. Two examples of PSD analysis are shown in Fig. 24.4. These analyses were performed with the NanoScope IIIA software, version 5.30r3.sr3, from Veeco. For most "real surfaces," the amplitude increases with morphological wavelength, as illustrated in the examples of Fig. 24.4. The PSD analysis presented in Fig. 24.4a was performed with a homogeneous and isotropic rough surface, generating a broad band. Figure 24.4b shows a PSD analysis with pronounced peaks; in this case the surface was anisotropic with parallel lines, defining specific morphological frequencies (or wavelengths) as displayed in the analysis.

In the following we present a number of studies for which quantitative morphological analyses were carried out using the parameters introduced.

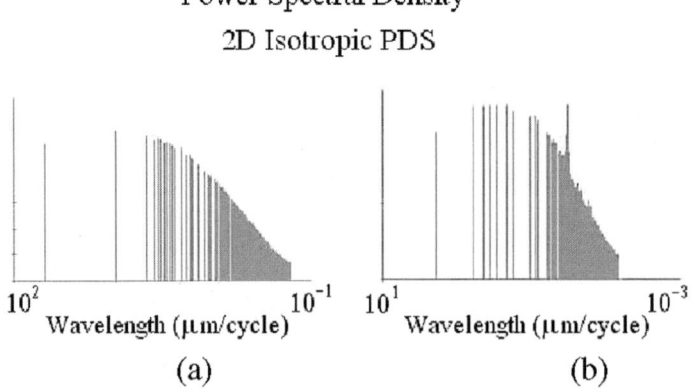

Fig. 24.4. Power Spectral Density (PSD) analysis for two different surfaces. (**a**) PSD for a homogeneous and isotropic surface, generating a broad band. (**b**) PSD with pronounced peaks, indicating the presence of specific morphological frequencies (or wavelengths)

24.3
Applications of Quantitative Morphological Surface Analysis

24.3.1
Dynamic Growth of Thin Films

Film growth by deposition is clearly of great technological importance. Fluctuations in the height $h(\mathbf{x}, t)$, surface location $\mathbf{x}$, and time t can be measured directly using a scanning probe microscope or indirectly by scattering. Analytical and numerical treatments of simple growth models suggest that, quite generally, the height fluctuations have a self-similar character and their average correlations exhibit a dynamic scaling form, named the Family-Vicsek scaling relationship [15]. The roughness ω and dynamic growth exponents α and z defined by this relationship are expected to be universal, depending only on the underlying mechanism that generates the self-similar scaling [16]. The determination of the growth exponents α and z is a fundamental problem of statistical mechanics. Considerable effort, both theoretical and experimental, has been made to investigate the surface growth process. Many references on the subject can be found in the excellent review of Barabási and Stanley [1]. Theoretical discrete models provided a substantial part of the driving force behind early investigations of the surface morphology. Discrete models, numerical simulations, and stochastic differential equations have been used to explain the growth mechanisms on d-dimensional substrates. Such equations typically describe the surface at large scales and times, which means that the short-range scale details are neglected and the focus is only on asymptotic coarse-grained (hydrodynamic) variables. In some sense, for $d = 1$ the growth phenomenon is reasonably well understood. Moreover, for $d > 1$, there are many challenging problems that seem insurmountable. Numerical simulations are generally impracticable or extremely difficult and numerical integrations are questionable, leading to somewhat inconclusive results [1–16]. On the other hand, more detailed and accurate measurements of the growth exponents of thin films are still lacking, and experimental confirmation of dynamic scaling is scarce [16, 17].

To exemplify a typical experimental approach for measuring the growth parameters of thin films, we describe a study of diamond films synthesized by microwave plasma-assisted chemical vapor deposition [18]. The roughness and dynamic exponents of these films were measured using atomic force microscopy.

The equipment used for the diamond film deposition is described in detail elsewhere [19, 20]. The substrate was silicon that had been scratched by 1-μm diamond powder and cleaned in an acetone ultrasonic bath. The following growth parameters were used: 300 sccm hydrogen flow rate (where sccm denotes cubic centimeter per minute at STP), 1.5 sccm methane flow rate (0.5-vol% methane in hydrogen), 70 torr chamber pressure, 820 °C substrate temperature, and nominal 850-W microwave power. The silicon substrate was divided into small pieces and eight films were synthesized with different growth times: 17, 20, 24, 26, 34, 48, 63, and 74 h.

The most economical way to characterize self-affine roughness is by a dynamic scaling form [15, 16]. In many situations, there is no information about the dynamics of the growth and it is not possible to produce surfaces with different sizes. Suppose that the only data available are collected at the final stage of the experiment, consisting of the values of the height $h(\mathbf{x}, t)$ of the surface, at different points $\mathbf{x}$ and times t. In this situation, the scaling of the "local roughness" $\omega_L(\ell, t)$ is studied, defined by [15]

$$\omega_L^2(\ell, t) = \left\langle [h(\mathbf{x}, t) - h_\ell(\mathbf{x}, t)]^2 \right\rangle_{\mathbf{x}}$$

where $L \times L$ is the system size, $\ell \times \ell$ is a window selected on the surface and $h_\ell(\mathbf{x}, t)$ is the average height in this window. The angular brackets $\langle \rangle_{\mathbf{x}}$ denote spatial (over $\mathbf{x}$) and ensemble averages. One can show that $\omega_L(\ell, t)$ obeys the scaling form [1,15,16]

$$\omega_L^2(\ell, t) \sim \ell^\alpha f(t/\ell^z), \tag{24.1}$$

where $f(u)$ is the scaling function of the argument $u = t/\ell^z$ and $z = \alpha/\beta$. The parameters α, β, and z are expected to be universal parameters, named growth exponents. For very small times, $u \ll 1$, we have $\omega_L(\ell, t) \sim t^\beta$ when the roughness grows as t^β and the different sites of the surface are practically independent. As time increases, different sites on the surface begin to be correlated. The typical distance over which the heights "know about" each other, the characteristic distance over which they are correlated, is called the correlation length and is denoted by $\xi(t)$, a parameter already mentioned, which increases as $\xi(t) \sim t^{1/z}$. When correlations are significant we have $\omega_L(\ell, t) \sim \ell^\alpha$ for $\ell \ll \xi$; for $\ell > \xi$ the roughness saturates, that is, $\omega_L(\ell, t) \cong \text{const} \sim \xi^\alpha$. For these conditions [1–12] the fractal dimension d_f of the film surface is given by $d_f = 3 - \alpha$. In the very long time limit, that is, as $u \to \infty$, it is expected that the roughness reaches its maximum (saturation) value $\omega_{\text{sat}}(L) \sim L^\alpha$. In what follows, omitting for simplicity the index L, the roughness will be indicated by $\omega(\ell, t)$; ℓ will be measured in micrometers and ω in nanometers.

The roughness and fractal dimensions of the seven diamond films were measured using an AFM, a Veeco NanoScope IIIA. Ten different regions of each sample were analyzed, each of size $160 \times 160 \,\mu\text{m}^2$. Using the AFM zoom facility, these $(160 \times 160) \,\mu\text{m}^2$ regions were divided into smaller regions (windows) of size $\ell \times \ell \,\mu\text{m}^2$, with ℓ ranging from 3 up to $160 \,\mu\text{m}$, and their local roughness $\omega(\ell, t)$ measured. It was not possible to analyze the 74-h sample due to the formation of large microcrystalline diamond grains, about $8.5 \,\mu\text{m}$ in diameter; it was found that $\omega = 600 \,\text{nm}$, exceeding the z limit of the AFM. The average grain sizes were about 2.8 and $7.0 \,\mu\text{m}$ for 17 and 63 h, respectively.

Figure 24.5 shows typical values of $\log_{10} \omega(\ell, t)$ as a function of $\log_{10}(\ell)$ for different growth times t. One can verify that for $t < 34 \,\text{h}$, the correlations between different sites of the surface are small and the α power law growth is not well defined. To determine the exponent α only the roughness of the 63-h sample was taken into account. Twelve different regions of this sample were analyzed. Figure 24.6 shows $\log_{10}[\omega(\ell, t = 63 \,\text{h})]$ as a function of $\log_{10}(\ell)$ for these 12 regions. It can clearly be seen that $\log_{10}(\omega)$ as a function of $\log_{10}(\ell)$ shows a power law growth, appearing as a straight line from $3 \,\mu\text{m}$ up to $\ell = \xi \cong 15 \,\mu\text{m}$, where ξ is the correlation length. For $\ell \geq \xi$, the roughness saturates, that is, $\omega(\ell, t) \cong \text{const}$ according to Eq. (24.1), and is given by $\omega \cong 400 \,\text{nm}$. The following α values were found for a straight line best fit [21] for the 12 different regions: 0.36 ± 0.06, 0.49 ± 0.04, 0.51 ± 0.04, 0.51 ± 0.08, 0.51 ± 0.03,

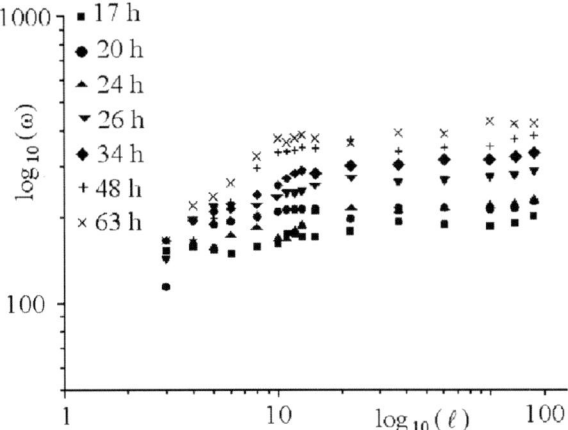

Fig. 24.5. Roughness $\omega(\ell, t)$ as a function of the length ℓ and of the growth time t; $\omega(\ell, t)$ is measured in nanometers, ℓ in micrometers, and t in hours

0.53 ± 0.03, 0.53 ± 0.06, 0.58 ± 0.07, 0.64 ± 0.08, 0.66 ± 0.11, 0.66 ± 0.08, and 0.70 ± 0.10. To analyze the goodness of this straight line best fit we used a standard statistical procedure [21] calculating the χ_v^2. The χ_v^2 values range from 0.97 up to 0.99, showing that the fit describes the data well. From the above α values we conclude that $\alpha = 0.56 \pm 0.09$.

The fractal dimension d_f of the 63-h sample was determined using the fractal algorithm of the AFM software, where the surface was analyzed as described below.

Let us now describe the surface fractal dimension [1–12]. A surface S can be measured by covering it with squares of linear dimension ℓ and area ℓ^2. $N(\ell)$ squares are needed to cover the surface, so $S(\ell) = N(\ell)\ell^2$. One might at first expect that for

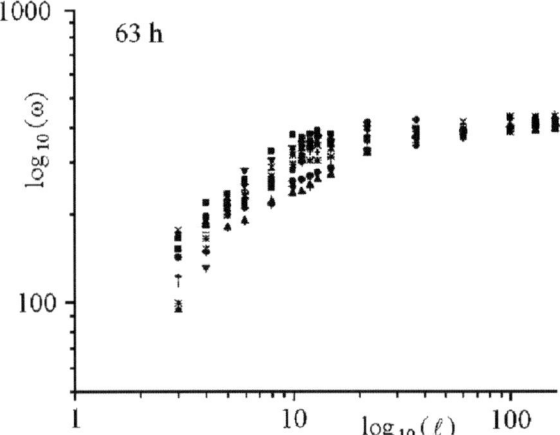

Fig. 24.6. Roughness $\omega(\ell, t = 63\,\text{h})$ as a function of the length ℓ. The 12 different symbols for each curve correspond to the 12 different regions analyzed

a surface $N(\ell) \sim \ell^{-2}$, since the area of the surface does not change if we change the unit of measurement of ℓ. But for fractal surfaces we have $N(\ell) \sim \ell^{d_f}$. Surfaces with $d_f > 2$ are fractal surfaces, where d_f is its fractal dimension.

Only the (160×160)-μm^2 images were considered. The following values were found for d_f: 2.43, 2.45, 2.46, 2.46, 2.47, 2.47, 2.48, 2.49, 2.49, 2.49, 2.50, and 2.50. So d_f would be given by $d_f = 2.48 \pm 0.02$ and consequently, as $\alpha = 3 - d_f$, α is given by $\alpha = 0.52 \pm 0.02$. This result is in good agreement with that obtained using the roughness measurement.

Figure 24.7 shows $\log_{10} \omega(t)$ as a function of $\log_{10}(t)$ for $t = 17, 20, 24, 26, 34,$ 48, and 63 h. Since for small growth times [15] we must have $\omega \sim t^\beta$, determination of the growth exponent β takes into account the roughness of the samples only for small growth times, that is, 17, 20, and 24 h: $\omega(17\,\text{h}) = 205.7 \pm 6.9$ nm, $\omega(20\,\text{h}) = 219.63 \pm 7.2$ nm and $\omega(24\,\text{h}) = 232.0 \pm 4.9$ nm. These values were estimated from the average $\omega(t) = [\omega(100\,\mu m, t) + \omega(130\,\mu m, t) + \omega(160\,\mu m, t)]/3$. With a straight-line best fit [21], the growth exponent is given by $\beta = 0.34 \pm 0.02$. Also in this case, estimating the variances and χ_v^2 indicates that $\chi_v^2 = 0.99$, showing that a straight-line fit provides a good description of the data.

Taking into account the above values for α and β, we obtain $z = \alpha/\beta = 1.65 \pm 0.28$. Consequently $\alpha + z = 2.21 \pm 0.30$, which satisfies the condition $\alpha + z = 2$ within experimental error. This could be an indication that the growth of diamond films is governed by the Kardar–Parisi–Zhang (KPZ) equation [1–12, 16, 22, 23]. However, according to the KPZ predictions, for $d = 2$ ballistic deposition [16], $\alpha \cong 0.38$ and $\beta \cong 0.24$. Thus $\alpha \cong 1/2$ and $\beta \cong 1/3$, which are the growth exponents predicted by the KPZ equation for $d = 1$. This could provide support for the superuniversality conjecture [24]. On the other hand, the above discrepancies could be a result of the formation of diamond microcrystallites, and in this case a somewhat different equation would govern the diamond film growth, as proposed by Hwa, Kardar, and Paczuski [25].

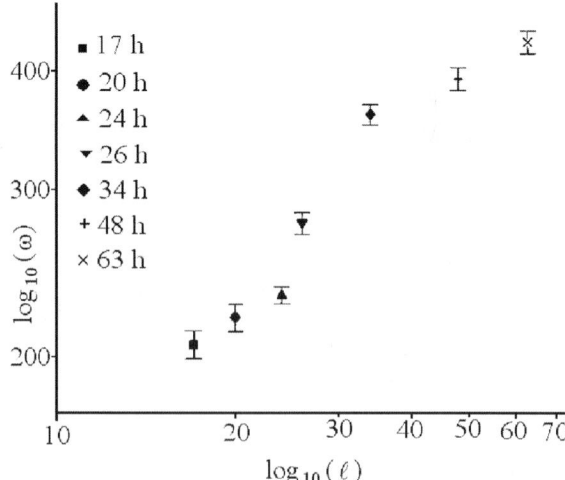

Fig. 24.7. Roughness $\omega(t)$ as a function of time t for $t = 17, 20, 24, 26, 34, 48,$ and 63 h

Summarizing, dynamic growth of thin films allows us to investigate different growth processes and associate to universality classes. With this approach one can find a number of studies in the literature about growth dynamics of thin films [2–12]. In addition, a dynamic growth study yields important parameters for describing the film surfaces quantitatively, such as correlation length ξ and fractal dimension d_f.

24.3.2
Periodicity Analysis of Lines Nanolithographed by AFM in PMMA

An AFM can be used for imaging and also for creating surface nanostructures; this technique is called scanning probe lithography (SPL). A number of SPL techniques have been developed, based on nanomanipulation [26–28], mechanical modification [29–36], thermomechanical writing [37, 38], local oxidation [39–43], electron exposure of resistance [44–46], dip-pen nanolithography (DPN) [47–49], electrochemical dip-pen nanolithography (E-DPN) [50,51] and rapid direct nanowriting of a conducting polymer by electrochemical oxidative nanolithography [52].

Polymethylmethacrylate (PMMA) has been widely used as a photo- or electropositive resist for nano- and microscale fabrications. Thin, uniform layers of PMMA are deposited on a substrate and selected areas exposed to ultraviolet light or to a scanning electron beam. The irradiation creates scissions in molecular chains in the exposed areas of the polymer, decreasing their average molecular weight and allowing them to be removed through a development process. For nanodevice fabrication, it can be advantageous to form desired nanostructures on the PMMA surface after it has been irradiated and developed.

In the following we describe the application of an AFM nanolithography technique to mechanically modify PMMA surfaces previously irradiated and developed [29]. A quantitative analysis using AFM images will be presented.

Glass microscope slides were cut into $14 \times 5\text{-mm}^2$ pieces, cleaned and baked at 150 °C for 10 min to remove residual humidity, and PMMA then deposited on the samples using a spin coater. The samples were then baked again at 180 °C for 20 min to evaporate the polymer solvent. PMMA ARP671.06 was used, with molecular weight of 950,000 -g/mol and concentration of 6% in chlorobenzene, from Allresist. The average PMMA film thickness was 500 nm.

A selected area of the PMMA was electron beam scanned in a scanning electron microscope (JEOL model JSM-6460 LV) using an e-beam nanolithography system (Nanometer Pattern Generation System, NPGS) to create micropatterns that allowed subsequent identification and location of smaller nanostructures to be formed by AFM. The e-beam lithography was performed with a 30 kV, 50-pA electron beam. The applied electron dose was $225\,\text{C/cm}^2$ with an exposure time of 362 μs at each scan point. The patterns consisted of two pads, each $20 \times 100\,\mu\text{m}^2$. The samples were then immersed in a developing solution of one-part methyl isobutyl ketone (MIBK): three-parts isopropyl alcohol (IPA) for 2 min and rinsed in IPA for 30 s.

The AFM used was a NanoScope IIIA with a tapping-mode probe. The procedure was to first obtain a tapping-mode image of a selected region, then to perform contact nanolithography, and finally to obtain a new tapping-mode image to visualize the modified surface. To image a surface using the NanoScope IIIA in AFM tapping

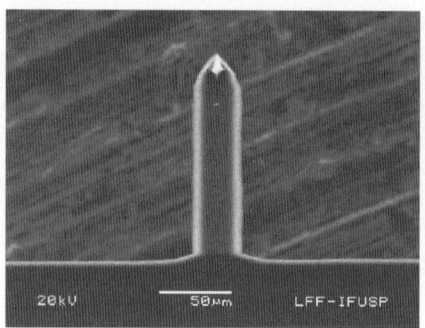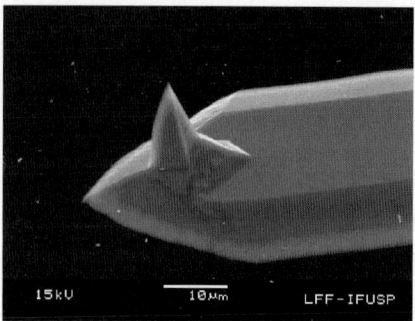

Fig. 24.8. SEM images of a typical AFM probe used for nanolithography. In the left image, low magnification, one can see the cantilever with a small pyramidal tip. In the right image, higher magnification, the pyramidal silicon tip can be seen

mode, the tip scans a maximum of 512 lines in a square area, interacting physically with 512 points along each of these lines. Importantly, note that the tip touches the surface only gently, and the polymer is not modified during image acquisition.

SEM images of a typical AFM probe used is shown in Fig. 24.8. The dimensions of each cantilever were measured (average dimensions were 120 μm long, 33 μm width, and 2.5 μm thick), allowing calculation of its elastic constant [53]. A force plot was then obtained so as to precisely determine the cantilever deflection to be used in the lithography process. A typical deflection used was about 70 nm. In this way it was possible to define the force between the tip and the polymer. The force range was 0.5–1.0 μN.

Following calibration, the tip was engaged on the surface in contact mode and nanolithography initiated. The mechanism is similar to image acquisition, but with the scanning tip maintained in constant contact with the surface so as to scribe each scan line. The distance between two scan lines is the ratio between the scan size and the number of scan lines. For example, using a 20-μm scan size and 512 scan lines, the distance between the scan lines is about 39 nm. Note that for a Nanoscope IIIa the number of lines in each scan can be 512 or 256 or 128. In this study, the region was scanned just once, using 1-Hz scan rate and 0° scan angle. Finally, a tapping-mode image of the nanolithographed area was acquired, using a 45°-scan angle for better visualization of the surface morphology. The scan size, number of lines, and distance between lines for the nanolithographed patterns are shown in Table 24.1.

A typical AFM image of a nanolithographed PMMA region is shown in Fig. 24.9. Specifically in this case, the scan size and number of lines were 20 μm and 512 lines, respectively. Consequently the distance between scan lines was 39.1 nm. The average morphological wavelengths present in this pattern were 39.2 nm and about half of this value, 19.4 nm. The reason for the occurrence of wavelengths of half the distance between scan lines is that the tip does not scan a single straight line, but technically follows a zigzag pattern. In other words, for each line the tip scans from left to right and from right to left, totaling two scratches for each line.

To characterize the morphological wavelengths present in each pattern formed by nanolithography we use the Power Spectral Density [54]. The PSD of the image

Table 24.1. Scan size, number of lines, and distance between lines for the nanolithographed patterns

Scan size (μm)	Number of lines	Distance between lines (nm)
10	512	19.5
20	512	39.1
20	256	78.1
30	512	58.6
40	512	78.1

shown in Fig. 24.9 is given in Fig. 24.10. The x-axis corresponds to the inverse of the wavelength (λ) and the y-axis is the PSD. The PSD is given in nm^2 and can be related to the roughness Δ (associated with each wavelength) and to the amplitude h of the surface Fourier transform as **PSD** $\propto \Delta^2 \propto h^2$ [55]. The wavelength $\lambda = 13.8$ nm in Fig. 24.10 is related to the tip radius due to the tip convolution, and will not be considered.

The sum of the PSD of nine different samples nanolithographed with 20-μm scan size and 512 lines is shown in Fig. 24.11. From this spectrum one can extract the average morphological wavelengths present for 20-μm scan size and 512 lines: 39.2 nm and 19.4 nm. The wavelength $\lambda = 12.9$ nm in Fig. 24.11 is related to the tip radius.

The mechanism of the process is to move material from the region where the tip scratches the surface to the side of this region. In this way, the pattern profile obtained is higher than that of the original surface, as illustrated in Fig. 24.12. Note that in this last figure, the z-scale is 30 nm/division and the x and y-scale are 200 nm/division, so the image is expanded in the vertical direction to emphasize the image profile.

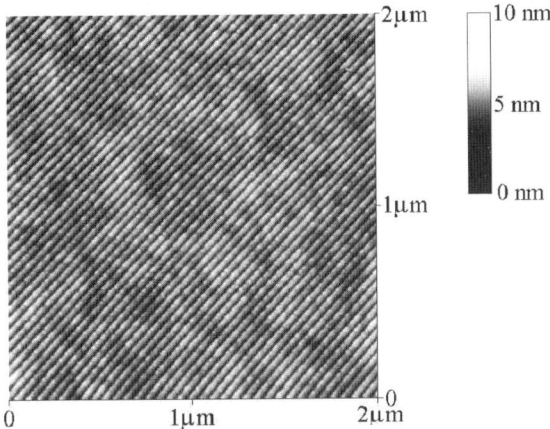

Fig. 24.9. Typical AFM image of a nanolithographed PMMA region. The scan size was 20 μm and the number of lines was 512

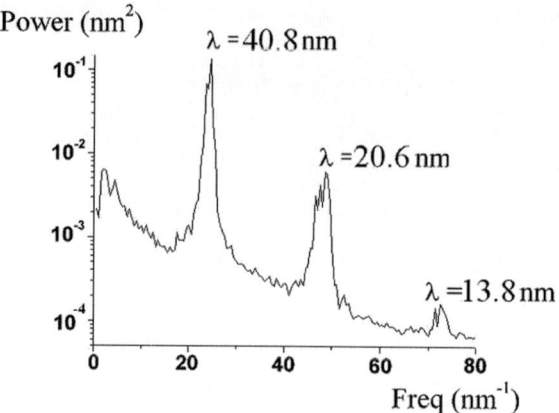

Fig. 24.10. Power spectral density of the image shown in Fig. 24.9

In some samples, the nanolithographed region shows a fine structure as can be seen in Fig. 24.13. In the AFM image of Fig. 24.13, aligned particles can be seen between the lines produced by the tip scratches. This image suggests that particles present in the developed PMMA [56, 57] are aligned during the nanolithography process.

Five samples were prepared nanolithographing the PMMA surface with 40-μm scan size and 512 lines and thus with distance between scan lines 78.1 nm. In all samples the morphological wavelengths obtained were: 78, 39, 26 and 20 nm (see Fig. 24.14). Note that 39 nm is half the distance between scan lines and 20 is about a quarter of the distance between scan lines. The 26-nm wavelength is probably related to fine structure of the developed PMMA.

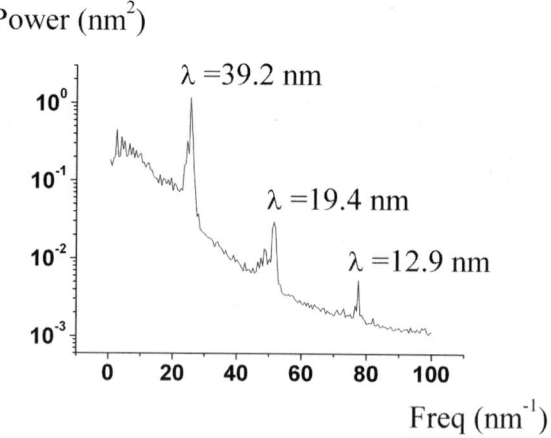

Fig. 24.11. PSD sum of nine different samples nanolithographed with 20-μm scan size and 512 lines

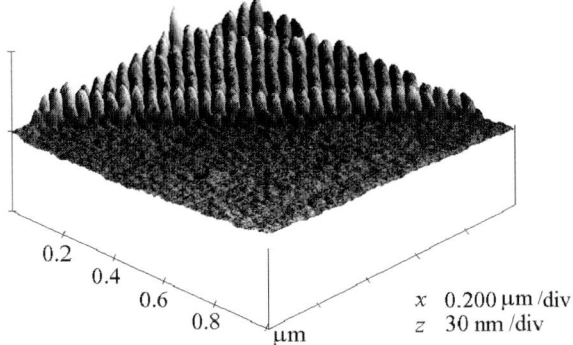

Fig. 24.12. AFM image of a border of the PMMA region nanolithographed using 20-μm scan size and 512 lines

In addition, five samples were nanolithographed with 30-μm scan size and 512 lines, generating distance between scan lines of 58.6 nm. In all samples a morphological wavelength of about 53 nm was present. In three of the samples a wavelength around 26 nm was observed, about half of the distance between scan lines.

Several more samples were prepared with different parameters, such as 10-μm scan size and 512 lines, and 20-μm scan size and 256 lines. Table 24.2 summarizes all the experimental results, showing the parameters used and the morphological wavelengths obtained.

Fig. 24.13. AFM image of a nanolithographed PMMA with 20-μm scan size and 512 lines. A fine structure is present, where aligned particles can be observed. The image width is 500 nm

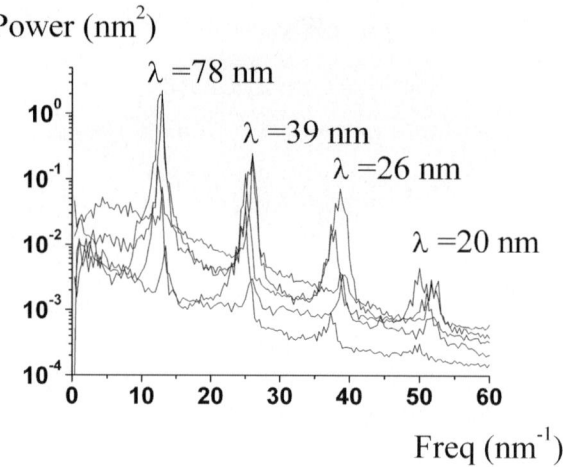

Fig. 24.14. PSD of five different samples nanolithographed with 40-μm scan size and 512 lines

A possible reason for morphological wavelengths of a quarter of the distance between scan lines is illustrated in Fig. 24.15. In the process of moving material from the region where the tip scratches the surface to the side of this region, depressions are created between the lines. In this way, the process introduces a new periodicity with wavelength of about one-fourth of the distance between lines (actually one-half of the distance between scratched lines).

Summarizing, nanolithography performed using an AFM generated repro-ducible and predictable periodicities, which were mainly dependent on the scan size and the number of scan lines. Power Spectral Density characterization was a powerful method for identifying the morphological wavelengths present in the nanolithography patterns.

Table 24.2. Summary of results. Scan size, number of lines, dis-tance between lines, and morphological wavelengths present in the patterns formed

Scan size (μm)	Number of lines	Distance between lines (nm)	Morphological wavelength (nm)
10	512	19.5	18.7
20	512	39.1	39.2, 19.4
20	256	78.1	76.9, 39.2, 26.0, 19.6
30	512	58.6	53, 26
40	512	78.1	78.7, 39.2, 25.9, 19.6

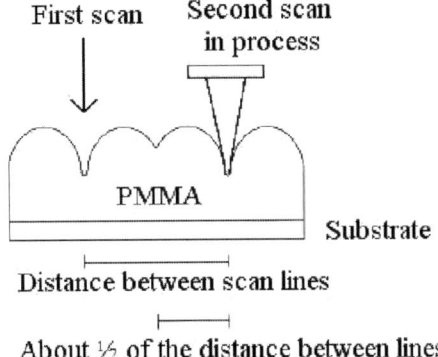

First scan

Second scan
in process

PMMA

Substrate

Distance between scan lines

About ½ of the distance between lines

Fig. 24.15. Schematic suggesting the origin of wavelengths of one-quarter the distance between scan lines. The figure is not in scale

24.3.3
Electrical Resistivity of Nanostructured Thin Films

The electrical resistivity (ρ) of thin films involves quantum effects when two conditions are satisfied. The first is that the film thickness (d) must be smaller than the electronic mean-free-path (ℓ_0), and the second condition is that the energy-level quantization must be enhanced in the direction along the film thickness d. This last condition occurs when we have a small number of Fermi subbands, given by

$N \cong d \left(\frac{3n}{\pi} \right)^{\frac{1}{3}}$, where n is the number of free electrons per unit volume.

For the specific case of platinum and gold films, these two conditions can be satisfied for very thin films [58], as shown in Table 24.3. The film thicknesses $d \leq \ell_0$ and the number of Fermi subbands, for the thickness used, was small (between 5 and 47). Thus, quantum effects are expected in determining the electrical resistivity of these thin films.

In the quantum model the calculation of the conductivity ($\sigma = 1/\rho$) as a function of the film thickness is done considering energy $E_v = \frac{\pi^2 \hbar^2 v^2}{2md^2}$ quantization of the conduction electrons in the direction along the film thickness d in the Boltzmann transport equation, taking into account the distribution function of the Fermi sub-

Table 24.3. Conditions for applicability of a quantum formalism for platinum and gold thin films

Metal	Conditions for quantum formalism			
	Condition (I)		Condition (II)	
	d (nm)	ℓ_0 (nm)	n (m^{-3})	N
Pt	$1.3 \leq d \leq 11.7$	~ 10	6.6×10^{28}	$5 < N < 47$
Au	$1.8 \leq d \leq 10.5$	~ 50	5.9×10^{28}	$7 < N < 41$

bands ν and the scattering potential $U(x, y)$ due to the film morphology. Then the conductivity generated by the surface is given by [58]:

$$\sigma_s(d) \approx \frac{e^2}{\hbar} \frac{d^5}{2\pi^6 \Delta^2 F_S} \frac{6}{N(N+1)(2N+1)} \sum_{\nu=1}^{N} \frac{k_\nu^2}{\nu^2}$$

where Δ is the roughness of the film surface, $k_\nu^2 = \frac{2m}{\hbar^2}(E_F - E_\nu)$ are the wave numbers associated with the quantization in the direction along the film thickness d, and F_S is a function that depends on the interaction between the conduction electrons and the film surface morphology.

As will be detailed below, F_S depends on a characteristic interaction distance (ℓ_s) and on the film morphology. Thus, in this case, the film surface morphology must be quantified. For this purpose the film surface is represented by height fluctuations given by $z(x, y) = \pm \frac{d}{2} + h(x, y)$.

The film surface height fluctuations $h(x, y)$ generate a scattering potential given by $U \approx \left(\frac{\partial E}{\partial d}\right) h(x, y) = \frac{\pi^2 \hbar^2 \nu^2}{m\, d^3} h(x, y)$ and the conduction electron scattering is calculated from the transition probability given by: $|<k\nu|U(x, y)|k'\mu>|^2$.

The conduction electron scattering by the surface is calculated [58] taking into account an average interaction distance (ℓ_s):

$$\ell_s(d) \approx v_{//} \tau_c,$$

where $v_{//}$ is the velocity parallel to the surface (x, y) and τ_c is the collision time, corresponding to the time that the electron interacts with the surface during the collision. It is possible to calculate $v_{//}$ using the conduction electron velocity in the z direction v_z and the Fermi velocity v_F:

$$v_{//}^2 = v_F^2 - v_z^2, \text{ where } v_F = \frac{\hbar}{m}\left(3\pi^2 n\right)^{1/3} \text{ and } v_z = \frac{\pi \hbar \nu}{md}.$$

The collision time is given by $\tau_c \approx \frac{d}{v_z}$. Then the average interaction distance can be written as:

$$\ell_s(d) = <\frac{v_{//}}{v_z}> d = \left(\frac{\pi}{3n}\right)^{1/3} \sum_\nu \left[\left(\frac{3n}{\pi}\right)^{2/3} \left(\frac{d}{\nu}\right)^2 - 1\right]^{1/2}. \qquad (24.2)$$

This quantity has been calculated for platinum and gold [58] thin films, showing an almost linear relationship between ℓ_s and d.

The morphology contribution to F_S is through the grain factor $g(d)$ that will be defined below.

At this point it is necessary to represent the film surface $h(x, y)$ as a Fourier expansion given by $h(x, y) = \Sigma_n h_n \sin(2\pi r/\lambda_n)$ [14,58–65], where r is the position vector modulus in the (x, y) plane given by $r = \sqrt{(x^2 + y^2)}$, λ_n is the morphological wavelength present in the surface, and h_n is the amplitude associated with each λ_n.

Specifically for the platinum and gold calculations, the morphological wavelengths λ_n were taken to be [14, 65] the film grain sizes D_n, so $\lambda_n \approx D_n$, and the wavelengths λ_n are given by an integer number of Fermi wavelengths $\lambda_n = n\, \lambda_F$.

With this approach, F_S is given by $F_S(g, \ell_s) = g(d)\,\ell_s(d)$, where $\ell_s(d)$ is given by Eq. (24.2) and the grain factor is given by [58–64]:

$$g(d) = \frac{1}{k_F} \sum_n (h_n/\Delta)^2 (\lambda_F/\lambda_n)^2 \qquad (24.3)$$

where k_F is the Fermi wavenumber and h_n and λ_n are associated with the Fourier expansion of the film surface height fluctuations.

For the case of metals, where $N \gg 1$, the total film resistivity is given by $\rho(d) = \rho_{bulk} + \rho_s(d)$, and is given in terms of the quantities defined above by [59]:

$$\frac{\rho(d)}{\rho_{bulk}} = 1 + \frac{C\Delta(d)^2\, g(d)\, \ell_s(d)}{d^2(1 - 0.15/d)}$$

where C is a constant that depends of the material, and specifically for platinum and gold is equal to $C_{Pt} = 6.261 \times 10^3\,\mathrm{nm}^{-2}$ and $C_{Au} = 28.072 \times 10^3\,\mathrm{nm}^{-2}$, $\Delta(d)$ is the roughness measured as a function of the thickness d, $g(d)$ is obtained from Eq. (24.3) measuring the morphological grain sizes as a function of the thickness d, and $\ell_s(d)$ is calculated as given in Eq. (24.2).

The distinction between morphological and crystallographic grain sizes will be discussed later in this text.

With this model, the platinum and gold resistivities were calculated for film thickness between 1 and 10 nm [60]. The calculated values were found to be in excellent agreement with the experimental data, as shown in Fig. 24.16.

On the basis of these results, it was further proposed that morphological anisotropy of the film surface should induce an associated anisotropy in the resistivity [61]. To estimate the metric scale needed for this effect to be manifested experimentally, the theory described above was used to calculate the resistivity of a simple surface with anisotropic morphology.

The geometry used for this estimate was similar to that shown in Fig. 24.1 of this chapter. The surface was defined by a sinusoidal profile in the x-direction given by $z = h \sin(2\pi x/L)$, where h is the amplitude of the sinusoidal profile and L the morphological wavelength. A granular profile was assumed in the y-direction instead

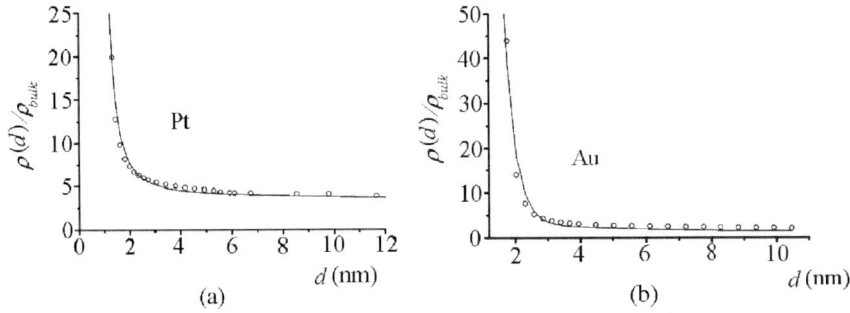

Fig. 24.16. Measured (*circles*) and calculated (*continuous line*) [60] resistivities for (**a**) platinum and (**b**) gold thin films. The fit is excellent

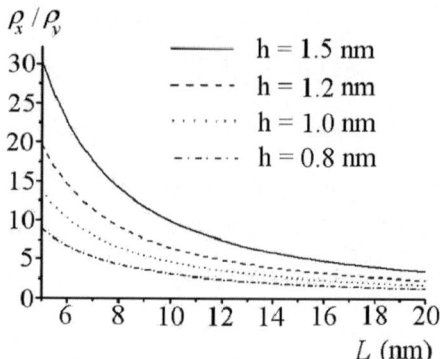

Fig. 24.17. Anisotropy factor ρ_x/ρ_y as a function of morphological surface wavelength L. Each curve corresponds to different amplitude of the sinusoidal surface profile

of a flat profile, as nanostructured thin films usually are when formed by filtered vacuum arc plasma deposition.

The results of the calculation are shown in Fig. 24.17, where the anisotropic factor ρ_x/ρ_y is given as a function of different morphological surface wavelengths L and each curve corresponds to different amplitudes of the sinusoidal surface profile between $h = 0.8$ nm and $h = 1.5$ nm.

In this calculation gold was taken as the film material and the average thickness was 5 nm.

As is clear from Fig. 24.17, the morphological wavelength L that is required in order for the anisotropic resistivity effect to be experimentally observable is too small (several nanometers) to be nanofabricated on the surface.

We have investigated an alternative approach for creating an anisotropic surface morphology. The substrate was a glass microscope slide scratched in one direction with $1/4$-μm diamond powder dispersed in water [61]. Figure 24.18 shows an AFM image of the glass surface with anisotropic morphology.

Two substrates were cut from a single scratched sample as indicated in Fig. 24.19. In this way substrate A has grooves in the longitudinal (y) direction and substrate B has grooves in the transverse (x) direction. Electrical contacts were formed on both ends of the substrates using silver glue followed by plasma deposition of relatively thick ($\sim$200 nm) platinum films onto the contacts, with a mask protecting the center of the sample, as shown in Fig. 24.19. Both substrates were then positioned in the plasma deposition vacuum chamber and their resistance measured throughout the platinum film deposition process. The film resistance was measured after the first six pulses of the repetitively pulsed plasma deposition process and subsequently after every three pulses, thus determining film resistance as a function of film thickness without removing the sample from vacuum for each individual measurement.

The films were formed by filtered vacuum arc plasma deposition [66–69] and the chosen material was platinum. The resistivity, ρ, was determined from the measured resistance R_m from $\rho = (R_m - R_c)dw/\ell$, where w and ℓ are the sample width and length, respectively, and R_c is an estimated contact resistance. The

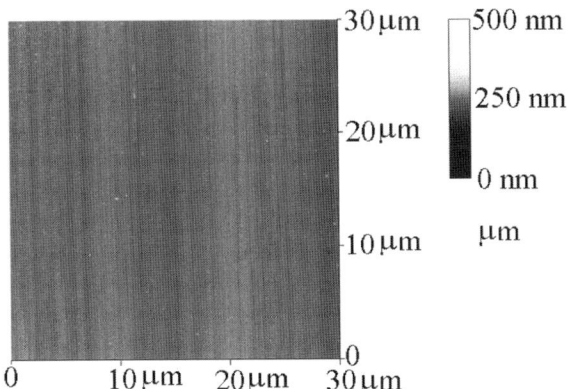

Fig. 24.18. AFM image of glass substrate surface after scratching with 1/4-μm diamond powder, showing the morphological anisotropy

measured resistance values varied over the range $2.5\,\text{M}\Omega$ to $15\ \Omega$ and the contact resistance was about 5Ω. Noting that the contact resistance is constant during the experimental measurements and that the highest resistances ($\sim\text{M}\Omega$) were measured for the thinnest films, it is clear that the contact resistance is irrelevant for these very thin films. The platinum film resistivity spanned the range 12 to $1.2\ \Omega.\,\text{m}$. The measured longitudinal and transverse resistivities, ρ_y and ρ_x respectively, are shown in Fig. 24.20 as a function of film thickness d. The resistivity of both samples increases with decreasing film thickness, as expected for very thin films [58–64].

The resistivity anisotropy ratio $\rho_x(d)/\rho_y(d)$ is shown in Fig. 24.21 as a function of film thickness, and is greater than unity and varies with thickness. The resistivity anisotropy increases significantly for film thickness less than about 2 nm, up to greater than a factor of 10 for the thinnest films investigated of thickness 0.4 nm. We point out parenthetically that the anisotropy ratio ρ_x/ρ_y is expected to reach near unity only for thicknesses d greater than about 70 nm [70, 71], which is consistent with an extrapolation of the Fig. 24.21 data that indicates about 90 nm.

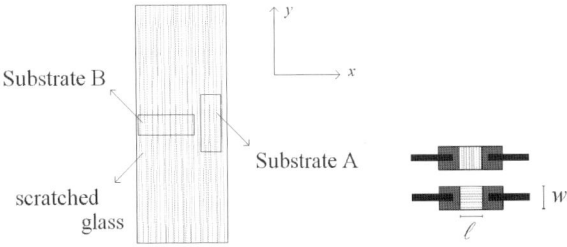

Fig. 24.19. Schematic showing preparation of samples A and B from the scratched glass substrate; substrate A is grooved in the longitudinal or y-direction, and substrate B is grooved in the transverse or x-direction. A schematic of the finished samples is shown on the right

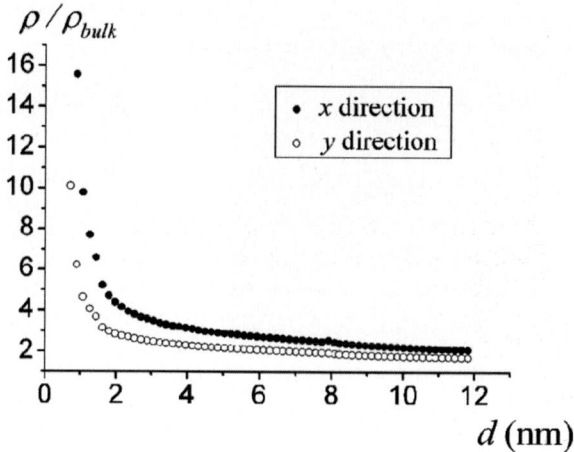

Fig. 24.20. Measured resistivity ratios ρ_x/ρ_{bulk} and ρ_y/ρ_{bulk} as a function of film thickness

These results indicate a significant resistivity anisotropy that is a consequence of the film morphological anisotropy, attesting to the importance of quantitative morphological analysis for this area.

Summarizing, according to our studies on platinum and gold thin film resistivities, we have shown [58–64] that the conduction electrons are scattered by the surface morphology. We have demonstrated that the quantitative morphological analysis of the films surface is a very important parameter and must be taken into account.

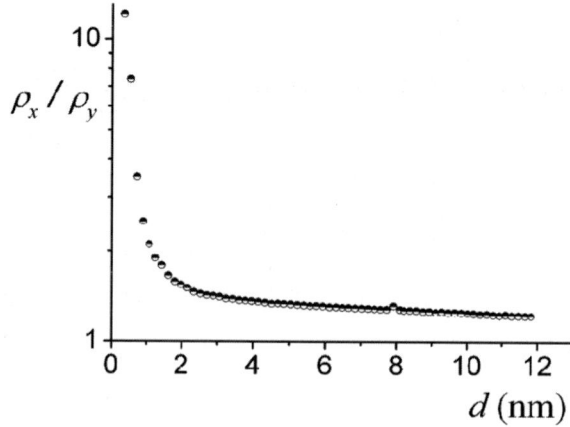

Fig. 24.21. Measured resistivity anisotropy ratio ρ_x/ρ_y as a function of film thickness d

24.3.4
Morphological and Crystallographic Grain Sizes

A number of applications of morphological surface analysis have been discussed in the above, including the importance of grain size determination for thin film resistivity.

A very common way to measure grain size is by using X-ray diffraction [72]. This technique allows measurement of an average grain size over the analyzed region, which can be two or three orders of magnitude larger than the area analyzed in one SPM image. In the following we discuss the distinction between morphological and crystallographic grain size and the importance of morphological grain analysis.

The crystallographic and morphological grain sizes for the platinum and gold thin films have been measured as a function of film thickness d using two different techniques: X-ray diffraction and STM [73].

Typical top-view STM micrographs of platinum and gold films are shown in Figs. 24.22 and 24.23. Figure 24.22 shows platinum films (a) 5 nm thick, and (b) 155 nm thick. Figure 24.23 shows gold films (a) 38 nm thick, and (b) 200 nm thick. These images clearly show the nanostructured nature of the films and how the granular structure changes with film thickness. It can also be seen that the surfaces are composed of different grain sizes. These surfaces have a characteristic fractal auto-affine symmetry [1, 4] with a "cauliflower morphology," where larger grains are composed of aggregates of smaller grains. The "grain diameter" or "grain size" D measured topographically on the film surface is the "morphological grain size." The grain dimension D_c measured by X-ray diffraction [72] is called the "crystallographic grain size." Thus, the morphological grain size is the grain dimension exposed in the film surface; and crystallographic grain size is the grain dimension imbedded in the film bulk.

The analyzed films were deposited by filtered vacuum arc plasma deposition [66–69]. Two kinds of substrates were used—monocrystalline silicon and an ordinary glass microscope slide. The second substrate, being amorphous, allows us to

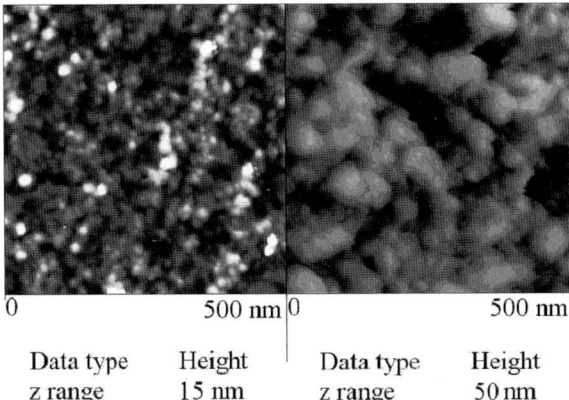

| 0 | 500 nm | 0 | 500 nm |

Data type	Height	Data type	Height
z range	15 nm	z range	50 nm

Fig. 24.22. STM images of Pt thin film: (**a**) 5 nm thick, and (**b**) 155 nm thick

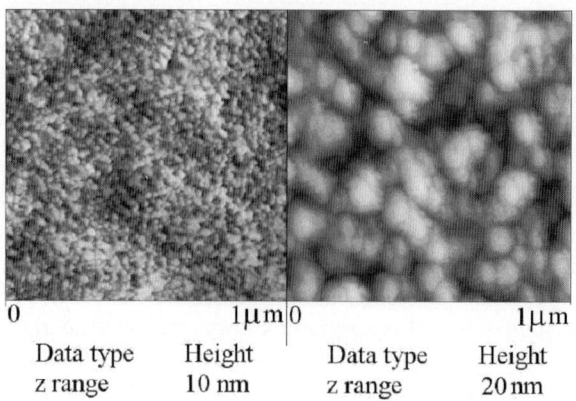

0	1μm 0 1μm
Data type Height	Data type Height
z range 10 nm	z range 20 nm

Fig. 24.23. STM images of Au thin film: (**a**) 38 nm thick, and (**b**) 200 nm thick

check whether or not any preferential film crystallographic orientation is induced by the substrate epitaxy.

The thickness d of the films was measured by placing a small piece of silicon close to the sample, with an ink mark that was removed after deposition and the step-height then measured by AFM.

The film thickness $d(t)$ was determined as a function of deposition time t for the platinum and gold films. The thickness increases linearly with time and the deposition rate was measured to be 7.6 nm/s for platinum and 29.6 nm/s for gold.

Characterization techniques used were STM for the morphological grain size and X-ray diffraction for crystallographic grain size and orientation.

The microscope used was a Scanning Probe Microscope, Veeco Nanoscope IIIA, in STM mode. Commercial platinum-iridium (STM) tips and homemade tungsten tips were used. About four different regions were imaged for each sample. The scan size was between 200 nm and 1 μm, with 512 × 512 pixels for the image resolution.

The X-ray diffraction measurements were carried out in a Rigaku diffractometer, with a 0.05°-step. The Cu-λK_α beam was produced by a conventional X-ray generator and monochromatized with a graphite crystal. The crystallographic grain size D_c was evaluated by the Scherrer equation [72].

Figures 24.22 and 24.23 show typical top view STM images of platinum and gold thin films, respectively. This kind of micrograph was used to measure the morphological grain sizes. Initially, the grain boundaries were defined and after their sizes D were measured. For each film with thickness d, the number of grains $N(D)$ of dimension D were plotted in histograms as a function of D. The D range, $D_{min} \leq D \leq D_{max}$ depends on the film thickness. The minimum grain size D_{min} was limited basically by the resolution of the micrographs, since the pixel size used was between 0.4 and 2 nm. Because of this limitation, we took $D_{min} \geq 3$–5 nm for platinum and gold films, independent of the thickness d. On the other hand, D_{max} increases as d increases. A typical $N(D)$ histogram is shown in Fig. 24.24 for a gold film with $d = 3$ nm. The continuous curve represents the histogram best fit. From these histograms were determined the average grain dimensions. Note that in the

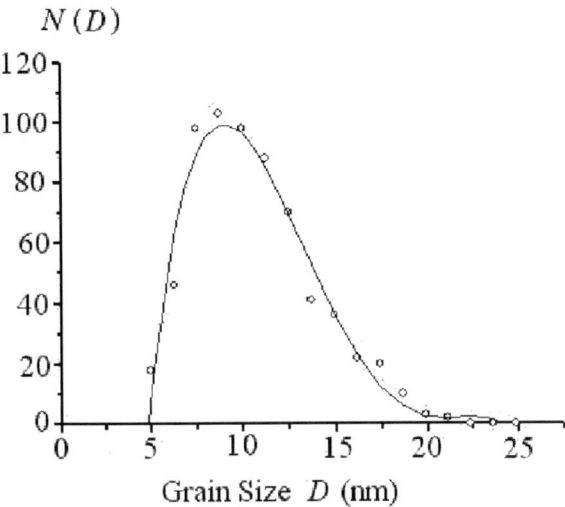

Fig. 24.24. Histogram of the morphological grain sizes $D(d)$ for a Au film 3-nm thick

histograms, there is always a sharp cutoff for smaller grains. This is due to the image resolution mentioned above.

The average values of the crystallographic and morphological grain sizes for each thickness d are indicated in the following by $D_c(d)$ and $D(d)$, respectively.

The morphological and crystallographic grain size analyses yield similar results for the silicon and glass substrates, within experimental error. Thus, in the following we do not distinguish between the substrate used for each sample analyzed.

In Fig. 24.25 the $D_c(d)$ values are plotted as a function of d. For platinum, the experimental results are indicated by open circles and for gold by solid circles. One can see from these figures that the grain size $D_c(d)$ saturates for large thickness [71]. For platinum films the saturation occurs at $d \approx 100$ nm, with a maximum grain size of around 23 nm. For gold films the saturation occurs at d ≈ 140 nm, with a maximum grain size of around 48 nm.

X-ray analysis confirmed that both platinum and gold films grow preferentially in the (111) direction. This effect was observed for both the silicon and glass substrates, showing that it is not an epitaxial effect. This implies that the films growth process is self-oriented.

The average morphological grain size $D(d)$, for platinum and gold, as a function of d is plotted in Fig. 24.26 in the form $\log[D(d)]$ vs. $\log(d)$. In both cases $D(d)$ increases rapidly in the region with thickness $d < 4$ nm. In the range $4 < d < 45$ nm, $D(d)$ is constant at $D(d) \approx 13 \sim 15$ nm. For thickness $2 < d < 45$ nm, $D(d)$ can be approximately described by the functions

$$D_{Pt}(d) \approx 15.5\{1 - 1.1\exp(-0.9d^{1.8}) - 2.3\exp[-0.001(d-45)^2]\} \text{ and}$$

$$D_{Au}(d) \approx 12.8\{1 - 3\exp(-0.7d^{1.8}) - 2.3\exp[-0.001(d-40)^2]\}.$$

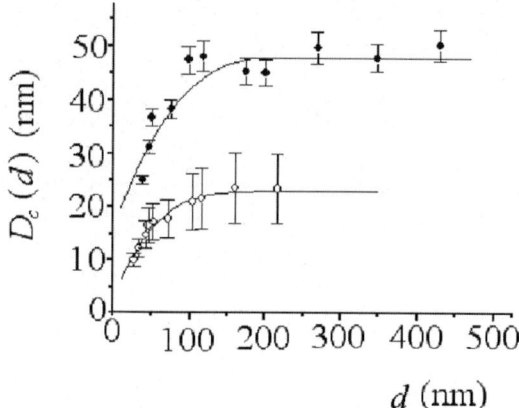

Fig. 24.25. Average crystallographic grain size $D_c(d)$ as a function of film thickness d, for Pt and Au films. The experimental results are indicated by open circles for Pt and by solid circles for Au

For $d > 45\,\text{nm}$, $D(d)$ increases with d and can be represented for both platinum and gold by $D(d) \approx 10.50 + 0.0334\,d + 0.0022\,d^2$. These results show that for $d > 45\,\text{nm}$ the morphological grain size $D(d)$ increases with d, as distinct from the crystallographic grain size $D_c(d)$ which saturates. This can be understood considering that for X-ray diffraction, crystalline defects are detected as grain boundaries, and thus the $D_c(d)$ saturation can be interpreted as high-density crystallographic defects. Figure 24.27 shows a three-dimensional micrograph view of a gold film 200-nm thick. This image suggests that the morphological grains are agglomerates of crystallographic grains, exemplifying the concept described above.

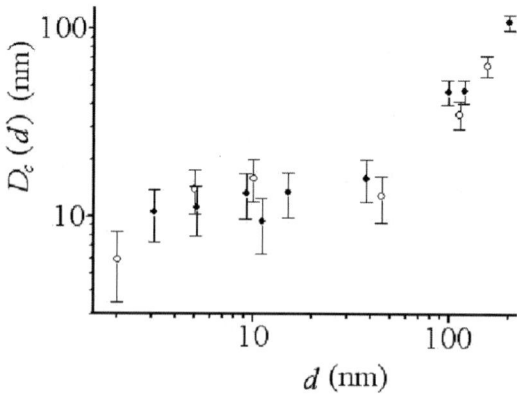

Fig. 24.26. Average morphological grain size $D(d)$ as a function of film thickness d, for Pt and Au films. The experimental results are indicated by open circles for Pt and by solid circles for Au

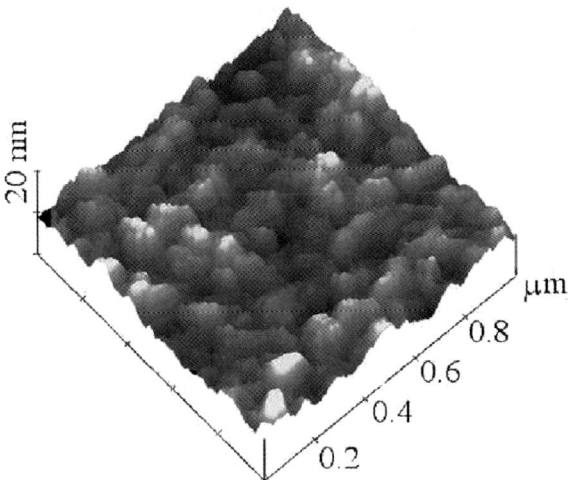

Fig. 24.27. Three-dimensional view of STM image of a gold film 200-nm thick

The early growth is generally subdivided into three main categories: (a) layer-by-layer growth, where a continuous monolayer is preferentially formed prior to the deposition of the subsequent layer; (b) island growth, where the deposition atoms tend to aggregate into island growth with thickness of several monolayers; and (c) mixed mode [66]. For the platinum and gold films described here, the early growth is in islands. Note that the thinner films analyzed here must be conductive enough to allow STM images to be obtained, this implies that the films have already coalesced.

Summarizing, nanostructured platinum and gold thin films with thickness between 2 and 430 nm have been fabricated by filtered vacuum arc plasma deposition. The films were analyzed measuring the morphological and crystallographic grain sizes as a function of film thickness. It was observed that for both platinum and gold films the crystallographic grain size saturates for large thickness and that they grow preferentially in the (111) direction. This effect was observed for both silicon and glass substrates, indicating that the film growth process is self-oriented. The morphological grain size, for platinum and gold films, increases monotonically except in the range of thickness between 4 and 45 nm, where the grain size saturates. The results show that for $d > 45$ nm the morphological grain size $D(d)$ increases with d, distinct from the crystallographic grain size $D_c(d)$ which saturates. This can be understood by considering that for X-ray diffraction, crystalline defects are detected as grain boundaries, so the $D_c(d)$ saturation could be interpreted as high-density crystallographic defects. Concerning the early growth of the films, the films grow as islands, but the thinner films analyzed must be conductive enough to allow STM images to be obtained; this implies that the films have already coalesced.

The explanation above defined "morphological grain size" as the grain dimension exposed on the film surface, and "crystallographic grain size" as the grain dimension imbedded in the film bulk. The importance of the morphological grain size has been pointed out by several workers.

As discussed in this text as the third application of quantitative analysis of surface morphology, the quantum theory applied to the electrical resistivity of nanostructured thin films [58–64] makes use of the morphological grain size as an important parameter.

Thermoelectric power in very thin films has been also studied [74,75], taking into account quantum size effects. In this study, the electrical resistivity theory [58–64] was successfully used, meaning that the morphological grain size was an important parameter in this case also.

24.4
Final Remarks

Quantitative analysis of surface morphology is an extensive field and it has not been our intension to discuss it in all possible facets. A number of significant parameters have been presented and four applications have been described where quantitative morphological analyses were required. Importantly, note that the SPM technique is fundamental for quantitative morphological analysis since it allows imaging surfaces with precise measurement in three dimensions.

Acknowledgments. The author would like to thank Professor Ian G. Brow, from Lawrence Berkeley National Laboratory, and Professor Mauro S. D. Cattani, from University of Sao Paulo, for the critical reading of this Chapter. This work was supported by Fundação de Amparo a Pesquisa do Estado de São Paulo (FAPESP), the Conselho Nacional de Desenvolvimento Científico e Tecnológico (CNPq) and the Coordenação de Aperfeiçoamento de Pessoal de Nível Superior (Capes), Brazil.

References

1. Barabási AL, Stanley HE (1995) Fractal concepts in surface growth. Cambridge University Press, Cambridge
2. Salvadori MC, Martins DR, Cattani M (2006) Surf Coat Technol 200:5119
3. Cattani M, Salvadori MC (2005) Surf Rev Lett 12:675
4. Melo LL, Salvadori MC, Cattani M (2003) Surf Rev Lett 10:903
5. Salvadori MC, Melo LL, Cattani M, Monteiro OR, Brown IG (2003) Surf Rev Lett 10:1
6. Salvadori MC, Melo LL, Martins DR, Vaz AR, Cattani M (2002) Surf Rev Lett 9:1409
7. Cattani M, Salvadori MC (2001) Surf Rev Lett 8:347
8. Salvadori MC, Pizzo Passaro AM, Cattani M (2001) Surf Rev Lett 8:291
9. Cattani M, Salvadori MC (2000) Thin Solid Films 376:264
10. Pizzo Passaro AM, Salvadori MC, Martins DR, Cattani M (2000) Thin Solid Films 377–378:285
11. Salvadori MC, Silveira MG, Cattani M (1999) Thin Solid Films 354:1
12. Salvadori MC, Silveira MG, Passaro AMP, Cattani M (1998) Acta Microscopica 7:465
13. Palasantzas G, Zhao YP, Wang GC, Lu TM (2000) Phys Rev B 61:11109
14. Feder J (1988) Fractals. Plenum Press

15. Family F, Vicsek T (1985) J Phys A 18:L75
16. Kardar M (1996) Physica B 221:60
17. Kim J, Palasantzas G (1995) Int J Mod Phys B 9:599
18. Salvadori MC, Silveira MG, Cattani M (1998) Phys Rev E 58:6814
19. Salvadori MC, Mammana VP, Martins OG, Degasperi FT (1995) Plasma Sources Sci Technol 4:489
20. Salvadori MC, Ager JW III, Brown IG, Krishman KM (1991) Appl Phys Lett 59:2386
21. Bevington PR (1969) Data reduction and error analysis for the physical sciences. McGraw-Hill, New York
22. Kardar M, Parisi G, Zhang YC (1986) Phys Rev Lett 56:889
23. Kardar M, Zhang YC (1987) Phys Rev Lett 58:2087
24. Medina E, Hwa T, Kardar M (1989) Phys Rev A 39:3053
25. Hwa T, Kardar M, Paczuski M (1991) Phys Rev Lett 66:441
26. Baur C, Bugacov A, Koel BE, Madhukar A, Montoya N, Ramachandran TR, Requicha AAG, Resch R, Will P (1998) Nanotechnology 9:360
27. Beton PH, Dunn AW, Moriarty P (1995) Appl Phys Lett 67:1075
28. Cuberes MT, Schlitter RR, Gimzewski JK (1996) Appl Phys Lett 69:3016
29. Teixeira FS, Mansano RD, Salvadori MC, Cattani M, Brown IG (2007) Rev Sci Instrum 78:053702
30. Chen J-M, Liao S-W, Tsai Y-C (2005) Synthetic Metals 155:11
31. Magno R, Bennett BR (1997) Appl Phys Lett 70:1855
32. Bouchiat V, Esteve D (1996) Appl Phys Lett 69:3098
33. Sohn LL, Willett RL (1995) Appl Phys Lett 67:1552
34. Hu S, Altmeyer S, Hamidi A, Spangenberg B, Kurz H (1998) J Vac Sci Technol B 16:1983
35. Kunze U, Klehn B (1999) Adv Mater 11:1473
36. Klehn B, Kunze U (1999) J Appl Phys 85:3897
37. Mamin HJ, Rugar D (1992) Appl Phys Lett 61:1003
38. Mamin HJ (1996) Appl Phys Lett 69:433
39. Snow ES, Campbell PM (1994) Appl Phys Lett 64:1932
40. Tsau L, Wang D, Wang KL (1994) Appl Phys Lett 64:2133
41. Campbell PM, Snow ES, McMarr PJ (1995) Appl Phys Lett 66:1388
42. Fontaine PA, Dubois E, Sti'evenard D (1998) J Appl Phys 84:1776
43. Dai H, Franklin N, Han J (1998) Appl Phys Lett 73:1508
44. Wilder K, Quate CF (1998) Appl Phys Lett 73:2527
45. Park SW, Soh HT, Quate CF, Park S-I (1995) Appl Phys Lett 67:2415
46. Tully DC, Wilder K, Fr'echet JMJ, Trimble AR, Quate CF (1999) Adv Mater 11:314
47. Piner RD, Zhu J, Xu F, Hong S, Mirkin CA (1999) Science 283:661
48. Hong S, Zhu J, Mirkin CA (1999) Science 286:523
49. Lee KB, Park S-J, Mirkin CA, Smith JC, Mrksich M (2002) Science 295:1702
50. Li Y, Maynor BW, Liu J (2001) J Am Chem Soc 123:2105
51. Maynor BW, Filocamo SF, Grinstaff MW, Liu J (2002) J Am Chem Soc 124:522
52. Jang SY, Marquez M, Sotzing GA (2004) J Am Chem Soc 126:9476
53. Sader JE (1995) Rev Sci Instrum 66:4583
54. Pratt WK (2001) Digital image processing: PIKS, 3rd edn John Wiley & Sons; New York
55. Command Reference Manual, Software version 5.12r3, Nanoscope IIIa, Veeco
56. Dobisz EA, Brandow SL, Snow E, Bass R (1997) J Vac Sci Technol B 15:2318
57. Dobisz EA, Brandow SL, Bass R, Shirey LM (1998) J Vac Sci Technol B 16:3695
58. Cattani M, Salvadori MC (2004) Surf Rev Lett 11:283
59. Cattani M, Vaz AR, Wiederkehr RS, Teixeira FS, Salvadori MC, Brown IG (2007) Surf Rev Lett 14:87
60. Salvadori MC, Vaz AR, Farias RJC, Cattani M (2004) Surf Rev Lett 11:223

61. Salvadori MC, Cattani M, Teixeira FS, Wiederkehr RS, Brown IG (2007) J Vac Sci Technol A 25:330
62. Cattani M, Salvadori MC (2004) Surf Rev Lett 11:463
63. Cattani M, Salvadori MC, Teixeira FS, Wiederkehr RS, Brown IG (2007) Surf Rev Lett 14:345
64. Cattani M, Salvadori MC, Filardo Bassalo JM (2005) Surf Rev Lett 12:221
65. Namba Y (1970) Jpn J Appl Phys 9:1326
66. Anders A (ed) (2000) Handbook of plasma immersion ion implantation and deposition. John Wiley & Sons, Inc, New York
67. nders A (1997) Surf Coat Technol 93:158
68. Monteiro OR (1999) Nucl Instrum Methods Phys Res B148:12
69. Brown IG, Anders A, Dickinson MR, MacGill RA, Monteiro OR (1999) Surf Coat Technol 112:271
70. Jalochowski M, Bauer E, Knoppe H, Lilienkamp G (1992) Phys Rev B 45:13607
71. Barnat EV, Nagakura D, Wang PI, Lu TM (2002) J Appl Phys 91:1667
72. Cullity BD (1978) Elements of X ray diffraction, 2nd edn. Addison-Wesley Publishing Company, Massachusetts
73. Salvadori MC, Melo LL, Vaz AR, Wiederkehr RS, Teixeira FS, Cattani M (2006) Surf Coat Technol 200:2965
74. Salvadori MC, Vaz AR, Teixeira FS, Cattani M, Brown IG (2006) Appl Phys Lett 88:133106
75. Cattani M, Salvadori MC, Vaz AR, Teixeira FS, Brown IG (2006) J Appl Phys 100:114905

25 Nanotribological Characterization of Carbonaceous Materials: Study of Diamond Coatings and Graphite

Marjorie Schmitt · Sophie Bistac

Abstract. Because of the development of nanotechnologies, it becomes essential to fully understand the properties of the materials used in these specific conditions, more particularly those of carbonaceous films. The mechanisms involved in the nanofriction of these coatings must be thoroughly studied; atomic force microscopy (AFM) seems to be a suitable technique to achieve this objective.

In this work, an inventory is first drawn up concerning the role of both experimental and intrinsic properties on the nanotribological behaviour of various carbonaceous films.

Then results obtained with AFM in contact mode on diamond coatings and graphite powders are presented: the effects of the scanning velocity, contact load and superficial chemistry on friction are more particularly studied.

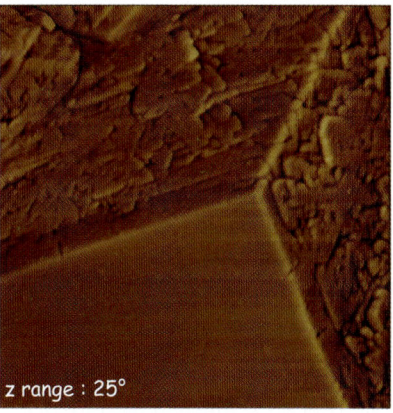

z range : 25°

Key words: Diamond, Graphite, Superficial chemistry, Nanofriction

Abbreviations

AFM	Atomic force microscopy
a-C	Amorphous carbon films
a-C:H	Hydrogenated amorphous carbon films
B−C−N	Boron carbon nitride films

C–N	Carbon nitride films
CN_x	Amorphous carbon nitride films
DLC	Diamond-like carbon films
DNP	Diamond nanoparticles
FFM	Friction force microscopy
HOPG	Highly oriented pyrolitic graphite
LFM	Lateral force microscopy
ns-C	Nanostructured carbon films
ta-C	Tetrahedral amorphous carbon films
VACNT	Vertically aligned carbon nanotubes

25.1
Introduction

With the development of micro-nanotechnologies, and more particularly of the micro- nanoelectromechanical systems (MEMS-NEMS), the understanding of the interfacial phenomena that occur in a contact is essential. Appropriate technology, not only in regard to the small size of the studied elements, but also to the specific properties that are investigated, had to be developed.

The surface force apparatus (SFA) was first realized in order to study the properties (both static and dynamic) of a molecularly thin film in between molecularly smooth surfaces. Then, the scanning tunnelling microscope (STM) was imagined to investigate, with an atomic resolution, electrically conducting surfaces. And finally, the atomic force microscope (AFM) was designed to test both electrically conducting and insulating surfaces at a nanoscopic scale; topographic analysis as well as adhesion and electrostatic force measurements can be carried out. The friction force microscope (FFM), stemmed from modifications of the AFM and led to the determination of friction. The conditions of use and operating parameters of these devices are described in [1].

Carbonaceous coatings are part of the materials used in the design of MEMS, and for other devices linked to nanotechnology; their friction and wear properties, at a macroscopic scale, have long been the subject of numerous studies [2–7]. But as it is well-known that friction is scale-dependent [8,9], the nanotribology of these materials is now under investigation.

Results concerning the influence of various parameters like the normal load, the scanning velocity or the environment on the nanotribological behaviour of carbonaceous materials are summarized in the first part of this paper. Then, after a description of the experimental details, nanofriction properties of diamond coatings and graphite powders are given.

25.2
Nanotribology of Carbonaceous Materials: State of the Art

The tribological behaviours of carbonaceous materials are varied, depending on the deposition method, the conditions of use; some results are presented below.

(111) and (001) H-terminated diamond surfaces are studied with diamond AFM tips: no difference in the friction is observed for the two surfaces (only differences in the friction images are noticed, due to the different lattice spacing and orientation) [10].

LFM tests realized on diamond-like carbon (DLC) coatings with Si_3N_4 tips led to a friction coefficient of 0.13 (even after 45 days of storage); this is in agreement with macroscopic results [11].

The wear debris particles of DLC films were found to be generated only for specific combinations of velocities, normal loads and number of sliding cycles. These results along with the corresponding friction and adhesion measurements are all indicative of a phase transformation mechanism of DLC samples at a nanoscale that is very similar to the ones predicted for the macro- and microscales [12]. The numerous studies dealing with DLC coatings indicate that their nanowear is strongly affected by many parameters, among which are the film thickness, the number of sliding cycles and the deposition techniques [13–16]. Moreover, the ultra low friction and wear of DLC coatings find their origin in the phase transformation of these films into a low shear strength graphite-like phase; this is the case not only at macro- and microscopic scales, but also on nanoscale [12].

Liu et al. demonstrate that for graphite and hydrogenated carbon films, the macrotribological behaviour is not directly comparable with the nanotribological one, due to the dependence of wear, plastic deformation and delamination on the contact area [17].

The friction on the Si_3N_4 tip against the basal plane of highly oriented pyrolitic graphite (HOPG) leads to a coefficient of friction of 0.06 [18].

Macro and nanotribological properties of HOPG are studied in parallel through fretting tests (macroscopic scale) and AFM/LFM experiments [17]. It appears that the nanoscopic friction coefficient remains relatively constant during the tests; this is not the case in macroscopic experiments where the coefficient of friction greatly depends on the tests duration. The initial coefficient of friction of graphite during fretting tests—corresponding to the contact between a Si_3N_4 ball and the atomically smooth surface of the graphite—is higher than the one obtained in LFM tests (with a Si_3N_4 tip). Moreover, the coefficient of friction of the fretting steady state can not be compared to those found in LFM measurements, because wear, elastoplastic deformation, delamination of atom layers and material transfer occurred during macroscopic tests, but this did not occur at a nanoscopic scale. A correlation between the nano- and macroscopic tribological properties of this material is consequently not obvious.

The concept of "superlubricity" was defined in 1993 by Shinjo and Hirano as the fact that friction can be softened to the point of being almost non-existent, even in the case of a dry contact between crystalline solid surfaces [19]. It has been shown that this superlubricity is the cause of the ultra-low friction obtained between two surfaces of graphite, during FFM tests [20]. These same authors also carried out nanoscopic friction tests by varying the rotational angle between a graphite surface and a tungsten tip: the origin of the ultra-low friction of graphite was then found to be linked to the incommensurability between this graphite surface and a graphite flake stuck (because of the sliding) on the tip [21].

Mate studied the friction force as a function of the load in air for graphite, diamond, a-C:H and buckminsterfullerene (C_{60}) [22]: the friction force clearly depends on the nature of the carbon, and the friction coefficient varies from 0.01 for the graphitic basal plane to 0.8 for C_{60}. The value obtained for graphite is smaller here than the typical macroscopic one (0.1) [23], it is however in good agreement with values obtained in previous studies [24, 25]. The intermediate value of 0.33 found for a-C:H films is not much greater than the one (0.24) obtained in [26] for unlubricated amorphous carbon films. The friction coefficient of diamond is initially equal to 0.3, but it drops to 0.05–0.15 for contact loads higher than 400 nN [22]: these values are similar to those identified in [27, 28] during macroscopic friction tests in air.

Extremely thin carbon-nitride (C−N) and boron and carbon nitride (B−C−N) of 1 and 3-mm thickness respectively are deposited on magnetic CoCrTa layers; indentation tests indicate that B−C−N films show better mechanical properties than C−N coatings. Lateral vibration wear tests are also conducted on these coatings with a LM-FFM apparatus: it appears that the wear resistance of the substrates are improved with B−C−N films. Moreover, the 1-mm B−C−N films show the highest hardness and the best wear resistance properties; this is linked first to the fact that these coatings are formed with graded composition and second because a strong interface is formed between the film and the plasma layer [29].

Riedo et al. observed a high microscopic friction for amorphous carbon thin films, medium friction for carbon nitride, and very low friction for graphite; these differences are neither related to the adhesion, nor to the surface morphology. The authors suggest that different phonon excitation in the amorphous carbon and CN_x films is at the origin of the friction behaviour seen [30].

FFM measurements can also be used to study the chemistry of carbon films; in the case of smooth a-C:H films deposited on silicon wafers, uniform friction maps indicate that the surface of these films is chemically uniform [22].

It was established that for a hard disk application, a carbon film thickness superior to 5 nm is necessary to obtain excellent nanofriction and wear properties [16].

The nanotribological properties of amorphous carbon films sputtered on Si(100) substrates, with different thickness (from 5 to 85 nm), are studied under the application of various normal loads (10–1, 200 μN) [31]: for N $<$ 50 μN (low-load range), the higher friction coefficient is observed for the thinner films, whereas in the high-load range (N $>$ 150 μN), the most important friction coefficient is noticed when using the thickest films.

FFM measurements performed on ns-C and a-C films led to the conclusion that the friction performances of the former are the best; it also appears that an increase of the size of the primeval clusters induces a higher effective coefficient of friction [32].

The FFM measurements realized on a-C:H films indicate that the topography and the friction force are strongly correlated: the low friction forces are observed on the summits of the roughness map, whereas the high friction forces are linked to the valleys of the topographical plot [22].

Grierson and Carpick present a review of a way to obtain hard carbon materials for AFM tips; it turns out that the performance of the carbon-coated tips is excellent when the film is well-adhered to the tips and is continuous [33].

25.2.1
Role of the Scanning Velocity

Prioli et al. have shown that scanning velocity up to $40\,\mu m\ s^{-1}$ does not influence the coefficient of friction of ns-C and ta-C films [34]; this independence in regard to velocity is in close agreement with the model [35] that suggests that the sliding friction forces are determined by two competitive processes, both with opposite logarithmic velocity dependence: the thermally activated cohesive forces between the two surfaces in contact, and the kinetics of capillary meniscus formation at the asperities at the tip–film surface interface. In the first process, an increase of friction is observed, while a decrease occurs with the second one. In the case of ns-C and ta-C films, both processes are present, and no velocity dependence is observed.

25.2.2
Role of the Environment

The friction properties of onion-like carbon (OLC) are studied with an ultra-high vacuum AFM, with and without oil addition [36]. Results obtained by lateral force microscopy (LFM) measurements indicate that in both experimental conditions, the friction coefficient remains low. It was also shown that an increase of the contact pressure, in the presence of oil, leads to a decrease of the friction coefficient: it seems that onion-like carbons create low friction material at the contact area; at high contact pressure, this reaction is accelerated. The formation of a smooth film consisting of deformed OLC is the strongest reason for low friction.

The influence of the relative humidity (RH) on the tribological properties of vertically aligned carbon nanotubes (VACNT), highly oriented pyrolitic graphite, diamond nanoparticles (DNP) and molecular beam epitaxially (MBE) grown C_{60} was tested through AFM/FFM measurements with a gold tip [37]; it appears that the maximum coefficient of friction for HOPG and DNP is obtained in the case of a relative humidity varying between 60 and 80%. There is no decrease of the friction for MBE C_{60} at 100% RH, probably because of the characteristic structure of the C_{60} clusters.

At the nanoscopic scale, the friction coefficient of ta-C remains constant in a relative humidity up to 70%, while a decrease of the friction is observed for ns-C films at humidity higher than 30%; in particular, at 40% RH, the friction coefficient of ns-C films, equal to 0.11 ± 0.02, is lower than that for ta-C films (0.13 ± 0.02) [34]. The hydrophobicity of a-C films is responsible for the independence with the humidity of the friction of ta-C, and the passivation of the dangling bonds of graphene planes (coupled to the presence of closed graphene planes) contributes to the decrease of the friction in the case of ns-C films.

The influence on friction of formation of capillary forces at the interface between a-C films and an AFM tip has been studied in [38]: its effect is not very pronounced for RH up to 75%. The contact at the tip–a-C film interface is influenced by the capillary condensation of water only for RH values that are close to saturation (effect of the condensation of water around the tip).

In the case of Si_3N_4 tips sliding against DLC films obtained by plasma source ion deposition, it appears that in the wearless regime, when the relative humidity increases from 5 to 60%, the friction is monotone and reversible, and increases with higher humidity. But adhesion is independent of RH: according to these results, the mechanism that occurs here is not linked to the formation of a meniscus (otherwise adhesion should increase); it could be the physisorption of water that increases the shear strength [39].

The friction between hydrocarbon-coated AFM tips against diamond and amorphous carbon has been studied in air (40–60% RH) and argon [9]: in dry argon, diamond showed the lowest interfacial shear strengths, followed by a-C in dry argon, then diamond in air and finally a-C in air. The presence of water on the surfaces samples could be at the origin of the higher friction in air.

25.2.3
Role of the Contact Load

The variation of the friction with load is accurately described by the Derjaguin–Müller–Toporov (DMT) model [9, 40–43]; however, a behaviour closer to the Johnson–Kendall–Roberts (JKR) model is observed in the case of a single crystal of diamond [44]. A strong sensitivity to defects can be at the origin of the variations that are observed from one location to the next.

When applying μN, the friction coefficient of VACNT dos not depend on the nature of the counterface [45]; whereas under high applied forces, these materials are both fractured and plastically deformed [37].

Wear of ns-C and ta-C has been studied through scratch tests [34]: the critical load to scratch ta-C films is higher than the one needed for ns-C films, in the micronewton range. The previous result is no longer valid for normal loads in the range of a hundred nanonewtons: in this case, ta-C films could be worn under the application of normal loads that are lower than the ones used in the case of ns-C films. The removal of a low density layer (due to the subplantation mechanism of the film growth) from the ta-C surface could be the origin of this phenomenon [46].

The influence of the normal load (from 2 to 20 mN) on the friction of amorphous carbon nitride (CN_x) films grown on Si(100) substrates by reactive ion sputtering and energetic ion bombardment during deposition (IBD) is studied in [47]. For sputtered films *without* energy IBD, three regimes of friction are observed, depending on the applied loads:

Regime I: the friction coefficient decreases to a minimum value of about 0.1 when the normal load increases; this corresponds to the initiation of the scratching.
Regime II: the friction coefficient increases to a maximum value of 0.4.
Regime III: the friction coefficient slightly increases (0.45).

The main friction mechanisms are adhesion and ploughing in regimes II and III, whereas only adhesion takes place in regime I.

For films sputtered *with* high-energy IBD, an increase of the normal load induces a rise of the friction coefficient to a maximal value; it remains then almost constant. The plastic flow seems to be the dominant deformation mode.

These three regimes are also observed in the case of amorphous carbon films tested under loads between 10 and 1, 200 μN [31].

A tungsten tip is used to study the nanofriction, in the ambient, of HOPG: the friction coefficient is found to vary between 0.005 and 0.015. Moreover, the dependence of the friction force on the normal load is shown to be almost linear on this material [25].

Various tips are also used for nanofriction tests against graphite, and again the link between friction force and normal load is proved to be linear, the friction coefficient still evolving in the same range of values as previously [9, 18, 35].

The connection that can link the friction force and the applied load during nanofriction tests on various carbon compounds is studied in [9]; well-defined single-asperity tips are used for these experiments carried out in ambient air and argon (the occurrence of plastic deformation or wear is avoided by applying low loads). On the basis of the model of the Hertzian-type tip–sample contact, the interpretation of the data shows that the friction is proportional to the contact area; and as the shear stress remains constant in the range of the applied pressures, the measured friction force (F_f) is proved to depend on the normal force (F_n) in the following manner: $F_f \approx F_n^{2/3}$. These authors have also showed that it is necessary to introduce an effective friction coefficient for point-contact-like single-asperity friction (the classical friction coefficient is not suitable for a comparison of the frictional behaviour of materials in the case of single-asperity friction); this coefficient is also helpful to classify the nanoscopic friction properties of carbonaceous materials. The highest friction is obtained for C_{60} thin film, a medium value is noticed for diamond and amorphous carbon, and a friction close to zero is observed for graphite.

25.2.4
Role of the Chemistry

The functionality of self-assembled monolayers (SAMs) of ω-functional n-alkanethiol has a great influence on their tribological properties ($-COOH$ and $-OH$ functionalities lead to SAMs with high superficial energy). The highest friction coefficient is observed for SAMs with $-COOH$ functionality, whereas the lowest friction corresponds to the use of SAMs with $-CH_3$ functionality: the higher the surface energy, the greater the friction coefficient [48].

Riedo et al. showed that on the partially hydrophobic surface of DLC films, the friction coefficient, at a nanoscopic scale, increases with the sliding velocity [35].

25.2.5
Role of the Chemistry – Bulk Chemistry

The properties of magnetron-sputtered DLC films with various H concentrations (2, 28, 40 at %, but sp^2/sp^3 ratio not given) are studied at a nanoscopic scale, with a diamond AFM tip: with higher H concentrations, the coefficient of friction increases, and both the hardness and wear performances decrease [49].

On the contrary, in the case of micro-wave plasma chemical vapour deposition DLC films, an increase of the H concentration in the plasma induces a higher sp^3

bonding fraction, and then a lower friction coupled with improved wear performance (tests carried out with a diamond tip) [50].

The friction of the pulsed laser-deposited DLC films (with 34 or 53% sp^3 bonding)/Si_3N_4 tip couple is not affected by an increase of the sp^3 content. In a second test series, the authors used CN_x films with various N contents ($0.2 < x < 0.3$ leads to sp^3 bonding varying between 36 and 53%): the resistance to sliding does not depend on the sp^2/sp^3 bonding ratio. The nanofriction of DLC is higher than the one of CN_x, and both DLC and CN_x films present higher friction properties than HOPG [30].

The hydrophobicity of DLC films is modified by the incorporation of F: the adhesion and friction properties measured by AFM in air decrease when the contact angle increases [51].

25.2.6
Role of the Chemistry—Superficial Chemistry

During the friction in ultra-high vacuum between a Si AFM tip and a (111) diamond surface (single crystal), the presence-absence of surface H is measured by Low Energy Electron Diffraction [52]: for loads up to 30 nN, a removal of the H from the surface induces an increase of the average friction coefficient by more than two-orders of magnitude (compared with the H-terminated surface). This shows that the presence of dangling bonds increases the contribution of adhesion to friction, whereas their passivation leads to a serious decrease of this force.

In the case of ultrananocrystalline diamond films (sliding contact with W and diamond AFM tips in ambient air), the passivation effect of H on these diamond films in air is clearly demonstrated since a total hydrogenation of the surface (that removes sp^2-bonded carbon and oxygen) leads to a decrease of adhesion and friction properties.

25.3
Experimental Details

25.3.1
Diamond Coatings: The Flame Process

The diamond coatings used in this study are obtained by the combustion flame process, a technique initially devised by Hirose and Kondoh [53] to remedy the financial and materials drawbacks of Chemical Vapour Deposition (CVD). This method is based on the fact that a flame can be considered as plasma in air, which means it can play the role of a reaction chamber which gives the heat necessary for the creation of the radicals involved in the formation and the growth of diamonds.

This device is not expensive as it only needs an oxy-acetylene torch coupled to a mobile mount, flowmeters controlling combustible and oxygen flows, a copper substrate holder equipped with a cooling system, and a pyrometer checking the temperature of the substrate surface.

Even if this apparatus is not sophisticated compared with a CVD device, it requires the mastery of many parameters to succeed in synthesizing diamond coatings satisfying the requisite conditions of purity and morphology.

From a structural point-of-view, the main difference between these diamonds and those synthesized by CVD lies in the fact that the dangling bonds are here mainly saturated by oxygen, instead of hydrogen terminations. The advantages and drawbacks of the flame process as well as the conditions of deposition are given in [54].

The diamond coatings used for the AFM measurements are a mix of {111} and {100}-oriented crystals, deposited on triangular tungsten carbide plates; their grain size varies between 2 and 8 μm.

25.3.2
Graphite Pins

In this study, various types of graphite powders are tested. The following is the specific designation of these powders:

Graphites A15 and A75: these powders are made of synthetic graphite and have two different particle sizes, i.e., 15 μm (A15) and 75 μm (A75).

Graphite B: this powder is also a synthetic one, but its synthesis process is different from the one of the previous graphite; the size of its particles is 75 μm.

Graphite C: made of natural graphite powder, with an average particles size of 75 μm.

All of the powders were compacted with a mechanical press into the shape of small graphite pins of 5 mm diameter and 3 mm height (H_V : 1.4 kg mm^{-2}).

25.3.3
Atomic Force Microscopy

The working principle of the AFM is the following: a sharp probe scans the surface of a sample, and during this displacement, a laser is focussed on the tip; the so-obtained beam of light is reflected to a photodiode detector. The deflections of the cantilever, due to the morphological or chemical modifications of the surface sample, induce changes in the laser beam intensity; these variations are detected by the photodiode and are finally recorded and analyzed by a computer. Two imaging modes can be realized with an AFM: the contact and the tapping mode. In the first case, mostly used for friction measurements, there is a continuous surface–tip contact, which leads to the formation of adhesive and shear forces between the gas layer adsorbed on the sample surface and the probe. This drawback can be avoided with the tapping mode, where the contact between the tip and the surface is only intermittent; there is then less damage in this imaging mode, and the measurements are of higher resolution.

In this work, measurements were carried out with a Nanoscope III (Digital Instruments) device in the ambient (20°C, 40% RH), in contact mode. The selected cantilever was a Si_3N_4 triangular one with a stiffness of 0.58 N m^{-1}. The size of the scanned zones varied from $800 \times 800\,\mathrm{nm}^2$ to $30 \times 30\,\mu\mathrm{m}^2$. The Trace-Minus-Retrace (TMR) values were recorded as a function of the scanned distance, for every test (scanning direction: from the top to the bottom). In this preliminary study, the observations were qualitative and not quantitative; no calibration of the cantilevers was made. Various scanning velocities and contact loads were used in order to study the influence of these two parameters on the friction properties of diamond coatings.

25.3.4
Tribometer for Macroscopic Tests

The graphite pins manufactured from the above-described powders were subjected to macroscopic sliding tests against silicon wafers (with different superficial energy); the applied normal load included 2, 5, 10, and 15 N and the corresponding friction experiments were carried out on a commercial CSM pin-on-disc tribometer.

All experiments were run at an ambient temperature of 25°C, and at a relative humidity of 30–40% for a duration of 40 min. The sliding speed was kept constant at a value of 10 rpm (which corresponds to a linear speed of $1.73\,\mathrm{cm\,s^{-1}}$). Each test was carried out three times to check the reproducibility of the results.

25.3.5
Modification of the Superficial Energy of Silicon Wafers

(110)-oriented silicon wafers, with one polished surface, were selected for the macroscopic experiments; they were first used just as they are (superficial energy: $47\,\mathrm{mJ\,m^{-2}}$) and then submitted to a hydroxylation (hydrophilic surface) and a grafting (hydrophobic surface).

Hydroxylation: the silicon wafers are immersed in a Piranha solution composed of 70% sulfuric acid (H_2SO_4, 96% purity) and 30% hydrogen peroxide (H_2O_2 30%), and heated at 50°C for 30 min. These substrates are then cleaned with bi-distilled and deionized water in an ultrasonic bath; they are finally dried with nitrogen. The so-obtained surfaces showed an important density of silanol groups ($5\ Si-OH/nm^2$) [55], and were hydrophilic (superficial energy: $73\,\mathrm{mJ\,m^{-2}}$).

Grafting: this process could only be realized on surfaces that were beforehand hydroxylated otherwise the strong bonds with the hydrolysable parts of the silanes can not be formed. The high density of silanol groups obtained at the end of the previous step will allow the adsorption of hexadecyltrichlorosilane (methyl ends). After being hydrolyzed with the Piranha solution, the silicon wafers were immersed in a 3:1 solution of carbon tetrachloride and HTS over 12 h; this step was realized in an ultrasonic bath in order to eliminate the silanes that were not grafted or physisorbed. The substrates were finally hydrophobic ones (superficial energy: $21\,\mathrm{mJ\,m^{-2}}$).

25.4
Nanofriction Results

25.4.1
Diamond Coatings (Obtained by Flame Process)

25.4.1.1
Influence of the Scanning Velocity

Friction tests were realized on a 10×10-μm^2 zone of the diamond coating, under a constant contact load (corresponding to the application of $\approx 0\,V$), at various scanning velocities: 0.2, 0.4, 0.5, 0.6, 0.8 and 1 Hz.

The scanning velocity seems to have little influence on the TMR, as the latter does not vary a lot when the speed is increased (Fig. 25.1).

Similar tests were carried out on smaller zones to evaluate if the size of the scanning area could play a role in the friction, in these experimental conditions: again, the variations of the velocity do not induce glaring modifications of the TMR [56].

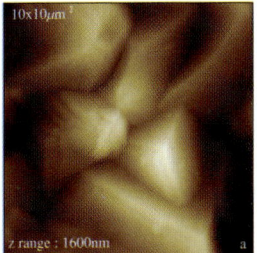

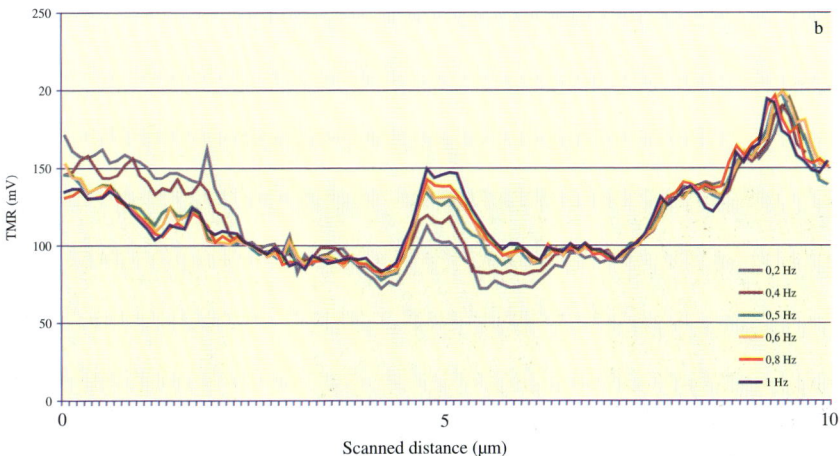

Fig. 25.1. AFM measurements on a $10 \times 10\,\mu m^2$ zone of the diamond coating. (**a**) Topography image; (**b**) TMR values as a function of the scanned distance, for different scanning velocities

It was shown in [34] that the scanning velocity up to $40\,\mu\mathrm{m\,s^{-1}}$ does not influence the coefficient of friction of ns-C and ta-C films; this independence with the velocity is in close agreement with the model according to which the sliding friction forces are defined as the result of the competition between two mechanisms, both with opposite logarithmic velocity dependence [35]: on the one hand, the thermally activated cohesive forces between the two surfaces in contact, and, on the other hand, the kinetics of capillary meniscus formation around the asperities at the tip–film interface. In the first case, an increase of friction is observed, while a decrease occurs with the second one. Both these processes could occur in the context of these measurements, since these tests were carried out in the ambient (sufficient moisture for the formation of a meniscus at the tip–surface contact); no velocity dependence is consequently observed.

25.4.1.2
Influence of the Contact Load

The measurements of these test series were realized on a *crystal facet* ($800 \times 800\,\mathrm{nm^2}$), on a *crystal* ($1 \times 1\,\mu\mathrm{m^2}$), and on *surfaces* of different sizes (from 20×20 to $30 \times 30\,\mu\mathrm{m^2}$), under various contact loads. The TMR values were again recorded as a function of the scanned distance, for contact loads corresponding to the application of 0, 1 and 2 V (scanning direction: from the top to the bottom).

Tests were first realized on small areas, which means on the *facet of a crystal* and on a *crystal*. It was observed, on these two surfaces, that the higher the contact, the greater the values of the TMR. Moreover, the variations of the TMR at 0, 1 and 2 V look the same: it seems that the topography of the surface is properly followed during the application of these loads. It was also noticed that the details of the topography on friction images are more visible when the contact load is bigger: the "sensitivity" of the TMR measurement seems to be more important at higher loads [56].

Measurements on surfaces of bigger size are carried out as follows: a 30×30-$\mu\mathrm{m^2}$ area is scanned (Fig. 25.2a), then a magnification of $25 \times 25\,\mu\mathrm{m^2}$ is made on the left bottom area of the previous surface (Fig. 25.2b); finally another magnification is realized on the right bottom area of the last surface, leading to a $20 \times 20\,\mu\mathrm{m^2}$-scanned zone (Fig. 25.2c). The TMR values are recorded on each

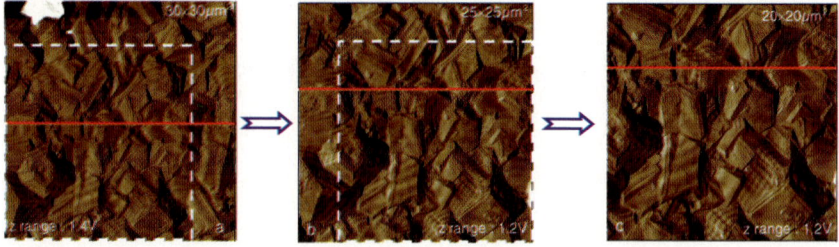

Fig. 25.2. AFM measurements on (**a**) $30 \times 30\,\mu\mathrm{m^2}$, (**b**) $25 \times 25\,\mu\mathrm{m^2}$ and (**c**) $20 \times 20\,\mu\mathrm{m^2}$ zones of the diamond coating (friction-retrace images); (**b**) magnification realized on the *dotted square* of image (**a**): (**c**) magnification realized on the *dotted square* of image (**b**)

surface, as a function of the scanned distance, under the application of 0, 1, 2, 3 and 4 V.

The phenomena that were observed on the surfaces of smaller sizes are still valid for these larger areas: the use of the highest contact loads again leads to greater values of the TMR [56]. Furthermore, it appears that the topography of the coating is carefully followed, even when applying the highest loads, in spite of the roughness that becomes more important here.

In order to try to analyse these results in a quantitative way, the values of the TMR, recorded on the same line, but at various contact loads (red line on Fig. 25.2) are studied; the evolution of the TMR as a function of the applied load is plotted for the three last scanned zones (20×20–25×25–$30 \times 30\,\mu m^2$) (Fig. 25.3).

When the contact load varies from 0 to 4 V, it clearly appears that the values of the TMR linearly increase; this variation is noticed for the three magnifications (the correlation factors are very good in the three cases). However, the values obtained in the case of the largest area ($30 \times 30\,\mu m^2$) are slightly higher than those noticed for the $25 \times 25\,\mu m^2$ and $20 \times 20\,\mu m^2$ zones (that are quite similar). This can be linked to the fact that the analyzed surfaces were scanned at the same frequency: an adequate combination of the scanning frequency and of the magnification (more particularly in the case of these rough coatings) could lead to optimal friction results.

The influence of the applied normal load on a macroscopic scale, has been studied for a variety of carbonaceous materials: in a case of the natural diamond/natural diamond couple, a rise of the normal load leads to an increase of the friction coefficient; the damage of the diamond surface is supposed to be the origin of this phenomenon [57]. However, for the same couple, Casey and Wilks observed no variation of its coefficient of friction when the normal load is modified [58]; these results are explained by taking into account the fact that the diamonds used as a counterface present neither the same crystal orientation nor the same direction of polishing [59]. Samuels and Wilks revealed a much more complex influence of the normal load:

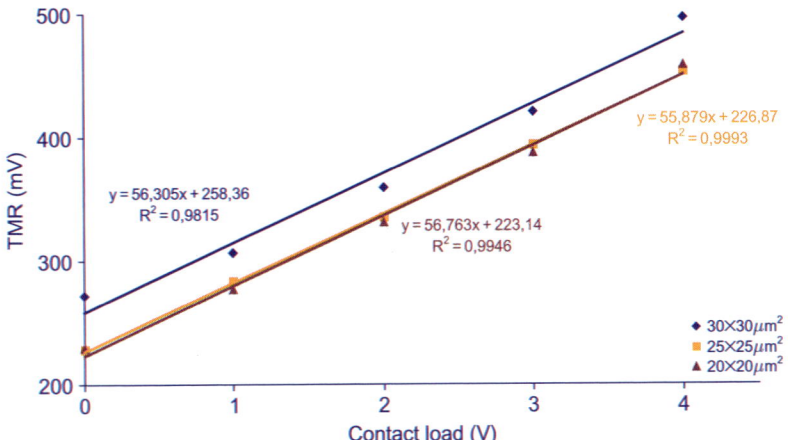

Fig. 25.3. Variation of the TMR as a function of the contact load (measurements along the *red line*) for scanned zones of different sizes

depending on the direction of sliding, and hence on the polishing direction of the diamond used as the counterface, an increase of the contact pressure induces either a decrease of the friction, or a decrease followed by an increase of the friction coefficient [28].

According to the great diversity of the tribological behaviour of diamond coatings that is observed at different scales, when the applied normal load is varied, it is very difficult to establish a correlation between their nano- and macroscopic friction properties in the present state of the research.

Moreover, in the specific case of the diamond coatings studied here, their significant roughness plays an important part; it was proved that a strong correlation exists between the topography of carbonaceous coatings and the friction force [22], and based on the previous measurements, it is clear that the "scale" on which tests were carried out (*facet* of a crystal, *crystal* or *part* of the coating) is of great importance. A correlation between the various measurements realized on the nanoscopic scale is not easy to determine, consequently, it is even more difficult to find a link between nano- and macroscopic tribological behaviours.

25.4.2
Graphite

25.4.2.1
Influence of the Contact Load

These preliminary tests were carried out on selected areas constituted of basal planes of similar size of graphites A75, B and C; the scanning frequency is equal to 1 Hz, and the contact load varied in the range of 0 to 3 V [60]. The TMR values were recorded as a function of the scanned distance, for the various contact loads; the example of graphite B is given in Fig. 25.4.

A precise quantitative study is not possible at this stage of the evaluation, but general trends can be however drawn: the friction of graphite B seems to be greater than that of graphites A75 and C; the values of the TMR observed for B are more than three-times higher than those of graphites A75 and C (the latest vary in the same range of values). The fact that the scanned surface for graphite B is $5 \times 5 \, \mu m^2$ (instead of $2 \times 2 \, \mu m^2$ for A75 and C) can play a part in these observed variations: the z range on this area is 600 nm (it is only 40 nm and 50 nm for graphites A75 and C), that indicates that the roughness is here more important; this may explain in part the higher friction of graphite B. However, even on the "flat" zone of graphite B, the TMR remains bigger than the values noticed on the smooth surfaces scanned on graphites A75 and C.

Graphite B is also the only one for which the increase of the TMR is clearly linked to the application of higher contact loads. For graphites A75 and C, the influence of the normal load is less markedly observable; it can be noticed, for example, that graphite A75 under a contact load corresponding to a voltage of 2 V presents TMR values which are not far from those observed at 3 V (sometimes they are even higher). The same phenomenon is noticed for graphite C.

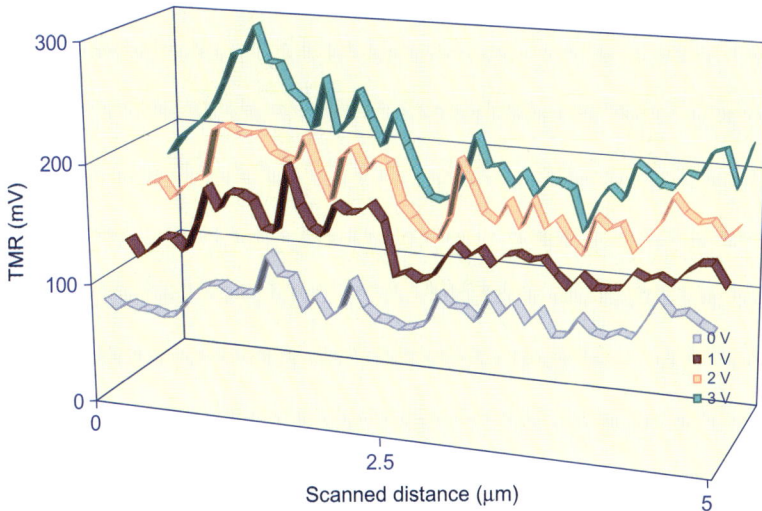

Fig. 25.4. TMR values as a function of the scanned distance under various contact loads, for graphite B

It appears then that graphite B is distinguished from the two others, whereas graphites A75 and C have quite similar behaviours.

25.4.2.2
Influence of the Tips Chemistry: Nanofriction of Grafted Tip/A15 Couples

In this part of the work, the superficial chemistry of Si_3N_4 AFM tips was modified in order to give them hydrophobic or hydrophilic characteristics (the same protocol as for the change in the chemistry of the silicon wafers was used here); measurements were then carried out with these tips against graphite A15.

The friction tests were realized on areas of different sizes (from 1×1 to $10 \times 10\,\mu m^2$); the possible influence of the size of the scanned areas will also be studied. The role of the experimental parameters is based on a variety of data: topography and friction images, and values of the TMR.

25.4.2.2.1
Role of the Scanning Velocity [61]

These tests were performed on the same area, with the following experimental parameters:

– scanning velocity: 0.5; 1; 2 Hz,
– contact loads: 0.5 and 2 V

Friction with a Hydrophobic Tip

The variations of the TMR, as a function of the scanned distance, were recorded for the three scanning velocities, at 0.5 and 2 V (Fig. 25.5).

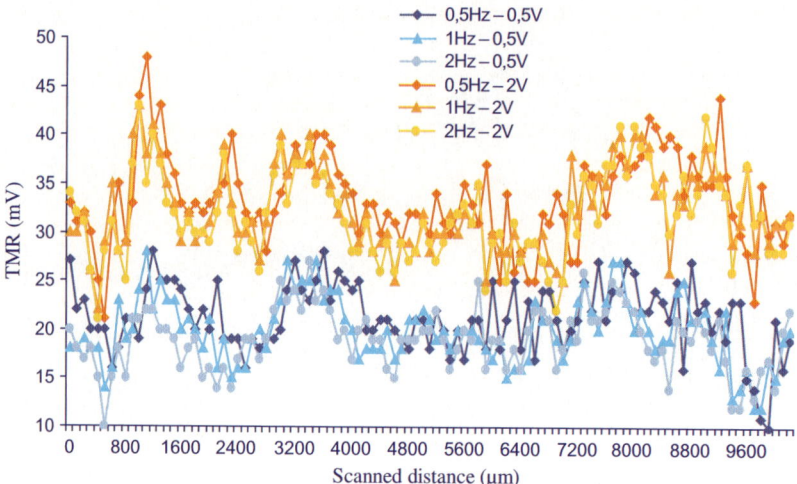

Fig. 25.5. TMR values as a function of the scanned distance for the A15/hydrophobic tip couple, for various scanning velocities

It appears that the scanning velocity has little influence on the tribological behaviour of graphite A15; for a given load, there is no significant difference in terms of TMR when the speed friction increases. However, for an applied contact load of 2 V, the values of the TMR seem to be higher than those found under 0.5 V.

A variation of the scanning velocity should have induced modifications of the interactions between the tip and the graphite at high and low speeds. But it turns out that when tested with the hydrophobic edge, this speed did not play a fundamental role on the tribological behaviour of the graphite.

Friction with a Hydrophilic Tip

The same tests as before were realized, but with a hydrophilic tip (Fig. 25.6).

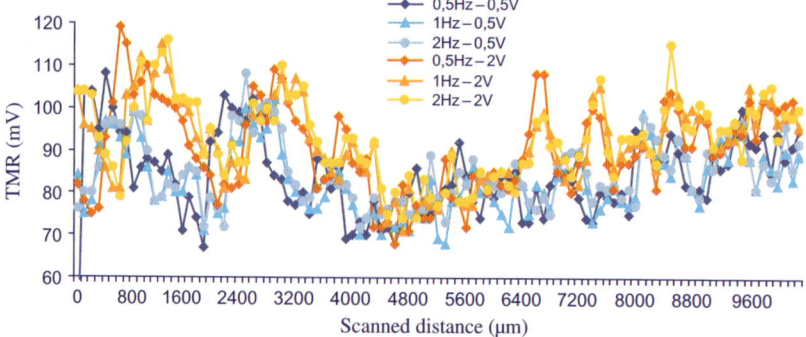

Fig. 25.6. TMR values as a function of the scanned distance for the A15/hydrophilic tip couple, for various scanning velocities

As was already the case for hydrophobic tips, it clearly appears that the scanning speed has no real influence on the values of TMR, since for a given load, the various curves intersect and no particular trends can be identified.

But it seems that even when the contact load increases to 2 V, the values of the TMR keep the same order of magnitude as for low load (except for a few areas).

Here again, the scanning velocity does not play an important part in the tribological behaviour of a hydrophilic tip rubbing against graphite A15.

For both hydrophilic and hydrophobic tips, the nanofriction properties do not seem to be influenced by the variations of the scanning speed.

During friction tests with the hydrophilic tip, it is important to take into account the presence of the capillary forces that occurred between the tip and the graphite surface; they induced the formation of a meniscus in the vicinity of the tip. Its presence could interfere with the motion of the tip, which would therefore require a larger force to continue the scanning of the surface. This might explain the higher values of the TMR during friction with a hydrophilic tip.

In the case of friction with hydrophobic tips, it is difficult to explain why the speed plays no role in the friction properties. In the case of ns-ta-C and C films, it was shown that the coefficient of friction remains constant when the scanning speed of the tip increases [34]. This behaviour has been attributed to the competition between two phenomena: adhesion between tip and the substrate, and capillary forces [35]. If neither of these two mechanisms was predominant, the coefficient of friction would remain constant.

Finally, having tested the influence of the scanning velocity through experiments under various contact loads, and as it was found that the TMR in the tests at high load (2 V) were in most cases higher than when tested at low load (0.5 V), it is therefore interesting to study the role of this parameter on the tribological behaviour of the previous couples.

25.4.2.2.2
Role of the Contact Load

Tests were carried out at a constant scanning velocity (1 Hz), various contact loads were used: 0, 1, 2 and 3 V on the hydrophilic tip/A15 and hydrophobic tip/A15 couples.

For both hydrophilic and hydrophobic tips, the variations of the TMR as a function of the scanned distance clearly reveals that an increase of the contact load induces a rise of the friction [61]. Moreover, experiments were conducted on areas of various sizes (from 2×2 to $10 \times 10\,\mu m^2$), and it appears that the friction is greater on zones of larger size; this could be attributed to the roughness that becomes higher on large areas.

At the end of these tests, it appears that the friction force becomes more important as the load applied to the contact increases: this could be explained by the fact that the tip must indeed continue its motion, following the topography of the scanned area, and overcome any indentation phenomena that might occur at the highest loads.

The behaviour observed here seems to be the trend for most of the couples studied in various works [34, 62–64]; the origin of such a behaviour could be attributed to parameters as varied as the low density layer on the surface of the film, the hardness of this film, etc.

25.4.2.2.3
Comparison Between Hydrophilic and Hydrophobic Tips

Various areas (from 2×2 to $10 \times 10\,\mu m^2$) of graphite A15 were scanned with both hydrophilic and hydrophobic tips, under different contact loads (0, 1, 2 and 3 V); the scanning velocity remained constant (1 Hz) for all the tests [61].

At the end of these tests, it clearly appeared that the TMR values measured with a hydrophilic tip are greater than those obtained with the hydrophobic tip (Fig. 25.7). However, at high loads (2 and 3 V), the values of the TMR recorded with a hydrophobic tip are sometimes close to those obtained with the hydrophilic one, but only on small parts of the area scanned.

It turns out then that in all the cases, for areas of the same topography, size and scale height (z range of the images), the friction due to the sliding of a hydrophilic tip is more important than that for the hydrophobic tip.

The same trend was observed in [48] in the study of the friction of grafted tips on glass substrates: nanofriction tests reveal that the SAMs functionalized with CH_3 have a lower coefficient of friction than the SAMs functionalized with COOH, which is explained by the formation of hydrogen bonds between the COOH terminations.

The phenomenon that can explain the observed difference between the hydrophilic tip/A15 and hydrophobic tip/A15 couples could be the creation of a meniscus between the tip and the sample, resulting in the formation of a "screen"

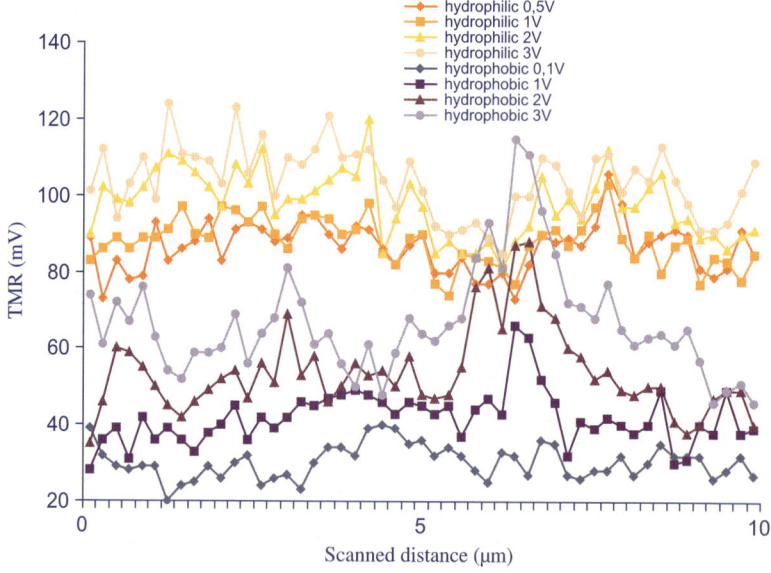

Fig. 25.7. TMR values as a function of the scanned distance for the A15/hydrophilic tip and A15/hydrophobic tip couples

that may cover interactions. These capillary forces are supposed to occur only in the case of hydrophilic tips: the tips with a hydrophobic CH_3 termination are indeed little or non-reactive, and are therefore less likely to oxidize, unlike hydrophilic ones.

At nanoscale, it is possible that, under the experimental conditions of this study, capillary forces are concentrated at the contact interface, and induce an increase of the friction force. Water vapour could then concentrate around the tip rather than in the tip–graphite interface; consequently, in order to move forward and continue the friction, the tip should increasingly defeat these capillary forces. The latter constitutes an additional obstacle to the advance of the tip, leading to an increase of the friction force.

25.4.2.3
Friction of the A15/Grafted Silicon Wafer Couples: A Possible Correlation Between Nano- and Macroscopic Behaviour?

The tribological properties of A15/Si couples, with Si wafers presenting different superficial energies, are evaluated through tests on a pion-disc tribometer, in order to provide a comparison between the results obtained at nano- and macroscopic scales [65].

Under these experimental conditions, it appears that the coefficient of friction of the A15/hydrophilic Si couple is smaller than the one obtained with a bulk substrate, whereas the use of a hydrophobic silicon leads to a friction which is nearly four-times higher compared with an untreated wafer: μ (bulk Si) $= 0.08$; μ (hydrophilic Si) $= 0.065$; μ (hydrophobic Si) $= 0.158$.

The extremely low coefficient of friction obtained in the case of hydrophilic wafers might seem surprising as a sliding on hydrophilic surfaces (that is to say strongly polar) is generally more dissipative than what is observed on hydrophobic (non-polar) surfaces. The specific behaviour of graphite in the presence of water (lubricating action) can explain these results and confirm the key role of the adsorbed moisture. After a Piranha treatment, the surface of the silicon wafer is covered with OH clusters; these polar groups are in favour of the adsorption of water vapour on the substrate. The amount of relative humidity at the level of the contact is, in the case of the hydrophilic substrate, more important than during friction against the bulk surface. Water acting as a lubricant for graphite, the friction coefficient in the presence of a hydrophilic surface is lower than in the case of a bulk surface, since the number of OH groups naturally present on the surface of bulk silicon is not sufficient for this self-lubricating process to be fully realized; the coefficient of friction of the graphite/bulk Si couple is consequently higher than that for the graphite/hydrophilic Si couple, but remains below the one obtained after sliding against the hydrophobic wafer.

It turns out that the higher the surface energy of the silicon wafers, the lower the coefficient of friction of the studied couples; this underlines the unusual behaviour of graphite rubbing. The friction on a non-polar surface will be more difficult than on a polar one which promotes the self-lubrication of graphite.

The SEM images realized after friction showed that the transfer on the hydrophilic wafer is composed of parallel tracks, themselves made up of clusters spread in length, and plastically deformed. By contrast, graphitic planes are still visible on the clusters that constitute the transfer on the bulk wafer; moreover, their ends look like "strips"; such morphology may suggest that the process of spreading of the material was underway, but that the "strips" were not sufficiently crushed to be integrated into the body of the cluster. These clusters with this typical morphology, are unfavourable to the friction, and might explain the higher friction coefficient noticed during tests against bulk wafers (compared to hydrophilic ones).

The transfer on hydrophobic wafers is made of relatively spread out and rounded clusters; such an aspect could let one suppose the formation of a "carrying" film, favourable to the friction. However, the highest stabilized coefficient of friction is observed during tests against this kind of substrate: it is then possible that, because of the hydrophobic nature of the wafer (relatively low superficial energy), the transfer is less important, limiting thus the formation of these small islands, which, in sufficient quantity, could have been favourable to the friction.

Transfer onto hydrophobic surfaces seems therefore to be more difficult, as if there was no (or little) affinity between graphite and the CH_3 terminations covering the hydrophobic wafers.

The surface chemistry plays then a fundamental role in the tribological behaviour of graphite; this parameter has not, however, the same consequences according to the scale on which the tests are conducted. The friction of graphite A15 against bulk, hydrophilic and hydrophobic silicon wafers was studied on a macroscopic scale, and it was found that the coefficient of friction of the A15/Si wafer couples decreases when the surface energy increases [65]: the use of a hydrophilic silicon wafer leads then to the lowest friction coefficient (macroscopic scale). On the contrary, it was observed that the friction due to the sliding of a hydrophilic tip against graphite A15 is clearly more important than during friction with a hydrophobic tip (nanoscopic scale).

The influence of surface chemistry on the tribological behaviour of A15 graphite is undeniable, but the nature of the contact (Si_3N_4 tip/graphite or graphite/silicon wafer), and the scale on which the tests are conducted also play an important role.

25.5
Conclusions

Atomic Force Microscopy is one of the most suitable techniques for determining the tribological characteristics of materials at the nanoscale. The friction properties of a wide variety of carbonaceous films have already been the subject of numerous studies; the results obtained through these works are summarized in a "state of the art" section.

Contact measurements carried out on diamond coatings (obtained by the flame process) have revealed that the friction became more important when the applied contact load increased; this was observed on a *facet of a crystal*, on a *crystal* and also on *larger zones* of the diamond coating. Moreover, the appearance of the TMR signal seemed to be similar to the topography of the coating.

Similar tests were realized on graphite powders; it appeared that if the scanning velocity did not seem to influence their friction properties at the nanoscopic scale, the contact load, on the other hand, played an important part: an increase of this parameter induced a rise of the friction. Moreover, the role of the superficial chemistry of the AFM tips was studied: a hydrophilic tip leads to a higher friction than in the case of a hydrophobic tip, under the same experimental conditions. These results are opposed to those obtained on a macroscopic scale: the nano-macro correlation remains at present a challenge.

References

1. Bhushan B (2005) Wear 259:1507
2. Andersson J, Erck R, Erdemir A (2003) Surf Coat Technol 163–164:535
3. Erdemir A (2004) Tribol Int 37:577
4. Ronkainen H, Varjus S, Koskinen J, Holmberg K (2001) Wear 249:260
5. Kennedy F, Lidhagen D, Erdemir A, Woodford J, Kato T (2003) Wear 255:854
6. Koskinen J, Ronkainen H, Varjus S, Muukkonen T, Holmberg K, Sajavaara T (2001) Diamond Relat Mater 10:1030
7. Erdemir A (2004) Tribol Int 37:1005
8. Homola A, Israelachvili J, McGuiggan P, Gee M (1990) Wear 136:65
9. Schwarz U, Zwörner O, Köster P, Wiesendanger R (1997) Phys Rev B 56:6987
10. Germann G, Cohen S, Neubauer G, Mc Cleeland G, Seki H, Coulman D (1993) J Appl Phys 73:163
11. Santos L, Trava-Airoldi V, Iha K, Corat E, Salvadori M (2001) Diamond Relat Mater 10:1049
12. Tambe N, Bhushan B (2005) Scripta Mater 52:751
13. Bhushan B, Koinkar V (1995) Surf Coat Technol 76–77:655
14. Gupta B, Bhushan B (1995) Thin Solid Films 270:391
15. Li X, Bhushan B (1999) J Mater Res 14:2328
16. Sundararajan S, Bhushan B (1999) Wear 225–229:678
17. Liu E, Blanpain B, Celis JP, Roos J (1998) J Appl Phys 84:4859
18. Ruan J, Bhushan B (1993) J Mater Lett 8:3019
19. Shinjo K, Hirano M (1993) Surf Sci 283:473
20. Dienwiebel M, Verhoeven G, Pradeep N, Frenken J, Heimberg J, Zandbergen H (2004) Phys Rev Lett 92:126101
21. Dienwiebel M, Pradeep N, Verhoeven G, Zandbergen H, Frenken J (2005) Surf Sci 576:197
22. Mate C (1993) Wear 168:17
23. Bowden F, Tabor D (1950) The friction and lubrication of solids. Oxford University Press, Oxford
24. Skinner J, Gane N, Tabor D (1974) Nat Phys Sci 232:195
25. Mate C, McClelland G, Erlandsson R, Chiang S (1987) Phys Rev Lett 19:1942
26. Miyamoto T, Kaneko R, Andoh Y (1991) Adv Inf Storage Syst 3:137
27. Tabor D (1974) The properties of diamond. Academic Press, London
28. Samuels B, Wilks J (1988) J Mater Res 23:2846
29. Wang M, Miyake S, Saito T (2005) Tribol Int 38:657
30. Riedo E, Chevrier J, Comin F, Brune H (2001) Surf Sci 477:25
31. Ma X, Komvopoulos K, Wan D, Bogy D, Kim Y (2003) Wear 254:1010
32. Buzio R, Gnecco E, Boragno C, Valbusa U (2002) Carbon 40:883
33. Grierson D, Carpick R (2007) Nanotoday 2:12
34. Prioli R, Chhowalla, Freire F (2003) Diamond Relat Mater 12:2195
35. Riedo E, Levy F, Brune H (2002) Phys Rev Lett 88:3793
36. Matsumoto N, Joly-Pottuz L, Kinoshita H, Ohmae N (2007) Diamond Relat Mater 16:1227

37. Ohmae N (2006) Tribol Int 39:1497
38. Binggeli M, Mate C (1994) Appl Phys Lett 65:415
39. Carpick R, Flater E, Sridharan K (2004) Polym Mater Sci Eng 90:197
40. Enachescu M, Van den Oetelaar R, Carpick R, Ogletree D, Flipse C, Salmeron M (1999) Tribol Lett 7:73
41. Buzio R, Boragno C, Valbusa U (2003) Wear 254:981
42. Carpick R, Flater E, Ogletree D, Salmeron M (2004) J Occup Med 56:48
43. Enachescu M, Van den Oetelaar R, Carpick R, Ogletree D, Flipse C, Salmeron M (1998) Phys Rev Lett 81:1877
44. Gao G, Cannara R, Carpick R, Harrisson J (2007) Langmuir 23:5394
45. Kinoshita H, Kume I, Tagawa M, Ohmae N (2004) Appl Phys Lett 85:2780
46. Davis C, Amaratunga G, Knowles K (1998) Phys Rev Lett 80:3280
47. Charitidis C, Logothetidis S (2005) Diamond Relat Mater 14:98
48. Ahn H, Cuong P, Park S, Kim Y, Lim J (2003) Wear 255:819
49. Jiang Z, Lu C, Bogy D, Bhatia C, Miyamoto T (1995) Thin Solid Films 258:75
50. Fang T, Weng C, Chang J, Hwang C (2001) Thin Solid Films 396:167
51. Prioli R, Jacobsohn L, Maia da Costa M, Freire F (2003) Tribol Lett 15:177
52. Van den Oetelaar, Flipse C (1997) Surf Sci 384:L828
53. Hirose Y, Kondoh N (1988) in Extended abstract of 35th Spring meeting of the Japanese Society of Applied Physics. Tokyo, p 343
54. Paulmier D, Schmitt M, Mermoux M (2001) in Proceedings of the Sixth Applied Diamond Conference (ADC/FCT 2001); NASA/CP-2001-210948, Auburn, p 293
55. Allara D, Patrikh A, Rondelez F (1995) Langmuir 11:2357
56. Boumaza S, Schmitt M, Bistac S, Jradi K (2007) Master thesis, Université de Haute-Alsace, France: "Etude du frottement du couple pion de graphite/pointe AFM à l'échelle nanoscopique par AFM"
57. Enomoto Y, Tabor D (1981) Proceedings of the Royal Society A 373:405
58. Casey M, Wilks J (1973) J Phys D6:1772
59. Samuels B, Wilks J (1988) J Mater Sci 20:213
60. Schmitt M, Bistac S, Jradi K (2007) J Phys Conf Ser 61:1032
61. Jradi K (2007) Ph.D. thesis, Université de Haute-Alsace, France: "Etude du comportement tribologique de poudres de graphite: transfert aux échelles nano et macroscopiques"
62. Riedo E, Chevrier J, Comin F, Brune H (2001) Surf Sci 477:25
63. Charitidis C, Logothetidis S, Gioti M (2000) Surf Coat Technol 125:201
64. Huang L, Xu K, Lu J, Guelorget B (2002) Surf Coat Technol 154:232
65. Jradi K, Boumaza S, Schmitt M, Bistac S (2008) Influence de la chimie de surface sur le comportement tribologique des couples graphites/silicium, accepted in Presses Romandes Universitaires

26 Atomic Force Microscopy Studies of Aging Mechanisms in Lithium-Ion Batteries

Shrikant C. Nagpure · Bharat Bhushan

Abstract. Lithium-ion batteries have been very popular in the past decade. They are now commonly used in portable devices such as mobile phones, laptop computers, and digital cameras due to their high energy density. Recently, they have been developed for applications requiring long life, such as electric vehicles (EV) and hybrid electric vehicles (HEV). Prolonged aging is one of the key attributes in such applications. Investigation of aging mechanisms in lithium-ion batteries becomes very challenging as aging does not occur due to a single process, but because of multiple processes occurring at the same time. Moreover, the anode and the cathode in lithium-ion batteries have different aging mechanisms. In this chapter, we review the recent studies conducted using the atomic force microscope (AFM) to understand the aging mechanisms in lithium-ion batteries. These include studies conducted to understand the surface film formation, the morphological changes, and the changes in surface properties of carbon-based anodes and $LiNi_{0.8}Co_{0.2}O_2$ and $LiNi_{0.8}Co_{0.15}Al_{0.05}O_2$ cathodes.

$$Li_{1-x}CoO_2 + Li_xC_6 \rightleftarrows Li_{1-x}C_6 + Li_xCoO_2$$

Key words: Batteries, Energy storage, Aging, Atomic force microscope, Hybrid electric vehicles, Electric vehicles

26.1
Introduction

Energy (from the Greek ενεργός, *energos*, "active, operation") is one of the most fundamental necessities for our existence. In recent years, "energy" has become the buzz word. Countries' economies revolve around their energy demands and supply. Countries try to achieve energy independence by identifying their energy sources and optimizing their uses. As shown in Table 26.1, energy sources can be broadly classified into two main categories; non-renewable and renewable sources [5]. Non-renewable energy sources take millions of years to be created and cannot be replenished in a short time. Non-renewable energy sources include oil, natural gas, coal, and nuclear. They are used extensively as compared to renewable energy sources, but their use produces greenhouse gases, and as such they have a detrimental impact on the environment. Renewable energy sources can be replenished in a short period of time. Renewable energy sources include solar, wind, geothermal, biomass, and hydro. Their use is clean and has much less impact on the environment.

Energy exists in the universe in different forms as described in Table 26.2 [5]. According to the law of conservation of energy, energy can neither be created, nor destroyed; it can only be changed from one form to the other. The two main forms of energy are kinetic energy and potential energy. Kinetic energy exists in a system by virtue of its motion, while potential energy is stored in a system by virtue of its position. Kinetic energy is further classified into different forms as motion, electrical, thermal, radiant, and sound. Similarly potential energy is classified as gravitational, mechanical, nuclear, and chemical. Out of these different forms of energy, electrical energy is the widely used form, and it is obtained from the various non-renewable and renewable energy sources mentioned earlier.

"Electricity is a basic part of nature, and it is one of the most widely used forms of energy" [5]. In 2007, according to the Energy Information Administration, the US generated 4.05 trillion kWh of electricity. In 2005, the US was the leader in the generation of electricity, generating 23% of the total electricity generated in the world [5]. Electrical energy helps in many ways; it lights cities, keeps houses warm or cool, one can play music, watch television, and it also energizes computers, refrigerators, washing machines, cooking ranges, etc. For some appliances, especially portable appliances, we need the electrical energy in a stored form. A battery is a device used to store electrical energy. Most notably, batteries are used in mobile phones and laptops. Batteries have also helped in developing electric cars, which have less impact on the environment.

A battery is a device that stores electrical energy in chemical form, and by the principle of a galvanic cell, it converts this chemical energy into electricity as and when needed. A galvanic cell is a device consisting of an anode and a cathode dipped in an electrolyte, and it produces electricity by spontaneous reduction-oxidation (redox) reaction. For example, consider a galvanic cell with a Zn anode dipped in $Zn(NO_3)_2$ solution, a Cu cathode dipped in $Cu(NO_3)_2$ solution, and a salt bridge of $NaNO_3$, as shown in Fig. 26.1 [11]. In the anode compartment, the Zn metal on the anode oxidizes, losing two electrons, and transforms to Zn^{2+} ions. The Zn^{2+} ions go into the solution while the two electrons travel through the external circuit, constituting the electric current. In the cathode compartment, the two electrons reduce Cu^{2+} and form Cu metal. This Cu metal is deposited on the cathode. The electrons can flow through the external circuit only if the two compartments are electrically neutral. The two compartments are separated from each other by a porous barrier, also called a salt bridge, which allows anions and cations to flow through, thus neutralizing the charge in each compartment. A battery consists of one or more galvanic cells connected in series or parallel. (Henceforth, in this chapter a battery stands for one galvanic cell.) Batteries are broadly classified as primary batteries and secondary batteries. Primary batteries are those which are cycled (fully discharged) only once and discarded. Secondary batteries, also known as rechargeable batteries, are those which can be cycled (charged and discharged) more than once.

Table 26.1. Different sources of energy (Adapted from [5])

Sources of Energy

Non-renewable	Renewable
These energy sources take million years to be created and so can not be replenished in short time. Their use leads to greenhouse gases.	These are the energy sources that can be replenished in short period of time. Their use is clean and has very less effect on the environment.

Oil

Over millions of years the buried remains of plants and animals are turned into oil, e.g.: gasoline. Though oil helps us in many ways, extracting, moving and burning oil causes greenhouse gases.

Natural Gas

Decay matter from plants and animals buried over millions of years turns to natural gas. It is used mostly as heating fuel. Though it burns cleaner than other fossil fuels, it still affects the environment.

Coal

Coal is sedimentary rock made of carbon and hydrocarbon. Most of the coal extracted in the US is used to generate electricity. Coal when burned generates greenhouse gases.

Nuclear

This is the energy contained in the nucleus of an atom, e.g. a uranium atom. Nuclear energy is used to generate electricity through nuclear fission reaction. It is clean, but the radioactive emissions are dangerous and as such needs special handling.

Solar

It is the energy from the rays of the sun. It is converted to thermal energy or electricity. The disadvantage of solar energy is inconsistent amounts of sunlight reaching earth and also a large surface area is needed to collect considerable amounts of sun rays.

Wind

Winds are created due to uneven heating of the earth's surface. These winds are converted to electricity using wind mills. It is not a continuous source of energy.

Geothermal

This is the energy from the heat within the earth. The two main ingredients are water and heat. The hot water can be directly used for heating. Electricity is also generated from geothermal resources. Geothermal plants have very low emissions.

Biomass

This is organic material from animals and plants, e.g. wood, crops, and some garbage. Biomass when burned can pollute the air. The other way of extracting energy is to convert it into methane or ethanol or biodiesel.

Hydro

In the US most of the electricity is generated by hydropower. Water is stored in dams and then released to drive the turbines that generate electricity. It is absolutely clean, though some may argue that it affects the natural habitat.

Table 26.2. Different forms of energy (Adapted from [5])

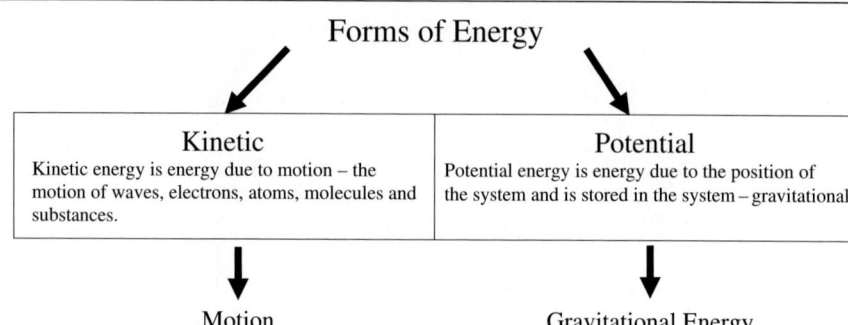

Forms of Energy

Kinetic	Potential
Kinetic energy is energy due to motion – the motion of waves, electrons, atoms, molecules and substances.	Potential energy is energy due to the position of the system and is stored in the system – gravitational

Motion
Objects move from one place to another when a force is applied according to Newton's law of motion. Energy due to this motion of objects is motion energy, e.g. flowing water (hydropower).

Electrical Energy
Electrical energy is due to the movement of electrons. Electrical charges moving through wire is called electricity. It is the most widely used form of energy, e.g. lighting.

Thermal (Heat) Energy
This is the internal energy of a substance due to the vibrations and movements of atoms and molecules within the substance, e.g. geothermal energy.

Radiant Energy
Radiant energy is electromagnetic energy that travels in transverse waves. It includes visible light, X-rays, gamma rays and radio waves, e.g. solar energy.

Sound
Sound is vibration of matter transferred through the matter as a wave. It is propagated through matter in longitudinal (compression/rarefaction) waves.

Gravitational Energy
Gravitational energy is energy by virtue of position of the substance, e.g. water stored behind a dam or in an over head reservoir.

Stored Mechanical Energy
Stored mechanical energy is energy stored in objects by the application of a force, e.g. compressed shock absorber.

Nuclear Energy
This is the energy contained in the nucleus of an atom, e.g. nucleus of a uranium atom.

Chemical Energy
Chemical energy is the energy by virtue of which atoms and molecules are held together, e.g. petroleum.

26.1.1
Battery Types

Batteries are classified according to their chemistries and shapes. Depending on the chemistry of the battery, different materials are used to make its components. Table 26.3 summarizes the properties of various battery chemistries, while Table 26.4 lists the advantages and limitations of these batteries. These different battery chemistries and their advantages and limitations are briefly discussed here [12].

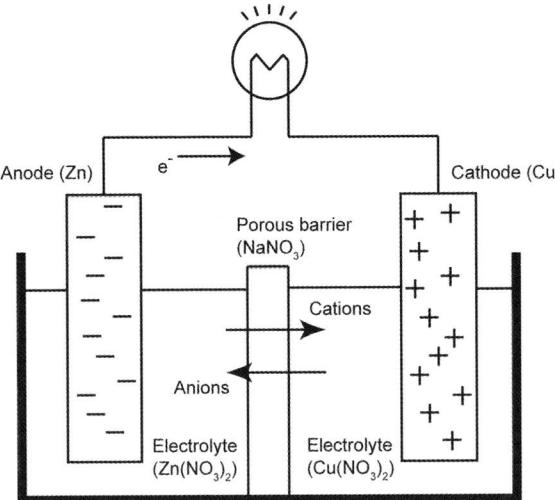

Fig. 26.1. Schematic of a galvanic cell. Electricity is generated in a galvanic cell by spontaneous oxidation of the anode and reduction of the cathode

26.1.1.1
Lead-Acid

French physician Gaston Planté introduced the lead acid battery in 1859. It was the first commercially available rechargeable battery. Though many battery chemistries were introduced later on, lead acid is still used in a lot of applications due its cost effectiveness and ruggedness. Because of gassing and water depletion, this battery can never be charged to its full capacity. Also, each time the battery is discharged completely, it tends to lose a small amount of its capacity. Because of the lead content, this battery is treated as non-friendly to the environment. They are used in wheelchairs, hospital equipment, automobiles, and uninterruptible power supply systems [12].

26.1.1.2
Nickel-Cadmium

The nickel-cadmium battery was invented in 1899 by a Swedish man, Waldmar Jungner. In 1947, Neumann was successful in introducing the completely sealed version of this battery. This battery has a moderate energy density and a long cycle life. The main drawback in this battery is the memory effect, in which the battery gradually loses its maximum energy capacity if it is repeatedly recharged after being only partially discharged. Though the metals used are toxic, it remains a favorite choice in areas where the most rigorous charge-discharge cycles are expected. Because of its ability to draw heavy load currents, it is also a favorite choice in applications like power tools [12].

Table 26.3. Characteristics of commonly used rechargeable batteries [12]

		Lead-acid (sealed)	Ni-Cd	Ni-MH	Li-ion cobalt oxide	Li-ion manganese	Li-ion phosphate
Commercial use since		1970	1950	1990	1991	1996	2000
Gravimetric energy density (Wh/kg)		30–50	45–80	60–120	150–190	100–135	90–120
Internal resistance (mΩ)		<100 12 V pack	100–200 6 V pack	200–300 6 V pack	150–300 pack 100–130 per cell	25–75 per cell	25–50 per cell
Cycle life (to 80% of initial capacity)		200–300	1500	300–500	300–500	Better than 300–500	> 1000
Fast charge time (h)		8–16	1	2–4	1.5–3	1 or less	
Overcharge tolerance		High	Moderate	Low	Low. Cannot tolerate trickle charge		
Self-discharge/month (at room temperature)		5%	20%	30%	< 10%		
Cell voltage (V)		2	1.25	1.25	Nominal 3.6 Average 3.7	Nominal 3.6 Average 3.8	3.3
Load current	Peak	5	20	5	< 3	> 30	> 30
	Best	0.2	1	0.5 or lower	1 or lower	10 or lower	10 or lower
Operating temperature (°C)		−20 to 60	−40 to 60	−20 to 60	−20 to 60	−20 to 60	−20 to 60
Maintenance requirement		3–6 months	30–60 days	60–90 days	Not required		

Table 26.3. (continued)

	Lead-acid (sealed)	Ni-Cd	Ni-MH	Li-ion cobalt oxide	Li-ion manganese	Li-ion phosphate
Safety	Thermally stable	Thermally stable Fuse recommended	Thermally stable Fuse recommended	Protection circuit mandatory Stable to 150 (°C)	Protection circuit mandatory Stable to 250 (°C)	Protection circuit mandatory Stable to 250 (°C)
Toxicity	Toxic lead and acids harmful to environment	Highly toxic Harmful to environment	Relatively low toxicity, should be recycled	Low toxicity, can be disposed of in small quantities		

Table 26.4. Advantages and limitations of commonly used rechargeable batteries (Adapted from [12])

	Advantages	Limitations
Lead-acid (sealed)	Very inexpensive and simple to manufacture	Low energy density – limits use to stationary and wheeled applications
	Well-developed technology	Voltage should never drop below 2.1 V
	Lowest self-discharge	Allows only limited number of full discharge cycles
	Low maintenance – no memory effect, no electrolyte to fill on sealed version	Environmentally unfriendly due to the lead content
	Capable of high discharge rates	Thermal runaway can occur due to improper charging
		Some versions can never be charged to their full potential
Nickel-cadmium	Fast and simple charge	Relatively low energy density
	High cycle life	Shows memory effect
	Good load performance	Environmentally unfriendly and so some countries restrict its use
	Long shelf life	
	Easy storage and transportation	Relatively high self-discharge, needs recharging after storage
	Good low temperature performance	
	One of the most rugged rechargeable batteries	
	Lowest in terms of cost per cycle	
	Available in a wide range of sizes and performance options	
Nickel-metal-hydride	30–40% higher capacity than standard nickel-cadmium	Limited service life of 200–300 cycles
		Relatively short storage life
	Less prone to memory effect than nickel-cadmium	Limited discharge current
	Simple transportation	More complex charge algorithm due to heat generated during charging

Table 26.4.   (continued)

	Advantages	Limitations
	Environmentally friendly – contains only mild toxins; profitable for recycling	Trickle charge settings are critical because the battery cannot absorb overcharge
		High self-discharge
		Should be stored in a cool place at 40 °C
		High maintenance
Lithium-ion	Highest gravimetric energy density (Wh/kg)	Requires protection circuit to maintain voltage and current within safe limits
	Does not need prolonged priming	Aging is a major issue
	Relatively low self-discharge	Expensive to manufacture
	Low Maintenance – no memory effect	
	Cells with high current capacity can be manufactured for power tool applications	

26.1.1.3
Nickel-Metal Hydride

Development of the nickel-metal hydride battery started in 1970. Only after stable metal hydride alloys were developed in 1980, was this battery available for consumers. The cathode in the Ni-MH battery is composed of nickel hydroxide. The anode is mostly composed of an intermetallic compound of type AB_5 where A is a rare earth mixture of lanthanum, cerium, and titanium, and B is nickel, cobalt, manganese and/or aluminum [16]. The materials used in this battery are non-toxic. Though it is better in some aspects than the nickel-cadmium battery, it shares some of its drawbacks with the nickel-cadmium battery due to the nickel technology. The cycle life of this battery is less than that of nickel-cadmium. It has a higher energy density compared to nickel-cadmium and shows no memory effect. Before the lithium-ion battery was introduced, the nickel-metal-hydride battery was used in mobile computing and wireless communications. It is often believed among researchers that nickel-metal-hydride led to the development of the lithium-based battery [12].

26.1.1.4
Lithium-Ion

The research for lithium-ion battery technology started with lithium batteries. In 1912, G. N. Lewis began working on lithium batteries. Lithium (atomic number 3, group 1, period 2) was used as an anode in lithium batteries. Lithium is an alkali metal, and being the lightest of the metals with greatest electrochemical potential, has the largest energy density for weight. But, lithium is very unstable, as it has only one valence electron. This made lithium batteries very unsafe for commercial use and so, research shifted from a lithium battery to a lithium-ion battery, which is a much safer option [12].

26.2
Lithium-Ion Batteries

Sony commercialized the lithium-ion battery in 1991, and since then it has been growing and gaining popularity at a rapid pace. The lithium-ion battery was initially used in mobile phones and laptop computers. This battery shows no memory effect and requires minimal maintenance. Its energy density is the highest among the other chemistries. Recently, due to its improved load characteristics and ability to draw high currents, it has also been used in power tools and medical devices. A protection circuit is often found in lithium-ion battery systems to limit charge and discharge, voltages and currents for safety reasons. Also, to avoid any thermal runaway and consequential fire hazard, the temperatures are continuously monitored in a lithium-ion battery system. In spite of requiring a protection circuit and a risk of thermal runaway, lithium-ion batteries have emerged as the single most favorite for portable applications, due to the other advantages it has over the alternate battery chemistries. The different types of lithium-ion batteries listed in Table 26.3 are discussed in detail later on.

26.2.1
Electrochemistry of Lithium-Ion Batteries

The anode in a lithium-ion battery is mostly made up of graphitic carbon [24], while different materials such as $LiCoO_2$, $LiMnO_2$, and $LiFePO_4$ are being tried for the cathode. The electrolyte in a lithium-ion battery is a lithium salt, such as $LiPF_6$, $LiBF_4$, and $LiClO_4$, soluble in an organic solvent, such as dimethyl carbonate (DMC) and/or ethylene carbonate (EC). The chemical reaction in the case of the lithium-ion battery is based on intercalation. In an intercalation reaction, guest ions are introduced in the host structure. The intercalation reaction is a reversible reaction and barely modifies the host structure. The host structure can be either two-dimensional or even three-dimensional. Figure 26.2 shows a schematic of an intercalation reaction between a carbon anode and a $LiCoO_2$ cathode [6]. As seen in the schematic, intercalation reaction occurs during charging, and Li^+ ions are transferred from the cathode into the anode. During discharging, a deintercalation reaction occurs, and Li^+ ions are transferred from the anode to the cathode while the electrons flow through the external circuit. The following chemical reaction occurs in a lithium-ion battery with a carbon anode and a $LiCoO_2$ cathode:

$$Li_{1-x}CoO_2 + Li_xC_6 \rightleftharpoons Li_{1-x}C_6 + Li_xCoO_2$$

The Li^+ ions are never oxidized in this chemical reaction but are only intercalated and deintercalated to and from the anode or cathode. The Co in $LiCoO_2$ undergoes

Anode (Carbon)   Cathode (LiCoO$_2$)

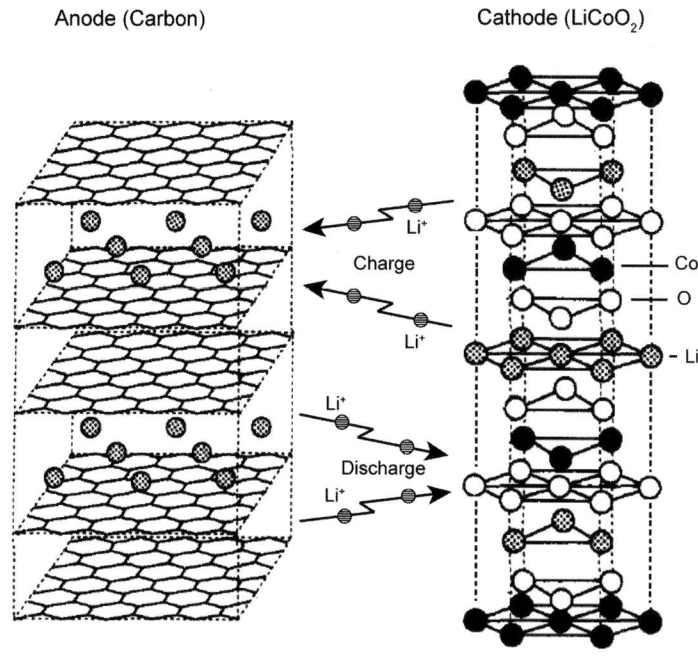

Courtesy of Panasonic

Fig. 26.2. Schematic of intercalation-deintercalation process. During charging Li$^+$ ions are inserted in the anode structure by virtue of intercalation and during discharging, these Li$^+$ ions are removed from the anode structure by virtue of deintercalation and transferred to the cathode [6]

oxidation from Co^{3+} to Co^{4+} during charging, and reduction from Co^{4+} to Co^{3+} during discharge. This is slightly different than the reaction in conventional batteries, where the anode undergoes oxidation and the cathode undergoes the reduction reaction.

26.2.2
Different Shapes of Lithium-Ion Batteries

Figure 26.3 shows three different shapes of commercially available lithium-ion batteries; cylindrical, prismatic, and pouch cell [12].

26.2.2.1
Cylindrical Cell

The cylindrical cell (Fig. 26.3a) is the most widely used among the available battery shapes. The components are rolled and packed in a cylindrical case. The cylindrical cell can hold more active material than any other battery shape. Hence, it has more energy density. Also, they are easy to manufacture and have high mechanical stability [12].

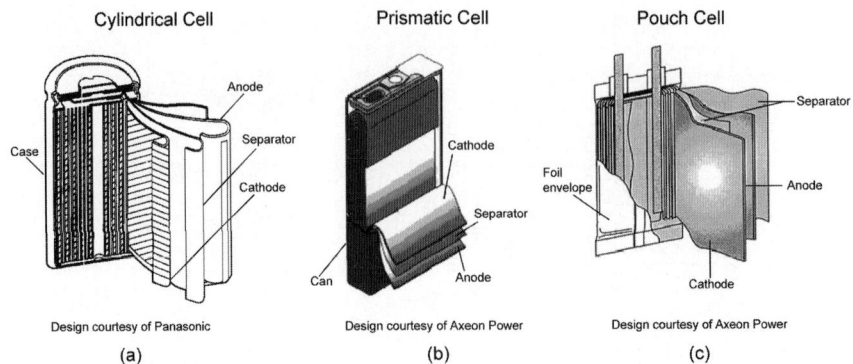

Fig. 26.3. (**a**) Schematic of a cylindrical lithium-ion battery. The components are rolled and packed in a cylindrical case [6]. (**b**) Schematic of a prismatic lithium-ion battery. The components are stacked in layers and enclosed in a rectangular casing [2]. (**c**) Schematic of a pouch cell. The components are stacked in layers and enclosed in a foil envelope [2]

26.2.2.2
Prismatic Cell

The prismatic cell (Fig. 26.3b) was developed in the early 1990s to serve the need of the semiconductor industry, where devices continue to get smaller and smaller. Because of its thinner geometry, it has lower energy density. Also, it becomes a manufacturing challenge to pack the components into a thinner geometry. They are mechanically less stable than cylindrical batteries and tend to bulge out due to a lack of proper venting for built up gases [12].

26.2.2.3
Pouch Cell

The introduction of the pouch cell (Fig. 26.3c) in 1995 drastically reduced the manufacturing cost of the battery systems. The components are stacked in a layer and enclosed in a foil instead of a metallic casing. It is very flexible and the most compact among the available battery shapes. It is lightest in weight as compared to any other battery shapes, but has a slightly lower energy density and load current. The drawbacks of the pouch cell are swelling due to the evolution of gases and high sensitivity to twist [12].

26.2.3
Types of Lithium-Ion Batteries

In lithium-ion batteries, as stated earlier, graphitic carbon is the most common material for the anode while newer materials are being tested for the cathode. The different types of lithium-ion batteries, summarized in Tables 26.3 and 26.5, come from the fact that cathodes can be composed of different materials [12]. All these types of lithium-ion batteries require a protection circuit for safe operation. They don't have any toxic material and can be disposed off very easily [12].

Table 26.5. Most common types of lithium-ion batteries [12]

Lithium-ion battery chemistry	Nominal voltage (V)	Charge limit (V)	Charge and discharge C-rates	Energy density (Wh/kg)	Applications	Note
Cobalt oxide	3.60 V	4.20 V	1C cont 1C pulse	110–190	Cell phone, cameras, laptops	Since the 1990s, most commonly used for portable devices; has high energy density.
Manganese (spinel)	3.7– 3.80 V	4.20 V	10C cont. 40C pulse	110–120	Power tools, medical equipment	Low internal resistance; offers high current rate and fast charging but lower energy density.
NCM (nickel–cobalt manganese)	3.70 V	4.10 V	~5C cont. 30C pulse	95–130	Power tools, medical equipment	Nickel, cobalt, manganese mix; provides compromise between high current rate and high capacity.
Phosphate (A123 System) Saphion®	3.2– 3.30 V	3.60 V	35C cont.	95–140	Power tools, medical equipment	New, high current rate, long cycle life. Higher charge V, increase capacity but shorten cycle life.

C rate is charge or discharge rate of the battery expressed in terms of its total storage capacity in Ah or mAh.

26.2.3.1
Cobalt Oxide

When Sony introduced the first lithium-ion battery they used cobalt oxide as a cathode. These batteries have an energy density of 110–190 Wh/kg and have a cycle life of 300–500 cycles. The drawback in these batteries is that, the charge-discharge rates can not exceed 1C, and as such they are only suitable for low current discharge applications. Another drawback in these batteries is that the internal resistance increases within 2–3 years of cycling [12].

26.2.3.2
Manganese Spinel

In 1996, scientists introduced manganese oxide spinel as a cathode material. The three-dimensional spinel structure of the cathode allows higher ion flow between the electrodes. Higher ion flow decreases the battery's internal resistance and increases its loading capacity. The manganese-spinel based battery also has less safety circuitry because of its thermal stability [12]. The high current discharging capability has made these types of batteries a favorite in current portable devices. The one drawback of a spinel-based lithium-ion battery is the low capacity, as compared to a cobalt-based lithium-ion battery. Its energy density is in the range of 110–120 Wh/kg. E-One Moli Energy of Canada is the leading manufacturer of this type of battery [12].

26.2.3.3
Nickel-Cobalt-Manganese

Nickel-cobalt-manganese cathodes are being developed by Sony. The cathode is a multi-metal oxide material consisting of nickel, cobalt, and magnesium to which lithium is added. This battery is a compromise between the high current rate of a spinel-based battery and the high capacity of cobalt-based batteries. It is very important to keep in mind that these two characteristics cannot be combined in one single battery system. Hence, various product ranges are available from Sony [12]. To state an example, if a nickel-cobalt-manganese-based battery is charged to 4.2 V/cell instead of its regular charging at 4.10 V/cell, which is 100 mV lower than cobalt and spinel-based batteries, the capacity increases but the cycle life reduces from 800 to 300 cycles [12].

26.2.3.4
Phosphate

A123 Systems has introduced a battery with a cathode made up of nano-phosphate materials. A123 Systems claims that their lithium-ion battery has the highest gravimetric energy density (Wh/kg) among all the other commercially available lithium-ion batteries. It can handle discharge rates 10-times the nominal rates and can be continuously discharged to 100% depth-of-discharge at 35°C. They also have a lower nominal and peak charge voltage than the other lithium-ion battery systems [12]. Valance Technology commercialized the phosphate-based lithium-ion battery systems and sold it under the brand name of Saphionâ [12].

26.2.4
Requirements in Modern Lithium-Ion Batteries

Lithium-ion batteries were introduced for devices like mobile phones and laptop computers. These devices undergo rapid development cycles, and newer versions arrive in a short period of time. As such, the age of the lithium-ion battery exceeded the age of these devices. Recently, lithium-ion batteries have been trying to acquire a market share in more durable goods, like the automobile industry. Lithium-ion batteries have found applications in electric vehicles (EV), hybrid electric vehicles (HEV), and temporary storage systems for renewable energy sources. EV or HEV not only reduce the burden on gasoline consumption and in turn on the non-renewable energy sources, but can also help protect the environment by lowering emissions. The US Advanced Battery Consortium (USABC, www.usabc.org) was established by the US Council for Automotive Research (USCAR). One of the objectives of USABC is to continue the development of battery technology for EV and HEV. According to USABC, a 42-V battery in a HEV should have a calendar life of 15 years [1]. Electric vehicles should have a battery system that can last for 10 years [3]. In terms of cycles, 1000 cycles at 80% depth-of discharge are expected in EV [3] and 300,000 cycles at 50 Wh are expected in a plug-in HEV [4]. This makes a prolonged life for a lithium-ion battery an absolute necessity, if it is to succeed in the automobile industry. End of life for the batteries used in automobiles is usually considered to be reached when the battery delivers only 80% of its rated ampere-hour capacity [4].

26.3
Aging of Lithium-Ion Batteries

Every secondary or rechargeable battery has a finite life, and the lithium-ion battery is no exception to this fact. Aging in a lithium-ion battery is a complex mechanism. Various interrelated processes occur at the same time during the cycles of a lithium-ion battery. Because of the interactions of these processes, it is difficult to study and predict the effect of any single process on the aging mechanism of the battery [24]. During each charge and discharge cycle, a battery undergoes certain chemical reactions which lead to its aging in subsequent cycles. Aging of a battery is not only dependent on its inherent chemistry but also depends on the charge-discharge rates, depth of discharge (DOD), and operating environment. Aging can be caused by changes in the electrode-electrolyte interface, changes in the electrolyte and/or changes in the composite electrodes [24]. In a lithium-ion battery, aging studies are mainly focused on the changes occurring at the anode and the electrolyte interface, and the cathode. As stated earlier, in almost all commercially available lithium-ion batteries, the anode is made up of carbon, especially graphite, and the cathode material varies as shown in Table 26.5. As such, many aging studies related to the graphitic anode are available in the literature. Less data is available about aging studies related to the cathode material. The majority of references found in the literature regarding the AFM studies of the cathodes in lithium-ion batteries come from the Advanced Technology Development Program (ATD), an initiative by the US Department of Energy (DOE).

Various destructive or non-destructive techniques are used to study the changes in the anode and the cathode as the battery ages. Electrochemical impedance spectroscopy (EIS), synchrotron infrared microscopy (SIM), Fourier-transform infrared spectroscopy (FTIR), X-ray diffraction (XRD), Raman spectroscopy, scanning electron microscopy (SEM), and atomic force microscopy (AFM) are a few techniques regularly used by researchers in this field. Recent studies have shown that although during aging the bulk properties change, such as a rise in impedance, capacity or power fade of the battery, the underlying phenomenon can be attributed to the interfacial processes. Study of surface properties and nanostructural changes, if any, along with morphological changes, can give better insight into the aging of a lithium-ion battery.

The aging mechanisms in the anode differ from the aging mechanisms in the cathode and so their aging effects are studied individually in the literature. While the aging in the anode is mostly related to the solid electrolyte interphase, aging in the cathode is related to nanostructural changes. We will divide our review of aging in the anode and the cathode into two separate sections.

26.3.1
Anode

Carbon is the most common anode material in all the lithium-ion batteries [24]. Several references are found in the literature regarding aging studies conducted on carbon anodes. Some studies were conducted on a carbon film deposited on a substrate, and some studies were conducted on the anode from a full-sized lithium-ion battery. Here we will review scanning probe experiments conducted on natural graphite, carbon film deposited on glass, and a graphite anode from a full-sized lithium-ion battery.

26.3.1.1
Literature Review of Anode Aging

An in-situ study was conducted by Jeong et al. [14] on composite graphite electrodes, prepared from graphite powder (NG-3), a caboxymethyl cellulose binder, and a styrene butadiene rubber binder in a weight ratio of 98:1:1: on copper foil (20 mm × 20 mm × 18 μm). In-situ AFM images (Fig. 26.4) showing surface morphological changes in a composite graphite electrode during constant current charging at 30 μA cm^{-2} in 1-M LiClO$_4$/ethylene carbonate (EC) – diethyl carbonate (DEC) (1:1) were obtained using Electrochemical Atomic Force Microscopy (ECAFM). Figure 26.4a shows a scan area of 20 × 20 μm, and Figs. 26.4b, c, d, e, and f show approximately the square area marked by the white border in Fig. 26.4a. At open circuit potentials of ∼3.3 V (Fig. 26.4b), and 1.4 V (Fig. 26.4c), there are no substantial changes in the surface morphology of the composite graphite anode. The surface morphology starts to show changes when the open circuit potential is reduced beyond 1.4 V. At the open circuit potential of 1.1 V (Fig. 26.4d) the particle edges start to curl up. This curling continues as the open circuit potential is further reduced to 1.0 V (Fig. 26.4e) and then to 0.8 V (Fig. 26.4f). Figure 26.4e, and f show

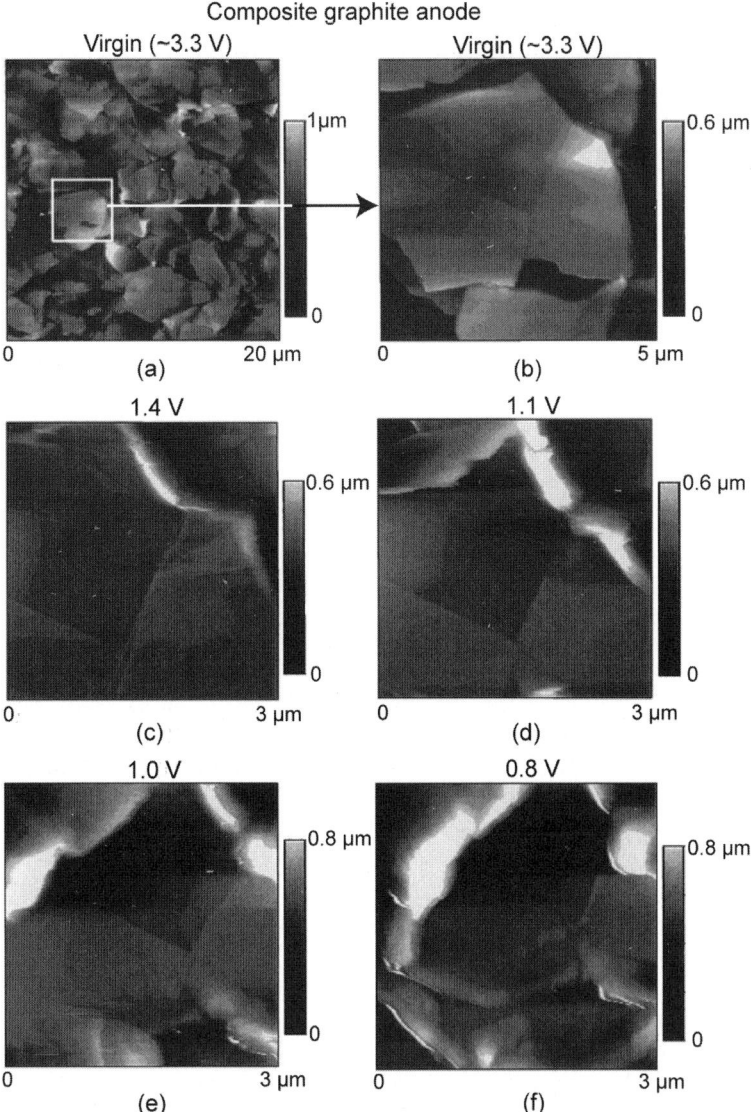

Fig. 26.4. Topographic AFM images of the composite graphite electrode surface obtained at the open circuit potential of (**a**) ~3.3 V, (**b**) ~3.3 V, (**c**) 1.4 V, (**d**), 1.1 V, (**e**) 1.0 V, and (**f**) 0.8 V during constant current charging at $30\,\mu A\ cm^{-2}$ in 1-M LiClO$_4$/EC – DEC (1:1). (**b**)–(**f**) is the square area shown in (**a**) [14]

the edges being swelled and also part of the edges being exfoliated. These changes in the edges of the composite graphite electrode are observed due to the intercalation of solvated lithium ions in the graphene layers.

Figure 26.5 [14] shows approximately the same area as in Fig. 26.4a but with a scan size of $10 \times 10\,\mu m$. Figure 26.5a shows the morphology of the graphite

Composite graphite anode

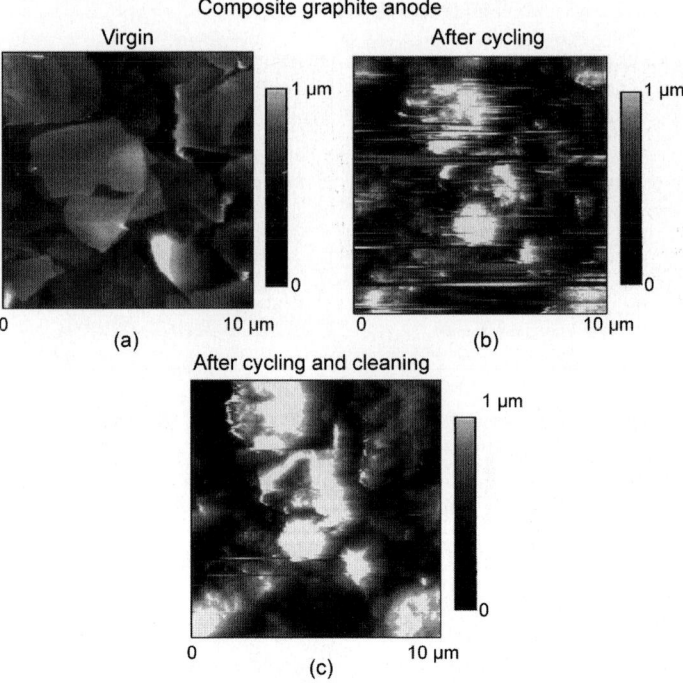

Fig. 26.5. Topographic AFM images of the composite graphite electrode surface obtained (**a**) before and (**b**) after cycling it from its open circuit potential of ~3. 3–0 V at $30\,\mu A\ cm^{-2}$ in 1-M LiClO₄/EC – DEC (1:1), (**c**) is obtained after cleaning the same sample by using a micro cantilever and AFM in contact mode [14]

anode at an open circuit potential of ~3. 3 V, while Figs. 26.5b and c show the morphology when the graphite anode is charged to the open circuit potential of 0 V in 1-M LiClO₄/EC – DEC (1:1). In Fig. 26.5b the entire surface is seen covered with precipitates that are considered to be the decomposition products of the electrolyte solution. Figure 26.5c is obtained after removing these precipitates by using an AFM microcantilever tip in contact mode. This reveals the morphological changes in the edges of the graphite flakes due to the process of intercalation.

Studies of other types of carbon samples exhibit a similar effect. Kong et al. [15] studied the topographical changes in evaporated carbon film on glass (Fig. 26.6a), and natural graphite (Fig. 26.6b). The evaporated carbon film sample was obtained by electron beam evaporation of carbon films on glass. For the natural graphite sample, natural graphite powder was compressed in a die without a binder by a hydraulic press into thin disks of ~1 mm thickness. The disks had a graphite density of 2 g cm⁻³ or about 88% theoretical density. Both samples were cycled from their open circuit potential of ~3. 0–0 V and then back to 3.5 V at 1–5 mV s⁻¹ in 1-M LiPF₆/EC – DEC (1:1). The virgin sample of evaporated carbon film (left-panel of Fig. 26.6a) shows smooth surface morphology with tightly packed globular grains. The virgin sample of natural graphite (left-panel of Fig. 26.6b) shows clearly aligned

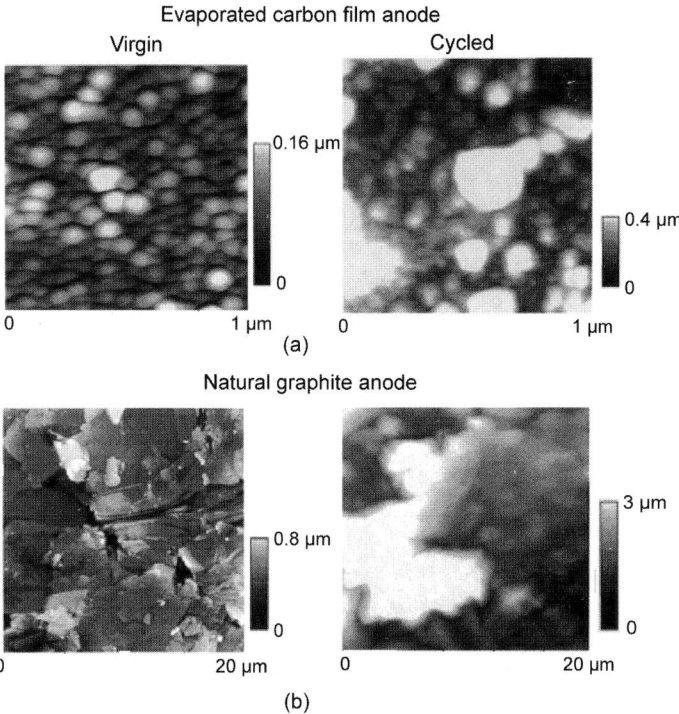

Fig. 26.6. Topographic AFM images of (**a**) evaporated carbon and (**b**) natural graphite electrodes before and after electrochemical cycling from their open circuit potential of $\sim$3.0–0 V and back to 3.5 V at 1–5 mV s^{-1} in 1-M LiPF$_6$/EC – DEC (1:1) [15]

graphite flakes with distinct edges. It shows the presence of visible cracks and an inhomogeneous surface due to the relatively unpolished piston faces of the press. The AFM images of both carbon samples after cycling show dramatic changes in the surface morphology and non-uniformity. The surface roughness increased by an order of magnitude after cycling. The size of some of the particles increased substantially, and they protruded from the surface. This is attributed to the local changes in carbon structure, mechanical breakdown of carbon particles, or the non-uniform surface layers.

An ex-situ study was conducted on the graphite anode of a full-sized lithium-ion battery by Kostecki and McLarnon [18]. A pouch cell was constructed for this study. The anode was composed of 92 wt.% MAG-10 graphite and 8 wt.% Polyvinylidene Fluoride (PVDF) on Cu foil. The cathode was made up of 84 wt.% LiNi$_{0.8}$Co$_{0.15}$Al$_{0.05}$O$_2$, 4 wt.% carbon black, 4 wt.% SFG-6 and 8 wt.% PVDF on Al foil. The electrolyte was 1-M LiPF$_6$/EC – DEC (1:1). The cell was subjected to a formation cycle by charging and then discharging at a low rate of C/25. The cell was then discharged at a constant current. During recharging the cell was charged to a voltage limit of 4.10 V, and was held at that voltage until the current dropped to C/20 or for a maximum of 2 h [23]. The cell was cycled at 60°C for 140 cycles after

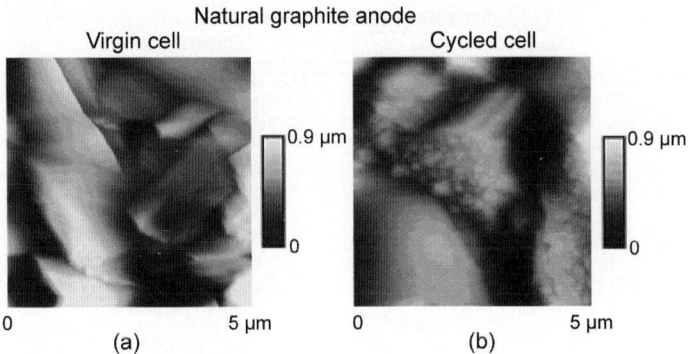

Fig. 26.7. Topographic AFM images of a graphite anode from the (**a**) virgin cell and the (**b**) cycled cell. The cell was cycled at C/2 rate, 100% DOD, between 3.0 and 4.1 V, at 60°C and lost 65% of its capacity after 140 cycles [18]

which it lost 65% of its capacity. The pouch cell was then fully discharged, disassembled in a He-atmosphere, and the graphite anode was washed in DMC for 24 h before being examined for morphological changes with an AFM. Figure 26.7 shows the AFM images from this study. The topography image of a virgin graphite anode (Fig. 26.7a) shows randomly oriented graphite flakes with distinct basal planes, steps, and edges. The topography image of the cycled anode (Fig. 26.7b) shows nanocrystalline deposits on graphite flake cross-section plane steps and edges but not on graphite planes. Thus, morphological changes in the graphite anode surface are observed due to intercalation/deintercalation in aged cells. Raman spectroscopy of this anode surface revealed spectral evidence of electrolyte decomposition products such as Li_2CO_3, $(PO_2)^-$, and $(PO_3)^-$ at locations of significantly disordered graphite surfaces [13, 18, 22].

26.3.1.2
Discussion of Aging Mechanisms in Anode Materials

The above review of AFM studies conducted on various carbon anodes suggests two basic changes occurring at the anode surface that can be related to the aging of the anode material. Firstly, due to the intercalation/deintercalation of Li^+ ions into the graphene layers during charge-discharge cycles, the anode surface undergoes morphological changes that lead to edge deformation, and also mechanical stresses being induced in the graphene layers. On the basis of their study (Fig. 26.4), Jeong et al. [14] attributed the morphological changes in the composite graphite anode to the intercalation of solvated lithium ions into graphene layers and the decomposition thereof. Figure 26.8 shows a schematic of the changes in the edges due to this intercalation of solvated lithium ions. The edges can swell, curl, and exfoliate due to the intercalation process [14]. The same results are observed in the case of highly oriented pyrolytic graphite (HOPG), except that the intercalation began at a more positive potential of ~1.1 V, and instead of curling edges, hill-like structures and their growth into the interior was observed [14]. The different behavior between the

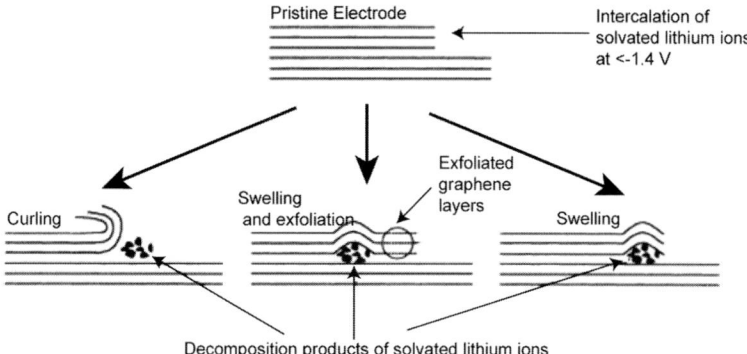

Fig. 26.8. Schematic model of three topographical changes observed in the anode due to intercalation of solvated lithium ions [14]

composite graphite and the HOPG anode was due to the difference in the edge plane structure, and that the intercalation reaction is greatly affected by the structure of the host material [14].

Secondly, a surface film is formed on the anode surface during charge-discharge cycles. Figures 26.5, 26.6, and 26.7 all show nanocrystalline deposits on cycled samples. This surface film is called a solid electrolyte interphase (SEI). It is formed from the products of the decomposition of the electrolyte at the anode surface during the first few initial charge-discharge cycles. This surface film plays a major role in the aging of an anode in lithium-ion batteries [24]. The SEI formation process is an irreversible reaction accompanied by the irreversible consumption of Li^+ ions resulting in a permanent loss of cell capacity, commonly referred to as an irreversible capacity loss (ICL) [15]. The SEI layer should be permeable to Li^+ ions to facilitate the intercalation and deintercalation reaction, and at the same time it should be impermeable to other electrolyte compounds. Thus, SEI has three major functions to perform: first, it should have high permeability for Li^+ ions so that they can reach the anode surface; second, it should not allow the electrolyte to permeate through and prevent it from further decomposing at the anode surface; and third, it should protect the anode surface from corrosion. The electrolyte components should be able to support the quick formation of SEI in the first few cycles without much loss of the electrolyte and the active material, but in subsequent cycles it should not attack and deteriorate the SEI. The mechanism and kinetics of the reduction of electrolytes in SEI formation is very complex and is a topic of on-going research [18]. Studies have shown that the chemical composition of SEI varies depending on the anode material and the electrolyte. According to Yang et al. [26] SEI contains $(CH_2OCO_2Li)_2$ in pure EC while $C_2H_5OCOOLi$ and Li_2CO_3 are formed in mixed solvents with DMC or DEC. Aurbach, in his studies [7, 8], showed that a very passivating agent such as $(CH_2OCO_2Li)_2$ is formed on carbon in EC – DMC. Studies are being conducted to further characterize the chemical composition and structure of SEI [14, 15, 18]. All the studies till now have indicated that the selection of electrolyte solvents, additives and other components becomes very critical for SEI properties, composition, and formation.

Studies also show the existence of two different kinds of SEI present on the same anode surface [24]. In the study conducted by BarTow et al. [9] on highly oriented pyrolytic graphite in 1-M LiAsF$_6$/EC – DMC, two different types of SEIs are formed. The SEI formed on the basal planes is thin, rich in organic compounds, and is mainly contributed by the reduction components of EC and DEC while the SEI formed on the edge sites is thicker, richer in inorganic compounds, and is mainly formed by the LiAsF$_6$ salt. Similar results were found by Kostecki and McLarnon [18] in their ex-situ study, in which SEI on the graphitic plane cross-section was dominated by inorganic products, and SEI on basal planes was formed from organic polymers. So SEI properties, formation, and composition are affected by the morphological changes in the anode surface. Thus, the cycling of a lithium-ion battery can cause disorders in graphene layers and lattice defects which can further lead to a non-uniform SEI layer.

Table 26.6, published by Vetter et al. [24], summarizes the cause and effects of aging in graphitic carbon anodes. Specific anode components can affect the aging mechanism. The summary of the dominant aging mechanism in anode material as given below is valid for most lithium-ion battery chemistries [24]:

- SEI is formed in the initial charge-discharge cycles. Though SEI is necessary to prevent the further reduction of electrolyte and corrosion of anode material, its growth beyond the initial cycles can increase the impedance in the battery. Increased impedance is directly related to capacity and/or power fade.
- Irreversible consumption of Li$^+$ ions can occur at the anode, which causes capacity fade due to loss of active material.
- Growth of SEI can reduce the permeability and prevent Li$^+$ ions from reaching the anode surface and reduce the intercalation/deintercalation capacity.
- Lithium metal plating can also occur due to uneven current and potential distributions over the anode surface. This causes Li$^+$ ion loss, leading to capacity fade.

26.3.2
Cathode

As indicated previously, cathodes can be made of different materials. Depending on the chemistry of the battery the cathode can be of cobalt oxide, manganese spinel, nickel-cobalt-manganese or phosphate. Here we shall review a study of the LiNi$_{0.8}$Co$_{0.2}$O$_2$ and LiNi$_{0.8}$Co$_{0.15}$Al$_{0.05}$O$_2$ cathodes used in the ATD program of DOE.

26.3.2.1
Literature Review of Cathode Aging

A study was conducted on a cell consisting of a carbon anode, LiNi$_{0.8}$Co$_{0.2}$O$_2$ cathode, and 1-M LiPF$_6$/EC-DEC (1:1) electrolyte [17]. Further details about cell construction can be found in Zhang et al. [27]. Figure 26.9a shows images of the LiNi$_{0.8}$Co$_{0.2}$O$_2$ cathode from the virgin cell while Fig. 26.9b shows the corresponding images from the cycled cell. The cell was subjected to two formation cycles and

Table 26.6.   Cause, effects, and influences of anode aging in lithium-ion battery [24]

Cause	Effect	Leads to	Reduced by	Enhanced by
Electrolyte decomposition ($\rightarrow$ SEI) (Continuous side reaction at low rate)	Loss of lithium Impedance rise	Capacity fade Power fade	Stable SEI (additives) Rate decreases with time	High temperatures High SOC (low potential)
Solvent co-intercalation, gas evolution and subsequent cracking formation in particles	Loss of active material (graphite exfoliation) Loss of lithium	Capacity fade	Stable SEI (additives) Carbon pre-treatment	Overcharge
Decrease of accessible surface area due to continuous SEI growth	Impedance rise	Power fade	Stable SEI (additives)	High temperatures High SOC (low potential)
Changes in porosity due to volume changes, SEI formation and growth	Impedance rise Overpotentials	Power fade	External pressure Stable SEI (additives)	High cycling rate High SOC (low potential)
Contact loss of active material particles due to volume changes during cycling	Loss of active material	Capacity face	External pressure	High cycling rate High DOD
Decomposition of binder	Loss of lithium Loss of mechanical stability	Capacity fade	Proper binder choice	High SOC (Low potential) High temperatures
Current collection corrosion	Overpotentials Impedance rise Inhomogeneous distribution of current and potential	Power fade Enhances other aging mechanisms	Current collector pre-treatment (?)	Overdischarge Low SOC (high potential)
Metallic lithium plating and subsequent electrolyte decomposition by metallic Li	Loss of lithium (loss of electrolyte)	Capacity fade (Power fade)	Narrow potential window	Low temperature High cycling rates Poor cell balance Geometric misfits

SEI – Solid electrolyte interphase, SOC – State of charge

one discharge cycle followed by a charge-neutral profile with 3% variation in state-of-charge (SOC) at 60% SOC and 70°C. The profile used for cycling the cell was specifically adopted to study the behavior of these batteries in HEV; and the details of the profile can be found in Zhang et al. [27]. After cycling, the cell lost power performance (W/kg) by 15–30%. The topographic image of the virgin cathode shows well-defined crystal planes and edges of the cathode material. In the topographic image of the cathode from the cycled cell, the entire cathode surface is covered by nanocrystalline deposits. The nanocrystalline deposits are seen in the intergranular spaces as well as across the crystal planes. The conductance images were obtained by holding the cathode sample at a positive potential of 1.0 V vs. the CSAFM tip. The black areas indicate areas of high electronic conductance while white areas indicate areas of low or zero conductivity. The magnitude of the current passing through the cathode is determined by the local contact properties of the cathode and the AFM tip, the local surface properties of the cathode and the AFM tip, and the AFM tip–cathode voltage difference. The virgin cathode mostly had areas of high electronic conductance. The conductance image of the cycled cathode showed areas mostly with high resistance and low conductivity. The areas of deep crevices and the intergranular spaces are the only areas seen to be conductive in the cycled cell [17].

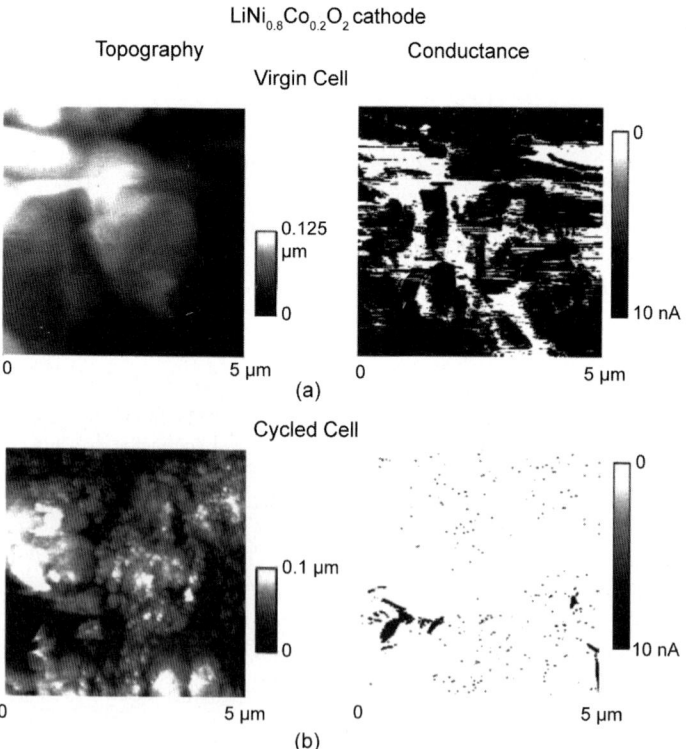

Fig. 26.9. Topographic (*left-hand panel*) and surface conductance images (*right-hand panel*) of $LiNi_{0.8}Co_{0.2}O_2$ cathodes from the (**a**) virgin cell and the (**b**) cell cycled at 70°C, 60% SOC, and 3% ΔSOC as per charge-neutral profile discussed in Zhang et al. [27] [17]

To establish a relationship between the surface morphology change and the cycling temperature, a plot of surface-average and root-mean-square (RMS) roughness vs. temperature was created for the $LiNi_{0.8}Co_{0.2}O_2$ cathode from the virgin cell, the above-mentioned cycled cell, and a similar cell cycled at 40°C (Fig. 26.10) [20, 27]. From the plot it can be clearly seen that both the surface-average and the RMS roughness of the cycled cathodes were lower than that of the virgin cathode. According to the authors, the roughness decreased in the cycled cathodes due to the nanocrystalline deposits in the intergranular spaces as well as across the crystal planes. The difference between the surface-roughness and the RMS roughness decreases as the cells are cycled at higher temperature (Fig. 26.10). The authors attributed this decrease in the difference to the formation of a more uniform surface with less protrusions and deep crevices [20, 27].

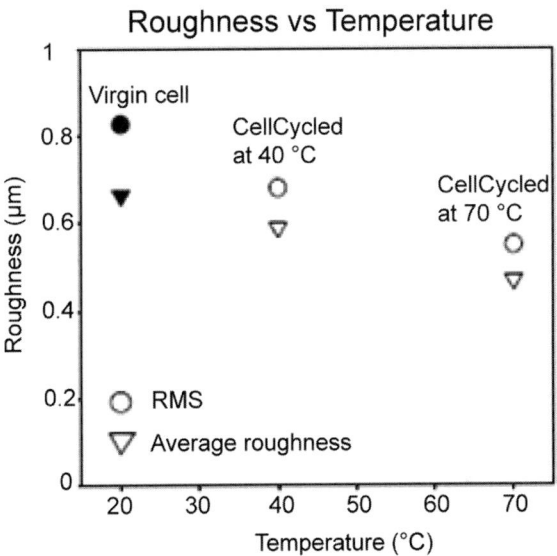

Fig. 26.10. Surface average and RMS roughness parameter of $LiNi_{0.8}Co_{0.2}O_2$ cathodes from the virgin cell, cell cycled at 40°C, 60% SOC, 3% ΔSOC and cell cycled at 70°C, 60% SOC, 3% ΔSOC as per charge-neutral profile (Adapted from [20, 27])

Another study was conducted on a cell with a $LiNi_{0.8}Co_{0.15}Al_{0.05}O_2$ cathode [19]. The cell had a synthetic graphite anode, and 1.2-M $LiPF_6$/EC-EMC (1:1) electrolyte. Figure 26.11a shows images of a $LiNi_{0.8}Co_{0.15}Al_{0.05}O_2$ cathode from the virgin cell while Fig. 26.11b shows the corresponding images from the cycled cell. The cell was cycled at C/2 rate, 100% DOD, between 3.0 and 4.1 V, and lost 34% of its power. The topographic images showed large polycrystalline agglomerates but showed no substantial change in surface morphology of the cycled cell [19, 21]. This contradicted the earlier results of the $LiNi_{0.8}Co_{0.2}O_2$ cathode as shown in Fig. 26.9 [17, 20]. Similar to the surface conductivity study of the $LiNi_{0.8}Co_{0.2}O_2$ cathode, the conductance images $LiNi_{0.8}Co_{0.15}Al_{0.05}O_2$ were obtained by holding the cathode sample at a positive potential of 1.0 V vs. the CSAFM tip. The black areas indicate areas of high electronic conductance while white areas indicate areas of low or zero conductivity. The magnitude of the current passing through the cathode is determined by the local contact properties of the cathode and the AFM tip, the local surface properties of the sample and tip, and the tip–cathode voltage difference. The virgin cathode mostly had areas of high electronic conductance. The conductance image of the cycled cathode shows areas mostly with high resistance and low conductivity.

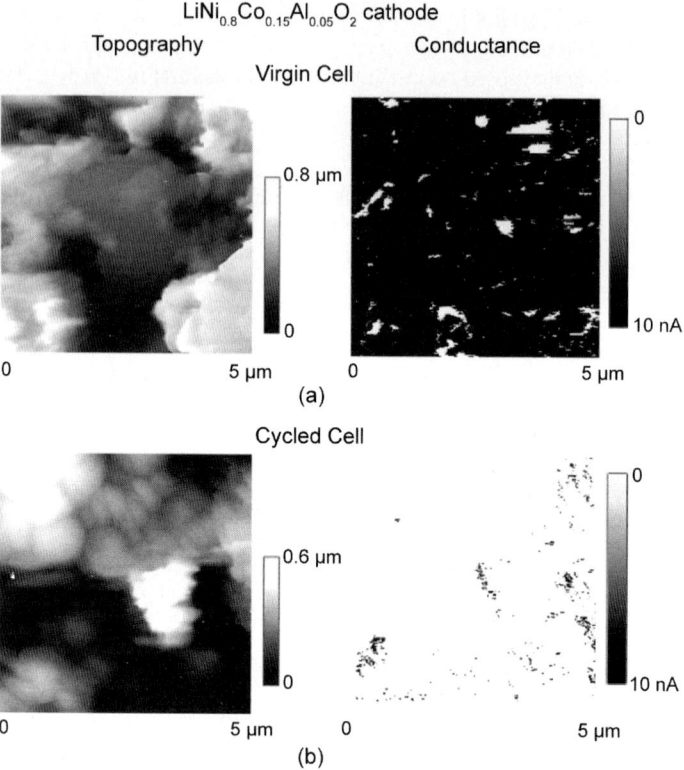

Fig. 26.11. Topographic (*left-hand panel*) and surface conductance images (*right- hand panel*) of LiNi$_{0.8}$Co$_{0.5}$Al$_{0.05}$O$_2$ cathodes from the (**a**) virgin cell, and the (**b**) cycled cell. The cell was cycled at C/2 rate, 100% DOD, between 3.0 and 4.1 V, and lost 34% of its power. The tip—sample voltage difference is at 1.0 V [19]

26.3.2.2
Discussion of Aging Mechanisms in Cathode Materials

The ATD study has identified the cathode as the principal cause of premature performance fade while the anode retained most of its original capacity [19]. In the CSAFM study of LiNi$_{0.8}$Co$_{0.2}$O$_2$ and LiNi$_{0.8}$Co$_{0.15}$Al$_{0.05}$O$_2$ cathodes the conductance of the cathode surface dropped but the surface morphology changes in LiNi$_{0.8}$Co$_{0.15}$Al$_{0.05}$O$_2$ and LiNi$_{0.8}$Co$_{0.2}$O$_2$ are different.

The topographic image (left-hand panel of Fig. 26.9) of LiNi$_{0.8}$Co$_{0.2}$O$_2$ shows that the surface of the cycled cathode is covered with nanocrystalline deposits while the topography image (left-hand panel of Fig. 26.11) of $LiNi_{0.8}Co_{0.15}Al_{0.05}O_2$ does not show any substantial change in the surface morphology. The nanocrystalline deposits were observed in the intergranular spaces as well as across the crystal planes. These nanocrystalline deposits in the case of the $LiNi_{0.8}Co_{0.2}O_2$ cathode decreased the roughness of the cycled cathode. The individual particle size was measured and was reported to be between 50 and 200 nm. These nanocrystalline deposits can have an effect on the surface properties of the cathode and, the result could be an increase in the overall cell impedance [20].

In a CSAFM study, both $LiNi_{0.8}Co_{0.2}O_2$ and $LiNi_{0.8}Co_{0.15}Al_{0.05}O_2$ cathodes showed a drop in surface conductivity. The virgin cathodes had high conductivity while the cycled cathodes showed low conductivity and high resistance. The high conductivity in the virgin samples was explained by highly conductive graphite and acetylene black additives. These are found in abundance in the composite cathodes [19]. Kostecki and McLarnon [19] considered the contribution from the active material of the composite cathodes to be negligible as compared to carbon. The drop in the surface conductance or rise in the resistance was attributed to the loose cathode particles, or the presence of a non-conductive film of polycarbonates [17, 19]. It was also suggested that the conductivity dropped due to the presence of a non-conductive PVDF binder, SEI layers, and deep cavities between the particles of the cathode. The reduced conductivity of the cathode surface was considered to be the reason for the drop in cathode capacity and the rise in the overall impedance of the battery. The SOC was observed to be non-uniform over the cathode surface. According to the authors, due to non-uniform SOC, the local load current distribution could be highly non-homogenous, leading to local overcharge or overdischarge. This could further cause effects such as non-uniform electrolyte decomposition, structural changes, etc.

The above discussion indicates that each cathode material can have an individual degradation mechanism leading to the aging of a lithium-ion battery and so independent study of each cathode material becomes evident. Both $LiNi_{0.8}Co_{0.2}O_2$ and $LiNi_{0.8}Co_{0.15}Al_{0.05}O_2$ cathodes showed reduction in surface conductance as the cell aged but there was no physical change in the $LiNi_{0.8}Co_{0.15}Al_{0.05}O_2$ cathode surface compared to nanocrystalline deposits observed in the case of the $LiNi_{0.8}Co_{0.2}O_2$ cathode. This indicates that more studies are needed to identify the interfacial changes that can uniquely identify the aging precursors in each of the cathode materials. As such, a single summary of cathode aging is hard to present. In general, aging of a cathode for all lithium-ion batteries is based on three basic principles [24]:

- Structural and surface property changes during cycling;
- Formation of a surface film or nanocrystalline deposits;
- Chemical decomposition/dissolution reaction.

Figure 26.12 gives a generalized summary of the cause and effect of the aging mechanism in the cathode materials. During Li^+ ion intercalation and deintercalation cycles, the volume of the cathode material changes, leading to structural disordering and mechanical stresses. Also, phase transformation of the cathode takes place during the intercalation and deintercalation cycles. In some cases, nanocrystalline deposits are formed during the intercalation and deintercalation cycles. Metal disso-lution and electrolyte decomposition takes place at the solid-electrolyte interphase. All these causes eventually lead to a drop in capacity and a rise in impedance of the battery.

26.4
Closure

Since Sony introduced the lithium-ion battery in 1991, it has become very popular for consumer appliances such as mobile phones and laptop computers. The advances in lithium-ion technology has furthered its use for durable applications such as elec-tric vehicles (EV) and/or hybrid electric vehicles (HEV). Along with high power capability, the batteries used in these applications need to meet the important crite-rion of a long life. According to the US Advanced Battery Consortium (USABC, www.uscar.org), a 42-V battery in a plug-in HEV should have a calendar life of 15 years [1]. Electric vehicles should have a battery system that can last for 10 years [3]. In terms of cycles, 1,000 cycles at 80% depth-of discharge are expected in EV [3] and 300,000 cycles at 50 Wh are expected in a plug-in HEV [4].

Lithium-ion battery technology has been shown to deliver high power for appli-cations such as power tools. It also has the potential to achieve higher energy den-

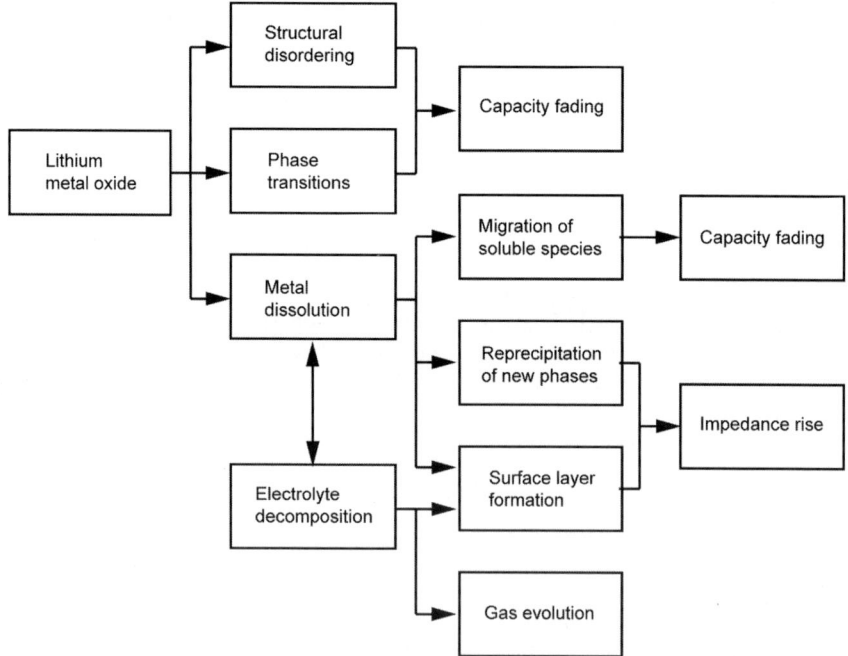

Fig. 26.12. Cause and effect of aging mechanism of cathode materials (Adapted from [25])

sities. The only hurdle in the success of the lithium-ion battery as the favorite for the EV and HEV is the shorter cycling life. Different nanostructured cathode materials are being tested to improve the life of the lithium-ion battery. In lithium-ion battery technology, graphitic carbon has been well-established as the anode material. Anode aging is known to be governed by the properties of solid electrolyte interphase. AFM studies of different cathode materials show different changes in interfacial properties as they age. As such it becomes necessary to study each cathode material individually and identify the unique aging precursors in each of them. A single technique would be insufficient to explain the complex aging mechanisms of anodes and cathodes, so various techniques, such as electrochemical impedance spectroscopy, synchrotron infrared microscopy, Fourier-transform infrared spectroscopy, X-ray diffraction, Raman spectroscopy, and scanning electron microscopy, are useful in studying the aging phenomena in lithium-ion batteries.

Along with these studies, atomic force microscopy topography study and current-sensing AFM (CSAFM) have proven useful in understanding the near surface or surface changes in anode and cathode materials. This review also shows that the in-situ electrochemical AFM (ECAFM) technique can be applied to study surface morphological changes in the active material and formation of SEI. Furthermore, techniques such as scanning spreading resistance microscopy (SSRM) and Kelvin probe microscopy (KPM) can be used to study the changes in the surface properties of the anode and cathode materials [10]. The results from scanning probe techniques, such as changes in surface morphology, electronic conductivity, and state of charge (SOC) over the material surface, can be correlated to the changes in the bulk properties of the lithium-ion battery such as rise in impedance, capacity fade, and power fade. These results collectively can help in identifying the difference in behavior of the cathode materials, the interfacial changes in the cathode materials at the onset of aging, and aging precursors. A theoretical model based on these results can help greatly in predicting the life of a lithium-ion battery. The experimental and theoretical results would help in selecting a better combination of nanostructured materials for the cathode in a lithium-ion battery and in predicting the aging mechanism for that cathode material. Thus, experimentation and theoretical modeling of lithium-ion battery materials can help to improve the lithium-ion battery technology beyond its current stature and meet the requirements of the USABC.

Acknowledgments. The authors would like to thank Dr. Giorgio Rizzoni and Dr. S.S. Babu of The Ohio State University for the opportunity to study the lithium-ion battery systems and some critical and thoughtful discussions during these studies.

References

1. Anonymous (2002) FreedomCAR 42 V energy storage system end-of-life performance goals. USABC, Southfield, MI (http://www.uscar.org/guest/view_team.php?teams_id=12).
2. Anonymous (2006a) Cell construction. Axeon Power Ltd, Dundee, Scotland. (http://www.axconpower.com/cell_construction.htm).

3. Anonymous (2006b) USABC goals for advanced batteries for EVs. USABC, Southfield, MI. (http://www.uscar.org/guest/view_team.php?teams_id=12).
4. Anonymous (2006c) USABC requirements of end of life energy storage systems for PHEVs. USABC, Southfield, MI (http://www.uscar.org/guest/view_team.php?teams_id=12).
5. Anonymous (2007a), Energy kid's page. Source: Energy Information Administration, Washington, DC. (http://www.eia.doe.gov/kids/).
6. Anonymous (2007b), Overview of lithium-ion batteries, Panasonic Corp., Secaucus, NJ. (http://www.panasonic.com/industrial/battery/oem/chem/lithion/index.html).
7. Aurbach D (2000) Review of selected electrode–solution interactions which determine the performance of Li and Li ion batteries. J Power Sources 89:206–218.
8. Aurbach D (2003) Electrode-solution interactions in Li-ion batteries: a short summary and new insights. J Power Sources 119–121:497–503.
9. BarTow D, Peled E, Burstein L (1999) A study of highly oriented pyrolytic graphite as a model for the graphite anode in Li-ion batteries. J Electrochem Soc 146:824–832.
10. Bhushan B (2006) Springer handbook of nanotechnology, 2nd edn. Springer-Verlag, Heidelberg, Germany.
11. Brown T, LeMay H Jr., Bursten B, Burdge J (2003) Chemistry the central science, 9th edn. Prentice Hall, New Jersey.
12. Buchmann I (2001) Batteries in a portable world: a handbook on rechargeable batteries for non-engineer, 2nd edn. Cadex Electronics Inc., Richmond.
13. Efimov (1999) Vibrational spectra, related properties, and structure of inorganic glasses. J Non-Cryst Solids 253:95–118.
14. Jeong SK, Inaba M, Iriyama Y, Abe T, Ogumi Z (2003) AFM study of surface film formation on a composite graphite electrode in lithium-ion batteries. J Power Sources 119–121:555–560.
15. Kong F, Kostecki R, Nadeau G, Song X, Zaghib K, Kinoshita K, McLarnon F (2001) In-situ studies of SEI formation. J. Power Sources 97–98:58–66.
16. Kopera J (2004), Inside the nickel-metal hydride battery. Cobasys, MI.
17. Kostecki R, McLarnon F (2002) Degradation of $LiNi_{0.8}Co_{0.2}O_2$ cathode surfaces in high-power lithium-ion batteries. Electrochem Solid State Lett 5(7):A164–A166.
18. Kostecki R, McLarnon F (2003) Microprobe study of the effect of Li intercalation on the structure of graphite. J. Power Sources 119–121:550–554.
19. Kostecki R, McLarnon F (2004), Local probe studies of degradation of composite $LiNi_{0.8}Co_{0.15}Al_{0.05}O_2$ cathodes in high-power lithium-ion batteries. Electrochem Solid State Lett 7(10):A380–A383.
20. Kostecki R, Zhang X, Ross PN Jr, Kong F, Sloop S, Kerr JB, Striebel K, Cairns E, McLarnon F (2001) Failure modes in high-power lithium-ion batteries for use in hybrid electric vehicles. Lawrence Berkeley National Laboratory, University of California, California, LBNL 48359,
21. Kostecki R, Lei J, McLarnon F, Shim J, Striebel K (2005) Diagnostic evaluation of detrimental phenomena in high-power lithium-ion batteries. Lawrence Berkeley National Laboratory, University of California, California, LBNL 59195.
22. Pasierb P, Komornicke S, Rokita M, Rękas M (2001) Structural properties of Li_2CO_3- $BaCO_3$ system derived from IR and Raman spectroscopy. J Mol Struct 596:151–156.
23. Shim J, Kostecki R, Richardson T, Song X, Striebel KA (2002) Electrochemical analysis for cycle performance and capacity fading of a lithium-ion battery cycled at elevated temperature. J Power Sources 112:222–230.
24. Vetter J, Novák P, Wagner MR, Veit C, Möller KC, Besenhard JO, Winter M, Wohlfahrt-Mehrens M, Vogler C, Hammouche A (2005) Ageing mechanisms in lithium-ion batteries. J Power Sources 147:269–281.

25. Wohlfahrt-Mehrens M, Volger C, Garche J (2004) Aging mechanisms of lithium cathode materials. JPower Sources, 127:58–64.
26. Yang R, Wang YY, Wan CC (1998) Composition analysis of the passive film on the carbon electrode of a lithium-ion battery with an EC-based electrolyte. J Power Sources 72:66–70.
27. Zhang X, Ross PN Jr, Kostecki R, Kong F, Sloop S, Kerr JB, Striebel K, Cairns EJ, McLarnon F (2001) Diagnostic characterization of high power lithium-ion batteries for use in hybrid electric vehicles. J Electrochem Soc 148:A463–A470.

Subject Index

Printing: Krips bv, Meppel, The Netherlands
Binding: Stürtz, Würzburg, Germany